OPEN SEA MARICULTURE

OPEN SEA MARICULTURE

PERSPECTIVES, PROBLEMS, AND PROSPECTS

Edited by Joe A. Hanson
Oceanic Foundation, Hawaii

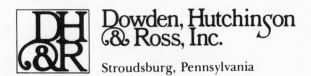

Dowden, Hutchinson
& Ross, Inc.
Stroudsburg, Pennsylvania

Library of Congress Cataloging in Publication Data
Main entry under title:

Open sea mariculture.

    1.   Mariculture.   I.   Hanson, Joe A., 1928-   ed.
II.   Oceanic Foundation.
SH138.06   639   74-13103
ISBN 0-87933-130-5

Manufactured in the United States of America.

Exclusive distributor outside the United States and Canada:
JOHN WILEY & SONS, INC.   -

74 75 76 5 4 3 2 1

PREFACE

This book is the culmination of a study begun in the fall of 1970 at the request of, and with financial support from, the National Oceanic and Atmospheric Administration's Sea Grant Office (Sea Grant No. 2-35285). Sea Grant supports tens of millions of dollars of oceanic research annually and has for some time had a particular interest in the promotion of aquaculture research. But space available for culturing marine organisms in inland and inshore waters is probably less extensive than space available for agriculture: so, if aquaculture ever is to develop grand-scale proportions, it eventually becomes appropriate to consider conducting it in offshore waters. Consequently, Sea Grant chose to devote a modest grant to an early consideration of the possibilities and problems that arise when culturing of marine organisms in the open seas is considered seriously . . . This preliminary analysis, to the degree that it could be successfully accomplished, could provide support for decisions on research thrusts and priorities.

The approach that Sea Grant considered appropriate for this somewhat futuristic study was oriented more toward systems engineering than toward biology. By 1970 the Oceanic Foundation had gained recognition for its marine systems engineering and aquaculture research, as well as some recognition of its willingness to tackle problems that wiser institutions might decline—hence the selection of the Oceanic Foundation to perform this initial feasibility analysis of open sea mariculture.

With no attempt at apologies, I must admit here that when I accepted Principal Investigator responsibility for the work, I did not fully appreciate the depth and scope of the problem. When the grant was awarded there was fair agreement on the topics to be covered; but the scope of material the topics implied and their predilection for interacting with each other was, I believe, incompletely recognized by all of us at the time. In short, one of the most difficult management tasks in this work was to keep its scope within practical limits.

Another difficult question was how to achieve acceptable authoritativeness throughout the breadth of material the work was to encompass. Stated frankly, this goal could not even have been approached had not many members of the Association of Sea Grant Institutions and numerous others contributed freely and enthusiastically of their time and resources. The strategy adopted was first to break the study down into major topics . . . and wherever an individual of recognized authority could be persuaded to accept responsibility for a given topic, he was gratefully awarded the task. Where no willing and able authority appeared, or where the nature of the topic precluded individual authoritativeness, an Oceanic Foundation staff member found himself responsible for a new research assignment. Although every attempt was made to match open sea mariculture research topics with individual realms of knowledgeability, it must be admitted that the matches were sometimes less than perfect. In order to compensate for questionable authoritativeness in some of the staff research assignments as well as to achieve maximal quality throughout, a nationwide review board was assembled: every chapter of this book has been reviewed and refined by several members of that board and frequently by members of their staffs as well. In fact, because many members of the review board enlisted assistance from still others, the reviewer participation shown here unfortunately is incomplete. Nonetheless, the names that do appear speak for themselves and, it bears repeating, the work could not have been done without the cooperation of these authorities.

We faced a third serious problem in the compilation and editing of the book itself. When a chapter is to appear under an individual author's name, the extent to which its content and format can be edited is ethically limited. And so arbitrary editorial control in a multiauthor book such as this must be much looser than in a publication involving only one or a very few authors. Although every attempt has been made to achieve consistency of presentation, a degree of reader indulgence must be requested.

Authors of individual chapters are listed in the table of contents and in the chapter headings. In several cases the authors are among the brighter young stars in marine science; other authors have well-established reputations. Edward D. Stroup (Chapter 4) is Associate Professor of Oceanography, University of Hawaii. Stephen V. Smith (Chapter 5) presently divides his time between the Hawaii Institute of Marine Biology, where he is Research Associate, and the Hawaii Laboratory of the Naval Undersea Center. Colin E. Nash (Chapter 8), internationally known for his work with the British Whitefish Authority, is Vice President and Research Director of the Oceanic Foundation. Lawrence P. Raymond (Chapter 7) is the Chief Scientist of the Environmental Studies Division of the Oceanic Foundation, and Paul K. Bienfang (Chapter 7) is Research Associate in that division. Guy N. Rothwell (Chapter 13) is Chief Engineer of the System Sciences Division of the Oceanic Foundation, where he has for some years been deeply involved in designing submersible, semisubmersible, and fixed marine structures. Stephen B. Ribakoff is Staff Engineer in the same division. Carol C. Hanson (Chapter 3) is a student of political science and ocean law at the University of Hawaii. John P. Craven (Chapter 3) is Dean of Marine Programs, University of Hawaii, and Marine Affairs Coordinator, State of Hawaii: his past ocean engineering accomplishments, including the U.S. Navy Polaris and Sealab programs largely mask the fact that he also has his J.D. degree and is an expert in matters of ocean law. George M. Sheets and Gerald Grimes (Chapter 3) both practice law in the State of Hawaii.

Jeanne M. Collier—given authorship credit for Chapters 3 and 11—more appropriately deserves coauthor credit for the entire volume: there is no chapter that has not been improved measurably by her gentle and not-so-gentle attentions. Ms. Collier, Senior Systems Analyst in the System Sciences Division of the Oceanic Foundation, brings two decades of experience in social and information systems research to her work. People of her sort are essential components of successful research groups, being not only competent researchers in their own right, but having the talent and determination required to construct coherent wholes from a multitude of diverse contributions.

To return to the review board, its membership is as follows:

- Orville W. Terry—Marine Sciences Research Center, State University of New York.
- Harvey R. Bulliss—Director, Southeast Fisheries Center, National Marine Fisheries Service
- Herbert F. Frolander—Coordinator, Marine Science and Technology Programs, Oregon State University
- Edward F. Klima—Plans and Policy Development, National Marine Fisheries Service
- Terrence Nosho—Aquaculture Agent, University of Washington
- Robert C. May—Assistant Marine Biologist, Hawaii Institute of Marine Biology
- Daniel B. Sass—Director, Environmental Studies Program, Alfred University
- George W. Klontz—Professor, Fisheries Management, University of Idaho
- E. Eugene Allmendinger—Associate Professor, Office of Marine Science and Technology, University of New Hampshire
- John Dermody—Assistant Director, Division of Marine Resources, University of Washington
- William J. D. Escher—Founding Associate, Escher Technology Associates
- Lawrence R. Glosten—President, L.R. Glosten & Associates, Incorporated, Naval Architects
- Jack C. Parker—Professor, Texas A & M University
- Willis H. Clark—Assistant Director, Center for Marine Resources, Texas A & M University
- Ira Dyer—Head, Department of Ocean Engineering, Masachusetts Institute of Technology
- John P. Harville—Executive Director, Pacific Marine Fisheries Commission
- Carl J. Sindermann—Director, Middle Atlantic Coastal Fisheries Center

Bonnie Rhodes and Dianne Henderson, two of the finest technical editors with whom I have had the pleasure of working, labored long hours working the manuscript into an acceptable form. Joyce Miller accomplished the major share of the manuscript typing, much of it over and over again.

Finally, a special acknowledgment must go to Harold Goodwin, Associate Director of the National Sea Grant Program, who was instrumental in initiating the work as well as in seeing it through.

And so it has taken two pages to list those who participated in the development of this work. I find myself uncomfortable expressing thanks, because such an expression implies that the work is mine and it is not: I have just listed those to whom it does belong. Nonetheless, ultimate responsibility for any errors of omission or commission remains mine.

Joe A. Hanson

Oceanic Foundation

CONTENTS

INTRODUCTION

Sea Grant

The mission is to perform a systems analysis of open sea mariculture. This means that the technical and economic feasibility of open sea mariculture must be assessed, the forms it may take must be derived, and assessments of the probable value of these forms as food-producing systems of the future must be made.

Systems analysis should result in a rigorous determination of fit between a system of interest and the larger system (environment) in which the system of interest is to function. A competent systems analysis exposes all the important facets of a system that interact with its external environment—that is, all the interfaces—and then determines what demands these interfaces impose on the structure and functioning of the system. Having accomplished this far-from-trivial task, the analysis may then proceed to determine how well an existing system does fit its environment, or how well a proposed system *will* fit its environment, or, if no system designs exist, to propose system designs that *should* fit the environment. Implicit in all this are requirements for (1) a relatively determinate environment and (2) a rather well-formed concept of the system that is to fit into that environment.

These points are made to alert the reader to the dilemma implicit in this work. On the one hand, the future form of society is most assuredly not determinate; on the other hand, open sea mariculture so far remains a term associated with few, if any, well-formed system concepts. Thus, if the task is not to be abandoned at its inception, it becomes necessary to postulate, first, future societies, and then, open sea mariculture concepts that might fit the societies postulated. Although a systems engineer devoted to dogmatic rigor might well choose to abandon the task, we feel that alternative future social orders can be (and, in fact, have been) logically derived—as can potential open sea mariculture concepts.

SOCIAL ORDERS REQUIRED BY OPEN SEA MARICULTURE

Alternative societal futures that appear possible from today's vantage point represent a broad spectrum of possibilities. At one end lie sparse, wandering, subsistence-culture tribes reduced to that state by global-scale thermonuclear war or famine. At the other end of the spectrum lies a global technological dictatorship in which all but the few humans in control serve, like the structures and machines that make up their environment, only as components of a world "technosocial system" that has self-perpetuation as its only

1

objective. But somewhere in the middle lie numerous possibilities for high-technology social orders that have evolved to a stage in which their own internal controls maintain an acceptable human quality of life within some range of equilibrium with available resources and the natural environment. Open sea mariculture, as a means of augmenting resources to maintain the health and vitality of a highly technological society, would serve either of the latter two possibilities; there seems little likelihood that it could come into being if the first postulation became reality. So, as a foundation for this work we necessarily postulate a world in which technological evolution proceeds apace but swings away from simple resource exploitation toward resource conservation, a world in which population growth rates taper off and then stabilize within the planet's ability to sustain them, within man's ability to tolerate and govern himself, and within his technological capability to provide for acceptable human quality of life over the long term. Because it is not necessary to the work, we ignore the question of whether the controls mandatory for the achievement of such a society are democratically or autocratically derived.

Such world orders entail some logical consequences. For example, the relative value of energy in all forms will almost certainly be greater than it is today, because there almost certainly will be a reduction in available energy per capita. Also, per capita environmental degradation allowances will have to be lower in more densely populated worlds, as must allowable per capita resource *consumptions*. As employed here, *consumption* means nonconservative resource exploitation, whereas resource *utilization* can be wholly or partly conservative. Clearly, also, resource consumption implies a higher per capita environmental degradation rate than does resource utilization, for utilization should involve recycling.

The fundamental logic underlying the kind of future world system for which open sea mariculture may become a food-producing subsystem is as follows: if populations become large enough to demand mariculture on the grand scale implied by open sea applications, this high ratio of humans to global resources will demand that societies be far more efficient in utilizing energy and other resources than are today's societies. In ecological terms, all forms of energy, including food, and structural materials as well will very likely become the "limiters" of human population. As limiters, they will be precious and, thus, relatively a great deal more expensive than they are today. If this is the likely social environment in which large-scale open sea mariculture will flourish, classical economic evaluations of its prospective profitability do not apply. Instead the cost effectiveness of open sea mariculture must be viewed within a much more fundamental context: the ability to produce a precious renewable resource with minimum consumption of nonrenewable resources and minimum environmental degradation.

OPEN SEA MARICULTURE DEFINED

The term "culture" is defined as the cultivation of plants or animals to serve human purposes by deliberate modification of their natural environment and by selective breeding of stock to improve cultured species. "Mariculture" is a term recently coined by Winston Menzel to designate aquaculture in a marine environment. Since the marine environment extends from bays and estuaries out into the open sea, *open sea* mariculture is employed here to distinguish culturing activity in unprotected waters, whether near to or far from the shoreline.

Where does fishing, per se, end and culturing begin? Controlled fishing, that is, controlled harvesting, might be considered the borderline; ideally, it allows natural processes to maintain populations at maximum sustained yield; yet it remains primarily a resource conservation measure. True culturing requires deliberate, production-oriented human intervention into that system of controls which, in the natural dynamics of a species, would be the sole dictators of reproduction, survival, growth, population sizes, and evolutionary directions.

However, mariculture need not be interpreted as absolute human control over *all* aspects of the cultured species' life cycle. Schemes that act only to ease the pressure of a primary limiting factor, such as nutrition, maturation rate, or reproduction rate, while not ideal culturing practices, may increase populations of desirable species to the point where the term "culturing" is appropriate.

Mariculture systems of the future may be *monocultures,* in which only one species is cultured intentionally, *polycultures,* in which several different organismic forms are intentionally cultured together so as to derive high overall system efficiencies, or *series monocultures,* in which products and by-products of one monoculture are used as human-controlled inputs to one or more other monocultures.

PROSPECTS FOR OPEN SEA MARICULTURE

In summation, the following major conclusions were derived from this study:

- As an engineering challenge, commercial-scale open sea mariculture systems lie within reach of present-day technology. Large-scale, environmentally benign systems of high overall efficiency could be achieved with the further development of present technology.
- As a biological challenge, mariculture in general suffers from a dearth of practical knowledge, particularly in the realms of controlled reproduction, nutrition, and diseases. These roadblocks must be at least partially removed

before significant funds are committed to developing commercial-scale systems.

- As a legal and political problem, open sea mariculture is likely to remain enigmatic for some time to come. Laws and agreements designed to protect and regulate mariculture intelligently in territorial and international waters are not likely to be developed until needed, owing to mariculture-stimulated conflicts. Yet the lack of such laws and international agreements is an inhibiting influence on the development of open sea mariculture.

- Commercial-scale open sea mariculture in America is probably at least two, but not more than four to five, decades away. In Japan, offshore mariculture experimentation is well along, and successful commercial-scale activities appear likely within the next few years.

- The most attractive evolutionary paths for open sea mariculture to follow involve either atoll basing or the synergistic development of mariculture systems with other forms of offshore industry. Offshore petroleum activities offer an excellent starting place, with offshore power production a likely successor.

- A national-level decade-long mariculture development program, if instituted, could result in initiation of commercial-scale open sea mariculture well before the end of the research and development decade, and would lay the necessary groundwork for the very large scale systems of the future. The cost of such a program would be in the neighborhood of $30 million per year. The short-term, as well as long-term, economic consequences appear attractive.

ABOUT THE WORK

The conclusions just stated are expanded and supported in Part Four; in Chapter 14 the potentials for initiating open sea mariculture systems are discussed, and in Chapter 15 a possible national-level open sea mariculture development program is outlined. The reader in a hurry may decide to satisfy himself initially with this introduction and the two chapters of Part Four, saving the remainder of the volume for reference purposes. But we feel that even the hurried and casual reader will find Part One worthwhile in that it provides interesting and valuable perspectives on the open sea mariculture concept. These range from a fundamental bioeconomic perspective, through several socioeconomic views, to an analysis of international law as it affects open sea mariculture.

Part Two contains summary descriptions of those aspects of the marine environment considered important to open sea mariculture prospects, along with the resulting conclusions and recommendations. Part Two is a basic summary essay on physical, chemical, and geological oceanography, which many will find to be an enlightening experience. Others, already conversant in these fields, may wish to skip over it or to skim lightly. Biological oceanography is not considered in any depth; to cover the entire field even in summary form would lengthen this volume unnecessarily, while contributing little to the study itself. Those aspects of marine biology important to mariculture are treated in Part Three.

Part Three is devoted to questions of science and technology. Its purpose is to determine where society stands today with respect to practical open sea mariculture and to expose tomorrow's prospects. The topics examined in Part Three encompass the basic questions of crop selection, nutrition, control of the biological environment, open sea operations bases, sources of energy, concentrating and harvesting crops, and processing the harvest at sea. The subject of physical and chemical environmental control (of critical importance to shorebound aquaculture) is not treated separately. It is our conclusion that the size and dynamic nature of the open sea environment at once preclude, and make unnecessary, tight human control over marine environmental quality in the open sea. Such quality, we judge, is better achieved by cautious site selection.

In summary, in Part One we attempt to place open sea mariculture in bioeconomic, social, technological, and legal perspective; in Part Two we analyze important environmental parameters; in Part Three we examine the tools in hand, and postulated, that could be employed by open sea mariculture; and in Part Four we attempt reasonable conclusions concerning the future.

It is appropriate to emphasize here that the purpose of this work is to assemble and analyze current information in order to achieve a logical assessment of the open sea mariculture concept. We believe that this construction and analysis of previously existing information is a unique effort that sheds new light on the question at hand. Beyond that, any new scientific or technological information that may be contained herein is the as-yet-unpublished result of the work of others or the product of speculation on our part.

PART I

PERSPECTIVES ON OPEN SEA MARICULTURE

1

OPEN SEA MARICULTURE IN BIOECONOMIC PERSPECTIVE

J. A. Hanson

Culturing in all its forms is, in the final analysis, the economic exploitation of biological phenomena: the value of the products derived from a culturing enterprise must, in the long run, equal or exceed the economic value of the energy and matter devoted to it, or the enterprise will not be feasible economically. Exceptional cases in which a culturing enterprise may be subsidized for one reason or another are irrelevant to the truth of the foregoing generality.

If economics and energetics were to equate exactly, no human (material) enterprises would be feasible economically, since, overall, they invariably consume more matter and energy than they produce for sale. The economic viability of human enterprises results from the pragmatic nature of the economic systems in which they operate—economic systems that permit ideas concerning free energy fluxes, free materials, free waste disposal, and so forth, and that place differential values on different forms of energy and matter. For culturing in general and, in the present case, open sea mariculture in particular, this sort of picture implies that attention must be directed to the relative economic value of the energy and materials involved, as well as to the bioenergetic efficiencies with which a culture system operates.

In terms of bioenergetics, culture systems perform three basic types of operations: storage, conversion, and control. With no important exceptions, these operations require matter and energy; and all have conversion efficiencies of less than 1. In this chapter an attempt is made to structure the necessary marriage of economics and bioenergetics so as to derive a fairly clear picture of open sea mariculture as a hybrid offspring of the two.

ELEMENTARY BIOECONOMIC MODEL

In the simplest possible expression, mariculture consists of feeding a marine crop until it has grown to harvestable size, and then harvesting and processing it for market. This nutrient, to crop, to harvest, to processing, to market chain can be diagrammed as in Figure 1.1, in which each box represents both storage and conversion and implies control. Nutrients, even at the elemental level, represent conversion of energy to matter, or conversion of matter in

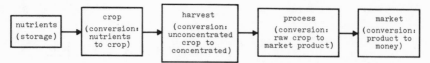

FIGURE 1.1 *Storage and conversion progression.*

one form to matter in another form. If nutrients are purchased, money is converted to nutrients. The crop converts nutrients to crop and stores them in that form. The harvest converts crop *in vivo* to crop in hand. And processing converts raw crop in hand to a market-ready product. The market converts product to money.

The nucleus of the crop must come from somewhere and conversion rates must be maximized and tuned to be compatible with one another. So, a complete mariculture picture will include controls and a hatchery operation. From the economic point of view, any storage or control operation is an economic negative unless it can be acquired free, in which case it will represent zero at best: storage and control operations cannot assume positive economic values directly. Conversion operations can have positive economic values only when the value of the second form is greater than the value of the first. Combining these statements and recognizing the inevitability of overhead expenses in any enterprise, we have the diagram in Figure 1.2.

Figure 1.2 is arranged to emphasize the fact that every operation in a mariculture enterprise represents an economic negative—or, at best, zero. The implications become clear if it is assumed that, from the market return, 20 percent is to be deducted for short-term net profit. This leaves 80 percent of the market return with which to fuel all eight basic consumers (negatives) shown in Figure 1.2, or 10 percent each on the average. Clearly, to the extent that some operations exceed the 10 percent average, other operations must be accomplished with less than 10 percent. The mariculturist, then, faces a classical problem in optimization; he must maximize market return while minimizing the costs of all necessary operations, as well as minimizing his overhead. Minimizing operation costs implies the clever exploitation of free

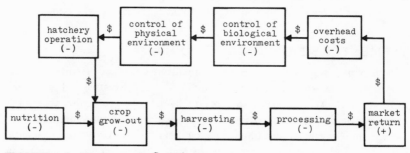

FIGURE 1.2 *Bioeconomic flow diagram.*

or low-cost energy and matter. Maximizing productivity implies maximizing conversion and control efficiencies.

In the interest of simplicity, such questions as transportation, warehousing, and marketing costs have been omitted by making the assumption that the product is wholesaled at its production site. For larger-scale enterprises such as those implied by open sea siting, it should be recognized that this assumption is almost certainly invalid.

Overhead

Overhead includes amortization of capital facilities, capital equipment, and interest paid, as well as operating expenses not charged directly to given operations. Considering the high cost of offshore facilities and operations, the open sea mariculturist will have to be either admirably clever in acquiring offshore sites, or find means in the offshore environment to nearly "zero out" the cost of other operations.

Biological Control

Control of the biological environment simply means protecting the crop from damaging levels of predation, disease, and parasitism. How to achieve such protection is not so simple; generally speaking, these three factors (or only one of them) can quickly devastate mariculture enterprises that otherwise would be profitable. In offshore mass culture, a combination of disease-resistant species, the avoidance of geographic problems, and inexpensive mass treatment is clearly indicated, if it can be achieved.

Physical Control

Control of the physical environment involves assuring that all nonbiological factors (light, chemistry, temperature, turbulence, etc.) in the culture environment remain within limits which are nonlethal and preferably optimum. In shoreside cultures, direct control over this category of parameters is usually possible and frequently economically attractive. In open sea operations, such control will seldom be possible and hardly ever economically attractive, except possibly in a few special circumstances. Achieving an acceptable physical environment probably will best be accomplished through careful site selection. Included in this category, too, are the problems of maintaining crop concentrations high enough to make other husbandry measures effective, and of temporarily achieving concentrations high enough for efficient harvesting; to do this in the open sea environment, the culturist probably will be required to make maximum use of natural phenomena if prohibitive capital and operating costs are to be avoided.

Hatchery Operation

In practice, hatchery operation may or may not be an integral part of an open sea mariculture enterprise. Typically, though, the juvenile crop ready for grow-out must be acquired at some cost to the culturist, whether or not he operates his own hatchery. In addition, controlled reproduction and larval rearing presently are, along with disease and parasitism control, major road-blocks to profitable culturing of many otherwise attractive species; so the hatchery function cannot be ignored in a bioeconomic perspective.

Harvesting and Processing

Harvesting and processing are necessarily technologically oriented, energy-consuming operations. In marine fisheries, these two operations together constitute the whole of the enterprise. In open sea mariculture, these two operations together must achieve a drastic reduction in cost as compared to fisheries; otherwise, considering the cost of culturing per se, the total cost of the cultured product will not be competitive with the caught product. Thus, the open sea mariculturist must make maximum use of the fortunate differences between his position and that of the fisherman. The advantages are that he (1) knows where his crop is, (2) has the opportunity to condition its behavior and location, (3) can more or less control his harvesting rates, and (4) can standardize the size of harvested individuals. These advantages must be employed in ways sufficiently effective to cover the costs of culturing in the open sea.

Nutrition

Nutrition can be a spectrum of operations ranging from simply locating the culture in an area where nutrients are abundant, to total, or provisional, feeding with artificially prepared foods. In open sea mariculture, nutrition may turn out to be a matter of finding inexpensive means to stimulate primary productivity in the culture area and then guiding the evolution of a multilevel trophic system from which several crops can be harvested. Until artificial feeds become extremely cheap, their use in mass open sea cultures seems economically doubtful; stimulation and the semicontrol of natural food chains appear the more probable avenues to economic feasibility.

THE TROPHIC CHAIN IN OPEN SEA MARICULTURE

Unless open sea cultures are very tightly controlled (at a very great cost), they seem unlikely to be pure monocultures (single-species cultures). Rather, the creation of an attractive ecological niche (the mariculture system) seems likely to result in the occupation of that niche by more than one species or, more correctly, more than one trophic level. Very likely the mariculturist will be well advised to manage the resulting multilevel trophic systems so as to achieve a polyculture of maximum productivity.

The trophic-level concept was first offered by Lindeman [1]. In it, bio-masses appear as levels of energy storage, and nutritional relationships are depicted as energy flows between storage, or trophic, levels. This view recognizes four basic trophic levels: (1) autotrophic plants (primary production), (2) herbivores, (3) carnivores, and (4) decomposers. Parenthetically, a slight but useful modification to Lindeman's original picture would be the addition of scavengers to the fourth category: decomposers (bacteria and fungi) at the microscopic level and scavengers at the macroscopic level both serve to break down dead organic matter and return it to inorganic forms usable by autotrophic plants. Scavengers are also important food animals, so there seems no reason to ignore them here. The general form of the trophic-level concept is shown in Figure 1.3.

The concept is open to criticism, however. For instance, quite a variety of marine species are omnivorous opportunists that consume plant and animal food, dead or alive, according to its availability. As species and as individuals, then, they occupy more than one trophic level during some or all stages of their life cycles. In other species the adults are carnivorous, and one or more larval stages are herbivorous; in still others the reverse is true. In these latter two categories, species may occupy only one trophic level at a time, even though they may occupy a total of several during various stages of the entire life cycle. Finally, many planktonic protozoans are both autotrophs (photosynthetic) and herbivores. In short, multiple trophic levels are occupied by some species either during a single stage or during successive stages of their life cycles; in contrast, other species are limited in their feeding to fairly narrow food-particle size ranges within a single trophic level. Therefore,

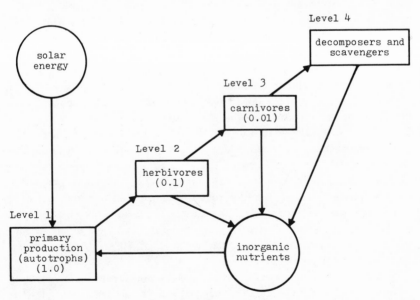

FIGURE 1.3 *Simplified overview: the trophic-level concept of nutritional interrelationships of aquatic organisms.*

although the trophic-level idea is a useful tool for conceptualizing marine food webs, it is seldom adequate for describing them completely.

This *feeding flexibility*, which tends to result in complex and variable marine food webs, undoubtedly has considerable survival value: a natural population capable of accepting only one source of food is inflexibly dependent on that source and vulnerable to variations in its availability, whereas a population capable of accepting a variety of food sources has a kind of "fail-safe" nutritional potential. It is important to recognize, though, that there are optimum natural foods for most species—foods which, if amply available, will produce the maximum attainable survival and growth rates. Thus, a multispecies mariculture system (i.e., a polyculture), to be most effective, very likely would conform rather closely to trophic levels. And the nutritional value that each potential species at one trophic level holds for prospective species at the next higher level would be of paramount importance to system efficiency. Therefore, the trophic concept is valuable in understanding and managing open sea polyculture systems.

"Pure" plants (with trivial exceptions) are obligatory autotrophs. They produce, rather than consume, organic matter. All other trophic levels may be viewed as arithmetic consumers in that the amount of organic material produced is less than the amount consumed. In fact, the ratio of consumption to production from one trophic level to the next usually is taken to be roughly 10 to 1. Thus, a *generalized productivity coefficient* for plants might be assigned as 1, with the resulting generalized coefficient for herbivores being 0.1, for primary carnivores, 0.01, and for secondary carnivores, 0.001. These are only order-of-magnitude approximations, however; actual coefficients for real species may vary widely from the 10 to 1 generality.

The relationship and importance of trophic position to mariculture becomes clear through a comparison of trophic chains in the Peru and California currents. In the California Current, the principal fishery product is the sardine, which occupies the third trophic level. The standing population size of *Sardinops* is estimated to be 750,000 tons, and the estimated yield is 500,000 tons. In the Peru Current, however, the major product is the herbivorous anchovy, *Engraulis,* which is a second-trophic-level organism. The estimated standing population size of *Engraulis* is 8 to 9 million tons and the annual yield is 8 to 9 million tons. In addition, this latter population contributes an estimated 4 million tons to guano production and another 4 million tons to tuna production. Note the significance of a decrease of one in trophic level; both the population size and the annual yield of the second-trophic-level organism are more than a power of 10 greater than for its third-trophic-level counterpart.

There is a subtle point in this that perhaps deserves further comment. Note in our example that the total annual yield of the second trophic level is greater than the size of the standing crop, whereas the third-level organism's annual yield is considerably lower than standing crop size. Standing crop size may be limited by a number of environmental pressures, of which fishing (harvesting) pressure is only one; food availability is another of prime importance. Thus, in a population limited in size by food availability, fishing pressure serves to reduce competition for food resources. If the population is one that can reproduce and grow rapidly, it can quickly replace losses due to fishing. Therefore, it is entirely possible for annual yield to exceed standing crop size when the crop is one whose size is limited by a factor other than fishing pressure and is one having high reproduction and growth rates. These criteria are more likely to be satisfied by lower-trophic-level organisms than by higher ones.

Trophic-Chain Significance

The significance of the trophic-chain concept to mariculture may be stated as follows:

1. Lindeman's trophic-level concept provides useful insight into general energy relationships and relative productivity in the natural marine food web, but is not appropriate as more than a very general basis for a functional model of the whole natural marine web.

2. However, controlled and simplified marine food chains (or polyculture systems) probably can be made to conform reasonably closely to the trophic-level model.

3. For open sea mariculture, overall energy efficiencies and order-of-magnitude productivity losses, as the trophic levels ascend, suggest simplified polycultures which minimize the number of trophic levels involved.

ADDITIONAL CONSTRAINTS ON MARICULTURE OPERATIONS

The measures that can be taken to control the bioeconomics of an open sea mariculture enterprise are constrained not only by the small share of the economic return that can be devoted to controlling trophic compositions and to each of the operations mentioned earlier, but by two other related factors as well: technology and society.

When a given technology has been reduced to standardized and common practice, its application usually tends to be comparatively inexpensive. When technology at, or beyond, the state-of-the-art is employed, its application tends to be comparatively expensive. Open sea mariculture is a new concept. Although much of the fundamental technology that it seems to require is already well advanced, new applications will be required. Thus, there seems little likelihood of inexpensively establishing open sea mariculture enterprises of any scale (in terms of technology expense versus market return), unless major components of the necessary technology can be acquired through means other than outright purchase for which the mariculture enterprise would bear the whole financial burden.

Moreover, one can no longer do entirely as he wishes on any significant scale, even in the open sea. Although one may have the requisite technology and the economic resources to employ it, he faces an array of legal and political constraints. The expense of conforming to these constraints will depend, of course, upon how much modification of the ideal operational design is necessary. Although some benefits may be derived from moving a mariculture enterprise offshore, where coastal-zone conflicts presumably are ameliorated, the blessing is likely to be mixed; for the farther one moves away from the overlapping jurisdictions of local, state, and federal governments, the farther one moves into jurisdictional vagaries of a more international nature. Thus, although a disproportionate amount of time and money may be required to satisfy the layers of regulations affecting the use of near-shore waters, it may be years before a mariculturist will be able to discern with any confidence what requirements he must satisfy in offshore waters and, possibly more important, what levels of legal protection his enterprise may be afforded (see Chapter 3).

CONCLUSIONS

The bioeconomic perspective on open sea mariculture reveals a difficult optimization problem. The bioenergetics of a multispecies marine ecosystem in most cases will require management for maximum efficiency and productivity; yet economic support for each management operation can be only a fraction of the total market return on productivity. At the same time, all operations must be conducted within a combination of legal, political, and technological constraints, which tend to increase their cost. The implications of this perspective are that open sea mariculture will be economically feasible only when some combination of two conditions prevails. The first condition is a high market value for the products. The second is a level of scientific and technological advancement that allows maximum exploitation of natural energy fluxes and other natural and human-generated phenomena in the offshore realm.

REFERENCES

1. Lindeman, R. L. 1964. The trophic–dynamic aspect of ecology. In W. Hazen, *Readings in population and community ecology*. W. B. Saunders Company, Philadelphia, pp. 206–266.

2

OPEN SEA MARICULTURE IN UTILITARIAN PERSPECTIVE

J. A. Hanson

Is it likely, by any stretch of the imagination, to be worthwhile accepting the bioeconomic challenge of open sea mariculture outlined in the preceding chapter? To answer this question, it is necessary to decide whether improved nutritional levels for existing and larger human populations are desirable, and, if so, to examine the prospects offered by the alternative systems of large-scale food production—agriculture and fisheries. (This is not to imply, of course, that the three systems are mutually exclusive, but rather to evaluate their relative capabilities to meet the total need.)

The first question is valid, even though superficially it appears cold-blooded, because humanity's basic disease is the rapid expansion of the human population: unmet human nutritional requirements in effect are only one of the many classes of symptoms of the population disease. Gregory Bateson [1] might contend that relieving the food-shortage symptom simply allows the population disease to continue expanding in chronic form until absolute limits are reached, and the entire civilization crosses one or more stress thresholds into a realm of critical instability. If this view is valid, any approach to increased nutritional capacity is undesirable in the long run.

But an opposing position is offered by the medical profession, which holds that symptomatic treatment is ethical when the disease is clearly terminal, when the disease will cure itself, or when symptomatic treatment assists in sustaining the patient long enough to allow other measures to deal with the disease. If this view prevails, improved nutritional capacities clearly are desirable. Rather than detour into this argument any further, suffice it to say that there is probably little real choice in the matter: so long as massive nutritional problems exist, societies are going to try to solve them

The remainder of this chapter treats facets of the second question: "How does open sea mariculture seem to rate against agriculture and fisheries as a supplier of food and as an economic enterprise for the future?"

AQUACULTURE'S RECORD TO DATE

Ancient Chinese literature mentions the management of fresh-water ponds as early as 1100 B.C. [2]. So it seems safe to conclude that the aquaculture

17

idea is not a new one. Yet aquaculture's impact on societies over the succeeding thousands of years appears to have been significantly less than that of the comparatively recent invention of the can and can opener.

Although societies have been culturing edible aquatic organisms for tens of centuries, aquaculture appears to have remained as an augmentation to agriculture rather than to have achieved major industry status; and this has occurred even in societies with traditional tastes for seafoods and seemingly undeniable mass nutritional needs. Probably this has occurred for two reasons. First, because beginning agriculture attempts were terrestrial and involved only seeding or stock restraint (or attraction) and feeding, they were very likely more successful than the first aquaculture attempts; and subsequently, in technological societies, agriculture fed on its own success so long as land and energy were available. Second, until recently the fishery resources of the seas were considered practically limitless. Thus, traditionalism and continual refinement carried fisheries and agriculture toward increasing productivity in the industrial nations, while the less technological nations, without external assistance, remained incapable of developing aquaculture on a significant scale.

In this vein, Bardach and Ryther [3] have mentioned the relative difficulty of working in the water, as opposed to land, and the fact that many marine organisms attractive for culture have larval stages that are difficult to rear. So it seems likely that early "technological" difficulties with aquaculture, combined with the heretofore high productivities of agriculture and fisheries, have given the latter two a head start that persists.

The thesis—that due to this head start by agriculture and fisheries aquaculture will not flourish until these two approach their productivity limits—seems to be supported by the recent surge of aquaculture activity in Japan. The Japanese are now the undisputed world leaders in aquaculture (not that competition is all that intense), producing yellowtail, *Seriola quinqueradiata;* the shrimp *Penaeus japonicus;* oysters, salmonids, other invertebrates, and freshwater forms at annual volumes running well into the millions of dollars. All this has been accomplished rather recently by combined government and private capital, stimulated by the existence of a well-established multimillion dollar annual market.

But if aquaculture has not flowered as a major industry throughout the world, neither has it expired; nor has it failed to show definite symptoms of development. The culture of mullet, milkfish, and others continues throughout Asia, at what might best be termed a subsistence farming level, to be sure, but it nevertheless continues. Great Britain is working mostly with marine flatfish and the indigenous oyster, *Ostrea edulis,* in order to improve techniques. A great deal of attention is being paid by the British to the engineering aspects of operating in the marine environment.

The United States has taken some advantage of its vast natural marine fauna

and a great deal of work has been accomplished with many species, including pompano, *Trachinotus carolinus* [4]; mullet, *Mugil cephalus* [5]; salmonids [6]; shrimp, *Penaeus* sp. [7]; oysters, *Crassostrea* spp. [8, 9]; clams, *Mercenaria* and *Mya* spp. [10–12]; and others. Nevertheless, with the exception of oysters and salmonids, most developments have not passed the laboratory level and, for many potentially commercial forms, little has yet been accomplished in the engineering and operation of facilities for hatchery production and impoundment on commercial scales. Pilot-scale tests are sorely needed, but government financing for these is limited, and the risk is often too high for the private sector.

The USSR has concentrated mostly on the sturgeon, *Acipenser* spp., and the carp, *Cyprinus carpio*, in fresh water and has ventured little into the marine field. The West European countries of Spain, Belgium, Holland, and Norway have preferred to rely on their existing shellfish interests, but France has recently launched a concentrated effort. Israel is the most aquaculturally advanced Middle Eastern nation, working with mullet and a variety of freshwater species.

One pattern that appears to emerge from this brief look backward emphasizes how critical brood stock and artificial breeding are. Where captive breeding has met with success (as with oysters, mussels, and salmonids), aquaculture shows strong growth signs. Where culture must rely on the capture of juveniles from natural stocks (as with yellowtail in Japan, mullet and milkfish in Asia, and pompano in the United States), the industry faces obvious growth limits, in addition to being in clear competition with natural-stock fisheries.

At least two general conclusions seem warranted by the foregoing: (1) aquaculture's economic performance so far has not been such that it should be expected to serve as an enticement to major investment; (2) an absolute prerequisite to commercial-scale aquaculture is controlled reproduction. No successful agriculture is based upon the capture of juveniles, nor is any truly successful aquaculture venture likely to be. The penaeid shrimp industry in the Gulf states (which has yet to prove profitable) is a possible exception; but as this industry expands, the price of captured juveniles and naturally gravid females is soaring. Clearly, even though the industry is not yet extensive, reliance on naturally occurring brood stocks is already becoming a limitation.

The conclusion that aquaculture in any form so far has not been competitive with either agriculture or the fishing industry in any significant way is inescapable, which leads to the supposition that aquaculture in general is likely to become competitive only when one or both of two conditions prevail. The first condition is a level of scientific and technological advancement adequate to support efficient and economical mariculture enterprises. The second is a decline in the efficiency and economy with which needed nutrition can be produced by other means, which would be accompanied by a rise in all food values and prices.

THE FUTURE OF AGRICULTURE

Medium projections by the United Nations indicate a world population slightly in excess of 6 billion people by the beginning of the 21st century [13]. Estimates of the total world terrestrial surface area are about 32 billion acres; this means a land-availability average of slightly more than 5 acres per capita in the year 2000. But only about one fourth of this, or 1.25 acres per capita, is considered potentially arable, with another fourth suitable for stock grazing. The rest is wasteland, mountains, forest, and tundra. Interestingly, this is about the per capita amount of land that feeds humanity today, and at least half the present world population lives in the shadow of starvation.

By 2050, the turn-of-the-century population could double again to 12 billion, and possibly rise even higher. This would leave only a little more than a half-acre each of arable and grazing land per capita, assuming that this burgeoning population did not cover agricultural and grazing lands as it grew. The mere existence of more people clearly reduces per capita agricultural land availability in a finite world; a world population of the size predicted would reduce it to alarmingly small numbers.

The purely arithmetic population effect is compounded by the tendencies of technological societies to usurp prime agricultural lands for urban industrial activities. When higher population levels are accompanied by increasing technological levels, per capita losses of agricultural lands tend to increase. Moreover, a high level of technological advancement is required for maximum agricultural productivity. Small acreages per capita cannot support human nutritional needs unless productivity per acre is maximized with technology. So the half-acre number is optimistic at best.

To compound the effect further, agriculture thus far is an industry based on fossil fuels. Odum [14] shows clearly that 19th- and 20th-century increases in unit area agricultural yield do not result so much from increases in the efficiency of solar energy utilization as from a fossil-fuel subsidy. The form of this subsidy is not only in the power used to construct and to drive cultivating, irrigating, fertilizing, pest control, harvesting, and processing machinery; it is also in the extensive use of petrochemicals in pesticides and fertilizers themselves.

As it stands today, then, modern agriculture is vitally dependent on two rapidly diminishing and nonrenewable resources—land and fossil fuels. As these commodities ever more rapidly grow scarce, their relative costs will rise exponentially, and the cost of land-based food production will rise even more sharply. The rise has already begun in spite of efforts to effect economies of scale through mass single-crop production employing highly advanced technology; and, because of the impending exhaustion of fossil-fuel supplies, it will continue even though populations may level off. In short, food derived

only from agriculture as we know it today appears insufficient to support expanding populations; it will continue to rise in real cost, and the rise is likely to be exponential.

THE FUTURE OF MARINE FISHERIES

By way of introduction, only marine fisheries are considered; freshwater fisheries are assumed to be of negligible importance in terms of the scale being considered here. Also, the figures given for marine fisheries are in terms of maximum sustainable yields.

Ricker [13] has stated that by the year 2000, total marine-fisheries yield will be on the order of 150 to 160 million metric tons. Ricker goes on to say that this yield would supply only about 30 percent of the expected world human protein requirement in 2000 A.D., and only about 3 percent of its caloric requirement. Chapman, in a later paper [15], estimates marine-fisheries yield at about 400 million metric tons in 2000 A.D. The disparity is not as significant as it appears since, as Chapman points out, the first figure includes only the larger (usually high-trophic-level) organisms, the "preferred" seafoods, whereas the second figure assumes the harvesting also of smaller (usually low-trophic-level) organisms that customarily are considered less desirable, but which are equally nutritious.

There is considerable debate concerning whether or not small organisms, scattered through the oceans, can be harvested economically, and, if so, whether they can be converted to a form that human populations will consume. Moreover, fishing, like agriculture, is fossil-fuel-dependent and so will doubtlessly become more expensive as time goes on, at least until it can convert to other fuel forms. Further, Chapman points out the problems of international politics that surround marine fisheries. Although he is optimistic over their eventual solution prior to sustained-yield population depletions, his data could easily lead to pessimism too. Nevertheless, for present purposes, Chapman's estimates are taken.

Applying Ricker's 30 percent and 3 percent proportions to Chapman's higher yield estimates leads to the conclusion that intensification of fisheries could lead to a situation in which fisheries alone could supply 75 percent of the world human protein requirement by 2000 A.D., and about 7.5 percent of the caloric requirement. However, Chapman implies that this is likely an upper limit, and many authorities think it far too high. Moreover, if human populations refuse to level off when this upper limit is reached, the ratios will gradually become less favorable. All this seems to imply an early emphasis on fisheries, but a longer-range view toward mariculture. If the optimistic estimates prove too optimistic, perhaps the mariculture emphasis should not be so long range after all.

SCIENTIFIC AND TECHNOLOGICAL REQUIREMENTS FOR MARICULTURE DEVELOPMENT

If open sea mariculture is ever to enter the realm of practicality, science and technology must be applied toward two basic objectives. The first is to maximize bioenergetic efficiencies in mariculture systems; the second is to maximize the efficiencies and cost effectiveness of mariculture operations per se. Included in the category of bioenergetic efficiencies are such research-and-development avenues as hatchery practice, nutrition, control of stock locations, and control of the biological environment. In the operational-efficiency and cost-effectiveness category lie such research-and-development avenues as conservative energy, offshore platforms, environmental prediction, harvesting, and processing.

Research and Development in Bioenergetic Efficiencies

Nutrition. Our analysis of nutrition indicates that pumping of deep nutrient-rich water (Intermediate Water) to the surface and holding it there by reducing its density or by mechanical means is, in the long run, the most attractive nutritional development avenue for open sea mariculture. Schemes for the utilization of urban wastes are second in priority, since this nutrient source frequently suffers from contamination problems. Perhaps urban wastes would best be used to replenish the nutrients in Intermediate Waters, where physical and biological processes could, over time, deal with contaminants while converting organic compounds to inorganic. Direct feeding of raw or blended feeds to high-trophic organisms is certainly a third-priority candidate. But efficiency losses through a chain which involves acquisition, processing, packaging, transportation, and dispersal indicate that this last approach is likely to be expensive and limited in the scope and scale of its application. The results of the nutrition analysis clearly imply open sea polyculture systems based upon primary production. Pilot experiments with this approach are underway, as discussed in Chapter 7, but commercial-scale application is several years in the future at best.

Spawning and Larval Growth. The development of practical hatchery technology is essentially only in its early stages. But results to date offer solid encouragement that any species can be induced to spawn artificially. Rearing of larvae to the postlarval stage, ready for grow-out, appears to be the major roadblock. Most marine species pass through several larval stages, each of which may have particular nutritional and environmental requirements. Research in this area, to date, indicates that hatchery problems will impact crop selection for some time to come; that the hatchery issue is of prime importance; and that future hatchery operations will be, at best, capital-intensive—and probably labor-intensive, too. In summary, hatchery research and development should be directed toward mass production of postlarval juveniles belonging to species that have the following characteristics: (1) favorable

energy-conversion rates (low trophic level), (2) high market value, and (3) practical to rear in an open sea enterprise.

Biological Control. Controls for disease, parasitism, and predation, as they could apply to large-scale open sea mariculture, are best characterized as being in early experimental stages. Preventions and treatments achieved to date have been oriented to marine aquaria and small-scale mariculture, and so tend to be labor-intensive, as is pointed out in Chapter 9. This problem appears to share first priority with the hatchery problems.

Physical Control. Locations of marine stocks in the open sea can be controlled through mechanical means with existing technology. But control is likely to be expensive with the tools at hand, except in the case of sessile animals. Where pelagic species are concerned, research and development to achieve inexpensive, survivable, and maintainable mechanical enclosures are indicated, as are efforts to exploit fortuitous behavioral characteristics of stock organisms as mechanisms for controlling their location and concentration.

Research and Development in Operational Efficiencies and Cost Effectiveness

Harvest Techniques. Thanks to the fishing industry, technology for harvesting marine organisms is well advanced. In some harvesting practices a relatively high degree of mechanization has been achieved; others remain labor-intensive. Nevertheless, for obvious reasons the technology has not yet been attuned to certain advantages (e.g., uniform size, set location, and behavioral conditioning) that mariculture can offer. The implications of the differences in harvested crops are explored in Chapters 10 and 11. Open sea mariculture might be able to utilize present-day harvesting technology as is; but improvements that would take advantage of the differences between the fishing and the mariculture situations would improve mariculture harvesting efficiencies significantly.

Offshore Platforms. Open sea platforms from which mariculture can be conducted exist in the form of natural atolls and islands as well as in man-made form. Man-made platforms can be either floating or bottom mounted, and may or may not pierce the surface. All man-made open sea platforms are expensive, although their costs will decline somewhat as technology improves and production rates increase. In spite of high platform costs, petroleum and power production are moving to sea, and other industry is likely to do so in the future as shoreside pressures and restrictions increase. Open sea mariculture should look to atolls, and to sharing man-made platforms with other industry, in order to acquire platforms at a price it can afford. There is little likelihood that the profits from open sea mariculture can support unilaterally the cost of man-made open sea platforms until the market values of the products reach very high levels compared with the cost of platform construction and maintenance.

Conservative Energy. The impending exhaustion of fossil fuel supplies is a major factor in estimates of the increasing costs of agriculture and fisheries products. It therefore would be unwise for open sea mariculture to develop a dependency on the same diminishing resources. This clearly suggests that the evolution of open sea mariculture should converge with the evolution of alternative energy sources, such as sea-based nuclear power plants (if radiation hazards can be minimized); geothermal power plants; various ocean-based, solar-energy power plant possibilities (i.e., wind, ocean thermal gradients, currents, and direct solar radiation); and, where appropriate, the use of hydrogen fuel as a long-range possibility. The potential for this avenue of research and development is expanded upon in Chapter 13. It is certainly true that early pilot projects in open sea mariculture might find it most economical to employ fossil fuels; but long-range design concepts should be associated with nuclear and conservative energy forms.

Environmental Prediction. The oceans are four-dimensional in the sense that all three of their geometric dimensions are of significant proportions and that they are, in addition, dynamic fluids. This four-dimensionality, coupled with sheer size, makes the oceans the most effective dissipators of matter and energy on earth. Nevertheless, localized contamination does occur and persist when the source is constant or repetitive. Furthermore, localized contamination tends to spread throughout the oceans—albeit with dilution factors that usually reduce contamination to innocuous levels. These same characteristics render control of the physical parameters within an open sea culture area well-nigh impossible. Consequently, open sea mariculture must find oceanic areas to its liking, . . . rather than attempt significant modifications to geographically convenient areas that are not to its liking.

Desirable site characteristics include freedom from extreme winds, waves, and currents, in addition to freedom from chemical contaminants, and adequate light and heat. All this implies that improvement is needed in the ability to predict marine environmental extremes and to predict evolutionary drifts in the characteristics of a localized area that is subjected to human-induced changes as well. Improvements in these capacities also will be important to other marine activities.

Recapitulation. Clearly, the science and technology picture reveals that most of the conceivable forms of open sea mariculture on a practical, commercial scale must await new developments and discoveries. However, for offshore work based on man-made platforms, it can be stated that sessile shellfish forms doubtlessly represent the brightest near-term prospect. Their location can be controlled with relative ease, they feed low on the trophic scale, hatchery technology is fairly well advanced for several species, and harvesting and processing do not pose major problems. Problems with disease, parasitism, and predation in the open sea realm are unassessed, but should be no worse, and might well be less severe, than in nearshore realms. Mariculture associated with atolls is a somewhat different short-range question. Although sessile organisms also appear to be the most attractive near-term candidates for large commercial cultures on atolls, smaller-scale commercial cultures of a variety of forms for local consumption by atoll societies and visitors are a distinct additional attraction.

THE LONG-RANGE ECONOMIC PICTURE

In earlier sections of this chapter we described why the costs of food produced by means other than mariculture are likely to continue to rise. Not so apparent is why power costs also are likely to continue rising. The reasons are simply that the inexpensive power sources quite naturally were exploited first by humanity—initially wood, then hydropower and fossil fuels. Although it is possible that nuclear energy may someday become a less expensive power resource than today's fossil fuels, this seems unlikely; increasingly restrictive safety and environmental protection regulations can be expected to offset any savings achieved by increased efficiencies in nuclear power production.

Since solar-energy sources represent "low-density" energy as compared with nuclear and fossil-fuel sources, it follows that very large capital expenditures will be required to build machines that can tap solar energy on large scales. Thus, it is unlikely that these energy sources will produce inexpensive power. In short, the power of the future will be more expensive than power today. Additionally, for the reasons explained in Chapter 13, it will be increasingly expensive to transport, and so will be least expensive near the source of its generation, just as power is today. In Chapter 13 we also present the thesis that tomorrow's power production plants are likely to be ocean-based; consequently, they will offer attractive ecological niches for open sea mariculture and could assist mariculture in bringing its production costs below those of agricultural and fishery products.

Also, over the long haul mariculture can expect to reduce its operating costs in proportion to its level of success. The more pervasive the practice of open sea mariculture becomes, the more routine its procedures and the production of its equipment and facilities will become. This will be in addition to cost reductions achieved through inevitable technological refinements and major improvements.

Therefore, the conceptual graph in Figure 2.1 may well be a real future possibility.

CONCLUSIONS

In this chapter we have explored the potential utility of open sea mariculture as a food producer for future societies. It has been shown that agriculture and fisheries are very likely nearing the limits of their practical

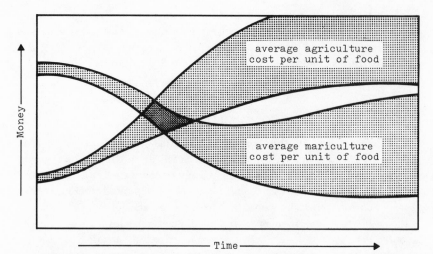

FIGURE 2.1 *Postulated dynamics of agriculture and mariculture costs. It is suggested that the costs of food derived from agriculture will continue to rise as land available for agriculture decreases with expanding populations and increasing industrialization, and as fossil-fuel power becomes more precious. If population levels off, costs of agriculturally derived food will still rise as a result of approaching fossil-fuel exhaustion and the accompanying necessity to convert to other, less economical, energy sources, sources of fertilizers, and means of pest control. It is also suggested that because of the arable space available mariculture costs, by comparison, can decline markedly if capital is directed toward developing the necessary technology—particularly with respect to utilization of conservative energy sources and waste recycling. Mariculture costs might rise subsequently with general inflation. But they might also continue to decline with technological refinements. The expanding width of the curves, of course, represents increasing uncertainty with time. Note that this graph is conceptual rather than quantitative and that its spatial relationships have been selected to illustrate the concept rather than to predict quantitative relationships.*

production capacities, that human populations, already inadequately fed on an overall worldwide basis, are continuing to increase their numbers, and that open sea mariculture is, on the whole, some years away from economic practicality. Mariculture now requires advances in scientific and technological realms to make the prices of its products competitive with the prices of products derived from agriculture and fisheries. However, as time goes on the prices of products derived from the two classical food sources show every indication of continuing their past exponential climb, and this, combined with developments in open sea mariculture that appear quite possible, could make mariculture economically competitive within a few decades.

REFERENCES

1. Bateson, G. 1972. *Steps to an ecology of mind.* Ballantine Books, Inc., New York.

2. Linn, S. Y. 1940. Fish culture in ponds in the New Territories of Hong Kong. *Fisheries Res. Sta.,* Vol. 1, No. 2.

3. Ryther, J. H., and J. E. Bardach. 1968. The status and potential of aquaculture, particularly invertebrate and algae culture. *U.S. Dept. Commerce Clearinghouse PBI 77-261.* 261 pp.

4. Finucane, J. H. 1970. Pompano mariculture in Florida. *Amer. Fish Farmer 1*(4):5–10.

5. Shehadeh, Z. H., and K. S. Norris. 1972. The grey mullet (*Mugil cephalus* L.): induced breeding and larval rearing research 1970–1972. *Oceanic Inst.* (Waimanalo, Hawaii) *Tech. Rept. OI-72-76-1.*

6. Huet, M. 1970. Breeding and cultivation of salmonids. In *Textbook of fish culture.* Fishing News (Books) Ltd., London, pp. 59–110.

7. Mock, C. R., and M. A. Murphy. 1971. Techniques for raising penaeid shrimp from the egg to postlarvae. In *Proc. First Ann. Workshop, World Mariculture Soc.* (Baton Rouge, La., Feb. 1970), pp. 143–156.

8. Matthiesen, G. C. 1970. A review of oyster culture and the oyster industry in North America. *Woods Hole Oceanographic Inst. Contr. 2528.*

9. Joyce, E. A., Jr. 1972. A partial bibliography of oysters, with annotations. *Florida Dept. Nat. Resources, Marine Res. Lab.* (St. Petersburg, Fla.) *Spec. Sci. Rept.* 34. 845 pp.

10. Loosanoff, V. L., and H. C. Davis. 1963. Rearing of bivalve mollusks. *Advan. Marine Biol. 1*:1–136.

11. Wallace, D. E. 1967. Shellfish studies in Maine. Paper delivered to the Natl. Shellfisheries Assoc. annual meeting, July 1967, Boston, Mass.

12. Zuraw, E. A., D. E. Leone, and C. A. Griscom. 1971. *Ecology of molluscs and the culture of* Mya arenaria. Marine Sci. Sec., Res. Develop. Dept., Electric Boat Div., General Dynamics, Inc., New London, Conn.

13. Ricker, W. E. 1969. Food from the sea. In Natl. Acad. Sci.–Natl. Res. Council, *Resources and man.* W. H. Freeman and Company, Publishers, San Francisco, Chap. 5.

14. Odum, H. T. 1971. *Environment, power and society.* Wiley-Interscience, New York.

15. Chapman, W. M. 1969. Some problems and prospects for the harvest of living marine resources to the year 2000. *J. Marine Biol. Assoc. India 11*(1–2):1–39.

Additional Readings

Avault, J. W., Jr., ed. 1971. *Proc. Second Ann. Workshop, World Mariculture Soc.* (Galveston, Texas). Louisiana State Univ., Baton Rouge, La.

Brown, L. R., and G. W. Finsterbusch. 1972. *Man and his environment.* Harper & Row, Publishers, New York.

Escher, W. J. D. 1972. *Nonfossil synthetic fuels.* Rept. to the Synthetic Fuels

Subgroup of the Comm. on Energy and Res. and Develop. Goals, Fed. Council on Sci. and Technol.

Idyll, C. P. 1973. The anchovy crisis. *Scientific American* (June), pp. 22–28.

Nash, C. E. 1970. Marine fish farming: I and II. *Marine Pollution Bull. (U.K.)* *1*(1):5–6; *1*(2):28–30.

White Fish Authority. 1970. *Annual Rept.* H.M. Stationery Office, London.

3

LEGAL AND POLITICAL PERSPECTIVES ON OPEN SEA MARICULTURE

C. C. Hanson and J. M. Collier
with
J. P. Craven, G. W. Grimes
and G.M. Sheets

Examining mariculture in the context of the legal and political environment as it has developed to date proves to be a frustrating exercise. A systems approach is incompatible with this essentially humanistic area. The evolution of the often ponderous legal structure is constantly molded by political realities and national interests, pressured by economic concerns, and seldom subject to the hard discipline of efficiency that biological evolution imposes. Laws evolve as a result of social customs and in cautious response to perceived need; they therefore tend to lag well behind technology, with an ever-increasing gap as the pace of technological evolution quickens. Where needs have not yet been perceived and where social custom has not kept pace with man's mushrooming technical capabilities, the law is a sleeping servant heedless of the inexorably advancing hours.

In this chapter we look at the existing legal framework as it may relate to our topic, point out the gray areas where it is still nebulous or nonexistent, and then discuss the practical realities of mariculture operation in the context of identifiable constraints.

ORIGINS OF MARITIME LAW

Let us set the stage for the current dilemma of legal uncertainty with a brief prologue reviewing the origins of maritime law governing the "high seas."

A loosely knit body of principles for controlling activities on and in the oceans developed over centuries of interaction between nations. Modern unification of the laws of the sea has been aided greatly by the existence of a number of resulting "maritime codes," which, although not promulgated by any sovereign authority, gradually assumed a binding character freely recognized by merchants and traders of all nations [1].

Early Maritime Codes

The earliest open-ocean expedition of record was ventured by an Egyptian king around 2900 B.C. The Phoenicians followed the Egyptians into the open seas, and the Greeks came next. The Phoenicians initiated the earliest known body of sea law, called the Rhodian Code, about 300 to 200 B.C. [2]. We can trace the development of maritime law through adoption of private customary codes by the following benchmarks.

- *Rhodian Sea Law.* The principles of the Rhodian Code of the Phoenicians were accepted by both Greeks and Romans, and held in memory for a thousand years. A compilation called the *Rhodian Sea Law* was prepared sometime during the later Roman Empire, between the 7th and 9th centuries A.D., and this was used in the Mediterranean for many years [3, 4].
- The *Basilika*. Belonging to the 7th century, the *Basilika* was a code of Byzantine law regulating the commerce of the Levant. Unlike succeeding codes, it contained positive law and derived its authority from the emperor rather than the consent of the merchants. The general principles in the *Basilika* were supplemented by special instances derived from *Rhodian Sea Law* [5].
- The *Tabula Amalfitana*. This "code of the Amalfi" dates from the 10th century and contains the maritime usages followed by the town of Amalfi, Italy. The decisions of the Consuls of the Sea of the City of Trani, on the shores of the Adriatic Gulf, also include the original sources of customary maritime law. These purport to have been compiled in 1063 by the Consuls of the Corporation of Navigators at Trani [5].
- *The Assizes of Jerusalem.* The sea law of the Levant (*Basilika*) was further developed by the Crusaders, who set up courts of their own to try disputes between the traders and seamen they found along the Mediterranean coasts. Their procedure of trial by battle for civil suits being inappropriate, they drew up codes known as the *Assizes of Jerusalem*. These were based on the customs of the merchants, and their application was entrusted to counsuls or sea magistrates, who were appointed in commercial ports such as Genoa, Venice, and Marseilles [6].
- *The Rolls of Oleron.* As an ordered code in a period devoid of maritime legislation on the Atlantic seaboard, the rise and spread of the *Rolls of Oleron* marked a distinct advance. Most probably promulgated in 1160 by Queen Eleanor of Aquitaine, the *Rolls* were so faithful a record of customs observed in the trade between the Atlantic ports of France that they were immediately successful and were gradually adopted by all nations of western Europe, influencing the customs of the Mediterranean as well. Richard I ordered the *Rolls* translated in England, where the English Admiralty Court regarded them as the edict of an English king. Soon becoming the recognized code in maritime matters, the *Rolls* are referenced in court records of the 14th century and the articles of the code appear in subsequent legal compilations of Flanders and Bruges [6].
- *The Wisby Sea Laws.* Published in the Saxon language at Copenhagen in 1505, the laws of Wisby, or sea laws of Gothland, further reflected the

influence of the *Rolls of Oleron*. The Wisby Code consisted of three parts: a transcription of the *Rolls*, extracts from the Laws of Amsterdam, and extracts from the statutes of the Hanseatic city of Lübeck [6].

- *The Hanseatic Code.* Another collection appearing on the Baltic was the statutory code enacted by the Hanseatic League at the beginning of the 17th century. A copy of the revised code of 1614, under the title *Jus Maritimum Hanseaticum*, was published at Hamburg in 1667 in the original German and a Latin translation. Even then this code was incomplete, consisting merely of the special customs and statutes of particular cities, and was largely based on the traditional background of Roman law [6].

- *The Black Book of the Admiralty.* In England, the most notable collection of sea law was contained in the ancient *Black Book of the Admiralty*, probably commenced during the reign of Edward III and continued under Richard II and Henry IV. This originated in 1375 as a collection of articles and rules concerning seafaring, compiled by a symposium of authorities, which was originally written in old French and later translated into Latin as *De officio admiralitatis angliae*. Several of these articles were based on the *Rolls of Oleron* and were subsequently included in the *Black Book of the Admiralty*, which was presumably intended to provide a sort of manual or collection of the rules of maritime law for use in the Admiral's Court [5, 7, 8].

- The *Consulato del Mare*. While England, France, and Germany were publishing versions of maritime codes in the 14th to 17th centuries, the Mediterranean trading centers were, in the 14th century, recovering from the effects of the Roman Empire's collapse and the ravages of Roman and Saracen pirates. Vigorous commerce was regulated by a great variety of local statutes written in Latin, originated by the consulandos, or self-governing organizations of merchants in the great ports, and administered in the courts by elected judge-consuls. In the 14th century, then, all the existing local statutes, customs, and usages of the Mediterranean were collected, translated into Catalan, and issued probably at Barcelona. This collection, titled the *Consulato del Mare*, came to be recognized by common consent as law throughout the Mediterranean, where it reigned supreme through the Middle Ages. With the advent of sovereign states, national legislation superseded the customary laws of the sea, although often incorporating many of the former rules. This same fate overtook the *Guidon de la Mer*, an early 17th-century collection, which was published at Rouen to assist the consuls in administering jurisdiction over insurance business, and which formed the basis of the French Ordnance enacted by Louis IV in 1681 [6].

Thus we see that the Romans adopted the sea laws of the Rhodians; the *Rolls of Oleron* passed from France to Spain, England, Holland, and the Baltic; and the laws laid down in the *Consulato del Mare* were followed throughout the Mediterranean. Taken altogether, these codes reveal the customary maritime law in an age of expanding commerce before the advent of the sovereign state. In spite of the fact that none of the codes per se became part of the English law, they contain many valuable principles and statements of marine practice that were used by the judges of the English Court of Admiralty in formulating the principles and practices of their court [1].

Sovereign Claims

Although the period that marked the development of maritime law by adoption of private customary codes did not end until the 17th century, when the age of national legislation began, the high seas have not always been free of sovereign claims. Real claims to sovereignty over parts of the open seas began in the second half of the Middle Ages, when advances in ship technology permitted Portugal and Spain to become maritime nations and passage to the New World was achieved in 1492. Portugal claimed sovereignty over the whole of the Indian Ocean and of the Atlantic south of Morocco, and Spain made the same claim over the Pacific and the Gulf of Mexico. These claims were incorporated in the Treaty of Tordesillas, 1494, and were based on two papal bulls promulgated by Alexander VI in 1493, which divided the New World between these two powers. Sweden and Denmark claimed sovereignty over the Baltic, and Great Britain over the Narrow Seas, the North Sea, and the Atlantic from the North Cape to Cape Finisterre.

Freedom of the Seas

The sovereign claims were more or less successfully asserted for several hundred years. However, eventually England's technological revolution provided her with many ships capable of distant voyages and powerful enough to defend themselves. When, in 1580, the Spanish ambassador Mendoza lodged a complaint with Queen Elizabeth against Sir Walter Drake for having made his famous voyage to the Pacific, the Queen reacted by declaring the seas rightfully free to all equipped to exploit them, by reason of nature and regard for the public use [9]. The British fleet proceeded to exercise that freedom of passage, often with armed might. Two decades later the Dutch trading interests challenged the legality of the Portuguese claim to exclusive use of a portion of the high seas. Their spokesman, Grotius, reflected Queen Elizabeth's thoughts on the subject in his short treatise, Mare Liberum, in 1609. Ironically, James I of England issued, in 1609, a proclamation excluding Dutch fishermen from operating off the shores of England, and Grotius' treatise prompted several English lawyers to reply and defend James I's proclamation [10].

Grotius went so far as to argue the right of free access to ports by foreign vessels, unhampered by local sovereign restraints. Free access to ports did not become a reality, but due to inability to police the ocean and lack of the establishment of a territorial limit around the coasts, freedom of the seas was eventually restored; witness: "The historic function of the law of the sea has long been recognized as that of protecting and balancing the common interest . . . of all the peoples in the use and enjoyment of the oceans, while rejecting all egocentric assertions of special interests in contravention of general community interest" [11]. The age of national legislation, then, began in the 17th century, creating an ever-increasing volume of statutory laws in all the principal maritime nations.

The highlights of nearly a century of international maritime agreements, such as they were, and their antecedents in the customary codes of ancient times, which we have just reviewed, illustrate the point that laws evolve in response to the needs and demands of society. The earliest maritime codes principally concerned navigation and commerce in coastal waters and the relatively protected Mediterranean Sea. Codification of maritime customs spread north to the western European countries of the Atlantic seaboard as these nations developed in the 12th to 14th centuries. Codes regulating piracy developed when advancing technology endowed ships with the capability to overtake and board one another and piracy became an economic and political weapon, as well as a depredation inimical to the interests of the merchants, in the 13th century. Again, as long-range navigation into distant waters grew technically and economically feasible at the end of the 15th century, various nations claimed sovereignty, enforced by treaty and regulation, over specific areas of the oceans and seas. In the course of a couple of centuries, then, the developing technical, political, and economic strength of first England and later other maritime nations forced legal restoration of the freedom of the high seas.

So we find that, in the course of events, concepts of rights and freedoms to exploit the organic and mineral resources of the seas were introduced relatively recently, as the technology for exploitation and the corresponding economic need built up pressure. These concepts assume importance as we proceed now to examine the present stages of legal evolution.

EVOLVING MARITIME LEGAL STRUCTURE

At this point in sketching the legal context for mariculture, the broad principle of freedom of the seas from sovereign claims undergoes modification. The modifiers, specifically, are the concepts of territorial waters, the continental shelf, and the living resources of the sea.

In the introduction to this book we spoke of the open sea environment as comprising all unprotected ocean waters without respect to their proximity to or remoteness from coastal land. (In this context, "unprotected" means exposed to the surge and full force of ocean currents and wave action.) It is important to keep this definition in mind as we explore the constraints upon the use of territorial waters and the difficulties inherent in enforcing regulations on the high seas.

World councils are once again in the process of dividing up the oceanic pie, in a manner of speaking. An Executive Order dealing with the continental shelves was issued by President Truman in 1945 [12]. The Truman Proclamation, in brief, stated that the United States regards the natural resources of the subsoil and seabed of the continental shelf contiguous to its coasts as subject to its jurisdiction and control. Where the continental shelf is shared with or extends to the shores of an adjacent state, the boundary would be negotiated. This does not affect the character of the waters above the shelf as high seas. The proclamation marked the beginning of a continuing controversy over territorial jurisdiction in the ocean. The order was not intrinsically the trigger; it serves, rather, as a benchmark for the accelerating development of oceanic law in response to technological advances that were outstripping it, thus creating a new situation with which existing law was unprepared to deal.

Many nations fear, with some justification, that they may have no voice in decisions affecting exploitation of one of the world's largest and richest natural resources—the ocean. Some are seizing the initiative by making diverse claims to sovereignty over territories and resources. It would be unrealistic not to recognize that some limits to use of the oceans, and some boundaries or claims of exclusive rights, will in the future almost certainly be imposed upon presently unclaimed high-seas areas. At this time, however, the validity, and even the extent, of these claims is highly ambiguous. Some nations claim exclusive or preemptive rights to fishing and mineral exploitation in waters adjacent to their coastlines for as far out as 200 nautical miles; others restrict their claims to 3, 6, 12, or more miles [13].

Let us attempt to introduce some sense of order into an essentially confusing and fluid legal picture by first identifying the known elements, and then reviewing the unknowns.

The Geneva Conventions

The present character of maritime law is embodied primarily in conventions negotiated in Geneva, Switzerland, in 1958 [14], and subsequently ratified by many nations. These comprise conventions on

1. The territorial sea and contiguous zone.
2. The continental shelf.
3. The high seas.
4. Fishing and conservation of the living resources of the high seas.

The following guidance is provided by the conventions:

Territorial Sea. A country's sovereignty extends beyond its land territory and internal waters to the adjacent territorial sea, its underlying bed and subsoil, and the airspace above it. A coastal nation must not hamper innocent passage through its territorial sea of foreign vessels proceeding to and from its ports and passing through straits used for international navigation that form a part of its territorial sea.

Continental Shelf. The continental shelf begins on the seabed at the seaward limit of the territorial sea. A coastal country is entitled to exercise exclusive rights to explore and exploit the natural resources and subsoil of the continental shelf adjacent to its territorial sea. These sovereign rights are limited to the sea floor; the coastal country must respect the legal status of the superadjacent waters and airspace, as defined in the next paragraph.

High Seas. The high seas begin at the seaward edge of the territorial sea. All countries enjoy the freedoms to navigate, fish, lay submarine cables and pipelines, and to fly over the high seas. All nations implicitly enjoy the freedom to conduct scientific research on the high seas and to explore and exploit the mineral resources of the bed and subsoil of the high seas lying beyond the continental shelf, in accordance with applicable principles of international law. These high-seas freedoms shall be exercised by all nations with reasonable regard for the interests of other countries in their own exercise of the freedom of the high seas.

Fishing. Nationals of all countries have the right to take fish on the high seas, subject to their treaty obligations, the interests and rights of coastal countries, and a series of rules related to the negotiation of conservation regulations, as specified in the Convention on Fishing and Conservation of the Living Resources of the High Seas.

The Geneva Conventions served to codify generally accepted international customs that had evolved over centuries; few provisions were agreed upon that had little or no precedent in customary international law. Unique elements dealing with new problems principally concerned exploration and exploitation of the ocean floor and conservation of living resources. However, the utility and enforceability of the conventions were severely handicapped by the many issues left unresolved. A review of the following list of deficiencies emphasizes the complexity of the problem.

1. The seaward limits of the territorial seas were not defined.

2. The precise limits of the continental shelf were not delineated.

3. The measures, if any, that coastal countries may take in the high-seas areas adjacent to their coasts to regulate the activities of foreign fishing fleets were not identified, nor was indication given as to what distance from the coastline such coastal-country rights would apply.

4. No substantive limitations were set on such coastal-country rights.

5. The measures, if any, that coastal countries may take in high-seas areas adjacent to their coasts to protect themselves against marine pollution caused by foreign nations or their nationals were undefined, as was the distance from the coastline to which such coastal-country rights would apply.

6. No substantive limitations were set on such coastal rights.

7. The means, if any, that coastal countries may employ in high-seas areas adjacent to their coasts to regulate the conduct by foreign nations of scientific research on the high seas and underlying seabed were not described, nor to what distance from the coastline such coastal-country rights would apply.

8. The substantial limitations of such coastal-country rights were not set.

9. The rights of individual countries to explore and exploit the natural resources of the ocean floor beyond the limits of the continental shelf were not described, nor were the rules and conditions defined under which such exploration and exploitation should be conducted.

10. Relative to item 9, the institutional and legal means for administering control of exploration and exploitation and for resolving disputes over activities were omitted.

It might seem that the Geneva Conventions are almost too vague to invite disagreement. A moment's reflection, however, reveals that they virtually ignore needs or rights of landlocked nations that have joined the world community over the last decade and a half and have, in recent years, been individually and collectively assessing their interests and objectives in the oceans. Some feel the 1958 Geneva Conventions contain rules of law that are detrimental to those interests and objectives. Consequently, they seek changes in the law by debate and negotiation; on occasion they deliberately ignore it.

International Law of the Sea

When an International Law of the Sea Conference in Geneva was planned for 1973 under the auspices of the United Nations, concerned nations, groups of nations, and organizations began concerted efforts in 1971 to draw up proposals and positions on prospective policy. These various proposals reflect the international dilemma over precisely delineating the territorial seas, continental shelves, and high seas. During pre-Geneva deliberations, reports on these efforts from groups as diverse as the Santo Domingo Conference of Caribbean States and the Yaounde seminar of African countries were discussed by subcommittees of the United Nations Committee on Peaceful Uses of the Seabed and the Ocean Floor Beyond the Limits of National Jurisdiction.

Patrimonial Sea. The Santo Domingo Conference promulgated a concept of a "patrimonial sea" for consideration, through the Declaration of Santo Domingo—approved, in form, by the participating ministers. The official text states:

1. The coastal state has sovereign rights over the renewable and non-renewable natural resources, which are found in the waters, in the seabed and in the subsoil of an area adjacent to the territorial sea called the patrimonial sea.

2. The coastal state has the duty to promote and the right to regulate the conduct of scientific research within the patrimonial sea, as well as the right to adopt the necessary measures to prevent marine pollution and to ensure its sovereignty over the resources of the area.
3. The breadth of this zone should be the subject of an international agreement, preferably of a worldwide scope. The whole of the area of both the territorial sea and the patrimonial sea, taking into account geographic circumstances, should not exceed a maximum of 200 nautical miles.
4. The delimitation of this zone between two or more states should be carried out in accordance with the peaceful procedures stipulated in the Charter of the United Nations.
5. In this zone ships and aircraft of all States, whether coastal or not, should enjoy the right of freedom of navigation and overflight with no restrictions other than those resulting from the exercise by the coastal State of its rights within the area. Subject only to these limitations, there will also be freedom for the laying of submarine cables and pipelines.

Implementation of the patrimonial-sea concept will require international agreement at least on the concept itself, the breadth of the zone in question, and the order of priorities to be observed in settling territorial disputes. At the risk of oversimplifying, it might be stated that this concept appears to support the position of those coastal nations now claiming sovereign rights in adjacent waters out to the 200-mile limit, while restricting sovereignty over the airspace to the area of the territorial seas. Cable and pipeline rights would also be free to all nations beyond the seaward limits of the territorial seas, a point not entirely clear in the Geneva Convention. In effect, coastal states would gain control of the waters superadjacent to the continental shelf, between the territorial sea boundary (still undefined) and the 200-mile limit; and the territorial sea would have relevance only in terms of airspace, undersea cables, and pipelines.

United States Position on Fisheries Regulation. The patrimonial-sea proposal has clear implications in three areas germane to this study: responsibility for promoting scientific research, protective controls against marine pollution, and control over the products of open sea mariculture operations. In the latter case, we may contrast the Santo Domingo concept of sovereign rights over renewable natural resources (i.e., biological forms) that lie within 200 miles of the coastline with the United States view, expressed by the U.S. alternate representative to the Committee on Peaceful Uses of the Seabed in 1972 in a revised draft of the proposed article on fisheries [15]. It is proposed therein that authority be granted to the coastal state to regulate the fish stocks inhabiting the coastal waters off its shores as well as its anadromous resources, and a right to reserve to itself all the available catch its vessels are capable of harvesting. Excess stocks would be allocated to the traditional fisheries of other nations by a formula negotiated by Subcommittee II of the International Committee on Peaceful Uses of the Seabed. Each state participating in this agreement would have priority rights to the fish passing through its territorial waters on a migrant course to the high seas. An annual harvest of fish would be allocated individually to each coastal and landlocked state acting as a party to this agreement, the size of the catch to be determined on the basis of conservation requirements, the capability of each state to harvest and utilize the available supply, and other relevant environmental and economic factors to be determined.

Two organizations in the United States, the New England Fisheries Steering Committee (NEFSC) and the National Federation of Fishermen (NFF), worked on proposals in 1971–1972 designed to give U.S. fishermen a voice in the pre-Geneva deliberations. Their proposals, divided into three sections— coastal species, anadromous species, and pelagic or high-seas species—generally support the U.S. position presented by McKernan, although the NEFSC and NFF differed in their views of where the boundaries of fisheries jurisdiction should be drawn. The NEFSC suggested that the United States should have a 12-mile territorial limit with fisheries jurisdiction to the outer edge of the continental slope. The NFF, on the other hand, proposed that coastal nations have ownership of all fish and shellfish resources which live on or above the continental shelf or slope adjacent to the coastal nation, or are dependent on the shelf and slope or the superadjacent waters for reproduction and/or survival during the major part of their life [16].

Mariculture Versus Fisheries Regulation. Although neither the Santo Domingo proposal nor the position stated by the United States acknowledges the possibility of mariculture operations, the patrimonial-sea concept would seem to impose no constraints on such an operation within the 200-mile limit. The United States fisheries article draft, on the other hand, could be inimical to open sea mariculture. For instance, were a state's mariculture stations to employ devices for attracting wild stock to the stations, such as feeding, artificial reefs, or the like, these would tend to increase the natural species' survival rates. The resulting increases in species populations would require constant monitoring, with appropriate adjustment of harvest allocations for the state conducting such mariculture operations, or else the mariculture operations would prove fruitless. A possible alternative solution, incorporating provisions in the proposal for excluding stock artificially supported by mariculture stations from the quotas imposed on "natural stock," would require positive techniques for distinguishing between the two categories. This aspect of mariculture will be discussed further in a succeeding section. Either alternative would have built-in lag times that might threaten the economic viability of the operation. The salient point here is that the United States, in this proposal barely 6 months old, ignored technology already developed in prototype today—a clear example of the phase lag in legal evolution.

Issues in Maritime Law. Why should there be such difficulty in developing a uniform body of maritime law acceptable to all nations? The answer is complex and deals not only with legal parameters, but also with intricate social and economic considerations. The process of arriving at international agreements, even between a limited number of nations, has historically proved ponderous and frustrating; what may seem fair in the eyes of some states frequently may strike others as grossly unfair and possibly even a threat to their very survival. Another basic factor complicating the situation is the inherent nature of the legal structure itself, which builds on a foundation of precedent. Obviously unique situations continually arise wherein there seems to be no directly applicable precedent upon which a court may base a ruling; nowhere does this occur more often than with new applications or expansions of technology. However, in these cases a ruling frequently is arrived at by stretching an established precedent to include the new situation.

Even in recent history we see that maritime law has been concerned principally with issues related to transportation and free passage, commerce, communication (undersea cables), logistics (pipelines and port usage), air-space rights, and—to a limited extent—with fishing rights as they pertain to controlling competition. Not until the 20th century did man presume to contemplate exercising controls over the natural habits of the wild creatures of the sea; it had been difficult enough to understand and track them. Neither did he envision the possibility that his careless and uncontrolled predation could reduce this apparently boundless resource to a point of diminishing supply, and eventually pose a threat even to himself. Some hint of this possibility entered his consciousness when the near or actual extinction of some species of marine mammals (whales, fur seals, etc.) began to arouse widespread alarm. Some precedent has now been set for conservation measures pertaining to the relatively visible mammal population. However, when we would seek authority to define the legal parameters of mariculture operation, we find ourselves turning to the land-based body of law, with its ample and well-defined precedent, to find rulings that can, by analogy, be applied to this new technology.

BASIC COMMON LAW

Bearing in mind that a mariculture operation is pointless without rights of ownership and control of the cultivated products, let us take a brief overview of the history of common law and precedents from other legal systems dealing with the ownership of animals. From these we can then seek to draw analogies.

Legal Basis for Property Rights over Animals and Fish

Since our interest in property rights is confined to biological entities in this overview, it would be useful at this point to define two descriptive terms that are frequently applied to animals in a legal context. These are: *ferae naturae,* meaning "of a wild nature or disposition," and *animus revertendi,* meaning "having the intention of returning." The latter term commonly refers to domesticated animals, which may in the future include fish—even on the high seas—if the predictions of some mariculturists prove true, although historically the legal view of fish has been as *ferae naturae.*

Going back to the earliest concepts of ownership, we find that

> in primitive society the first appropriator was regarded as the owner of all things. Whether the article possessed was real or personal property, corporeal or incorporeal, movable or immovable, was immaterial. By the original grant of the Creator, all mankind was regarded as having the right to pursue and take that which the Romans called *res nullius.* This is but the instinct of unorganized society, for the sense of property inheres in the human beast [17].

In other words, unless someone specifically laid claim to a wild animal, that animal belonged to no one, and anyone could take possession of it. As Sir John W. Salmond wrote, "the fish of the sea . . . belong by an absolute title to him who first succeeds in obtaining possession of them" [18].

During the gold-rush days in California, there were no laws governing the acquisition of mining claims and water rights, so a few customs evolved from the experiences of relatively unlearned men. The customs were based on the concept that " . . . he who first possesses is entitled to ownership" [17]. "Possession is nine points of the law," a phrase familiar to most of us, evolved from similar circumstances. When we are dealing with fixed property or water rights, these rules may be relatively simple to apply. But what do we find when we encounter property that is alive, ambulatory, and given to moving about under its own volition? Then the question of what constitutes "possession" of the property gains complexity and fairly frequently becomes the subject of contention under the law, particularly in instances where animals are not clearly members of domesticated species.

Historically, in English common law the crown owned all wild animals. Current practice in the United States seems to combine both the Romans' *res nullius* and the English approach; for example, " . . . the ownership of wild animals and fish, not reduced to *actual possession* [emphasis supplied] by private persons is in the people of the state in their collective sovereign capacity, or in the state as representing all people" [19].[1] This view may be

---

[1] Graves v. Dunlop et al., (87 Wash. 648). "While animals *ferae naturae* belong to the state, as indicated, yet, when they are reclaimed by the art and power of man, they are the subject of property, and a property right thereto may be acquired."

related to the present doctrines of *public trust,* that is, property held for the benefit of the general public.

In the case of animals *ferae naturae,* legal possession need not mean that the owner must be in physical proximity to his possessions at all times, or that escape be absolutely barred. Most writers, when dealing with this subject, quote from Blackstone's commentaries. Blackstone recognized three bases upon which animals *ferae naturae* could be considered the qualified property of man:

1. *Per industriam honisis:* by means of man's reclaiming and taming the animals by industry, art, and education or by confining them within his own immediate power so that they cannot escape and regain their liberty.[1]

2. *Ratione impotentiate:* on account of the animals' inherent immobility, as in the case of animal infants or, more pertinent to this study, the case of sessile organisms such as oysters and clams.[2]

3. *Propter privilegium:* (on account of privilege) meaning the special privilege of hunting animals on a park preserve or personally owned land, to the exclusion of other hunters.[3]

Let us look at the applications and implications of these three principles and their possible import to a mariculture operation. The first one, *per industriam honisis,* would seem to be the one most open to discretionary interpretation under the law. A review of the record bears this out, as shown by the sample cases cited in the following paragraphs.

Principle of Per Industriam Honisis. In the Bering Sea Controversy[4] in 1895, the common law rules of ownership of wild animals by individual persons were made a part of international law. In this case, the United States claimed ownership of fur seals that spent part of their time in the U.S. territorial waters and the rest on the high seas. The United States felt that Great Britian was seriously depleting the stock by hunting them on the high seas. In the absence of any previous international dispute over wild animals, the American legal staff turned to common law for precedent to sustain the U.S. claim. An analogy was drawn between the seal herd and animals such as bees or carrier pigeons, which, according to Blackstone,[5] continue to be the property of their owner even when flying great distances away from their homes, owing to their fixed intention to return, *[animus revertendi].* As spokesman for Great Britain, Sir Charles Russell asserted that " ... this *animus revertendi* conferred the right of property in wild animals at the common law only when it was induced by *artificial* [emphasis supplied]

[1] 2 Blk Comm. 391.
[2] 2 Blk 3, 4.
[3] 2 Blk 394.
[4] U.S. v. Great Britain, Arbitration Decision 32 Am. L. Reg. and Rev. 901.
[5] 4 Bl Comm 225.

means, such as taming them or offering them food." He asserted that the fur seals migrated to U.S. territorial waters through natural causes, not because the United States took any measures to attract them. Therefore, he asserted, they were to be considered like any other wild animal in this respect, and the law of animals remaining the property of their owner only so long as they continue to be in his domain applied. The decision exonerated the defendant, Great Britain, and the record reads:

> The decision of the Arbitrator's supports the further reference that there is no such thing in international law as a national right of property in a herd or body of wild animals as a whole, apart from ordinary right of property in each individual animal inherent in its custodian during the time that this possession of it lasts. [20].

At the time this case was decided and entered in the legal records, no consideration was given to the possibility of branding the fur seals, feeding them, and treating them as an owned herd. With increased knowledge of animal and fish habits, and with technological advances that seem to make systematic marking and feeding of fish a possibility, the international law may well prove inadequate to deal with new situations, and thereby may require substantial updating.

Moving now from herds to individual animals, we find that although right of property may be acquired in animals *ferae naturae* an absolute property is not usually acquired.[1] For example, a branded cow on an open range, although allowed to roam free and not given to returning by its own volition, would normally be considered an absolute property. The ownership is easily identified by a brand, and for many years cattle have been considered domesticated private property. However, if animals *ferae naturae* are captured and held in private ownership and subsequently escape to resume their natural liberty, they become the property of those who recapture them [21]. The rule was applied in a dispute[2] concerning a Pacific Ocean sea lion that had been captured on the West Coast by a dealer in wild animals and later transported to New York State with a number of other sea lions to be sold. The animal, unsold because of an imperfection, escaped into the Atlantic Ocean. A fisherman recaptured the sea lion over 70 miles from the escape point, and the original owner laid claim to it about a year later. The court denied recovery to the erstwhile owner because it conceded sea lions to be animals *ferae naturae,* wherefore the owner had only a qualified property right that ceased when the animal regained its natural liberty and evaded his control.

[1] Bl Comm.
[2] Mullett v. Bradley (24 Misc. 695, 53 N.Y.S. 781, 1898).

In another court case concerning a wild animal[1] the decision went the other way, owing to more complex factors. The animal in question was a silver fox, of a second generation born in captivity and valued at $750. The fox, colorfully named McKenzie Duncan, was caged to prevent escape, but tame enough to hand-feed. Escaping through an unlocked gate, the fox eluded his owner's pursuit and was shot on the following evening, 6 miles away, by a rancher who found McKenzie prowling around his chicken house. The rancher, presumably unaware of the fox's identity, value, nature, or ownership, sold the pelt to a trapper for resale on commission. The defendant in the suit bought the pelt for $75. Having learned of the fox's fate and the possessor of the pelt, the plaintiff—owner successfully brought action to recover its value in the courts of the state of Colorado. The defendant then appealed on the premise that McKenzie was a wild animal, so possession was essential to ownership; that is, when pursuit was abandoned, the plaintiff lost title, which the rancher obtained by killing the fox and transferred to the defendant for material consideration. He further claimed that the question of whether an animal is wild or domestic is determined by species and not by individuals. The plaintiff won the appeal decision on three points: (1) that the animal was domesticated and his disposition of *animus revertendi* must be presumed; (2) that irrespective of other facts, since a tax on foxes was proposed in Colorado, the common law rules as to domesticated animals should be applied; and (3) that his fox possessed intrinsic value, meaning it was a part of the plaintiff's livelihood and "such as contribute to the support of a family or the wealth of a community" (a consideration not raised in the previous case). The written opinion of J. Burke gave the following reasons for the court's decision:

1. The defendant was knowledgeable that the pelt was the product of a large, legitimate, generally known industry.

2. The pelt had an easily ascertainable value that was considerable.

3. The pelt bore indications of ownership.

4. The seller was not the owner.

5. No right of innocent purchase had intervened.

6. The pelt was taken from a locality where silver foxes did not normally run wild and large numbers were kept in captivity.

This case and the legal points involved have been covered in some detail because these points well may be pertinent to control over, and rights to, the products of a mariculture operation.

Another look at the concept of constructive possession in the physical absence of an owner is provided by the example of a beekeeper. [18]. The owner is not required to bar the entrance of the beehive, the bees are free to

[1] E.A. Stephens & Co. v. Albers, (Supreme Court of Colorado, 81 Colo. 488, 256 p. 15, 52 A.L.R. 1056, 1927).

come and go (as indeed they must), and they have repeatedly been the objects of larceny. In one case in point[1] the plaintiff sued a defendant for taking and destroying a swarm of bees and their honey. The swarm had left the plaintiff's hives and flown to a tree located on lands belonging to the Lenox Iron Company. Plaintiff kept track of the bees, marking the tree that served as their new home. Two months later the defendant cut the tree down, killed the bees, and took the honey. The decision of the lower court was in plaintiff's favor, and defendant appealed to the New York State Supreme Court, which affirmed the judgment. The decision was based partially on the premise that, although bees commonly are considered animals *ferae naturae,* in a circumstance where they have been reclaimed by man, the reclaimer has acquired a qualified property in them and retains a constructive possession. The decision, written by J. Nelson, states:

> They are now a common species of property, and an article of trade, and the wildness of their nature by experience and practice has become essentially subjected to the art and power of man.

Thus, even when this particular swarm left the plaintiff's hive it was still in his constructive possession because he was able to keep track of it, and therefore his property right remained enforceable. Nelson further states

> If a domestic or tame animal of one person should stray to the enclosures of another ... the absolute right of property notwithstanding would still continue in him. Of this there can be no doubt. So in this respect the qualified property in bees.

He continues on to comment that because of the

> institutions of civil society, and the regulations of the right of property, the law of nature where prior occupancy alone gave right ... has given way to the establishment of rights and property better defined and of a more desirable character. ...

So in this situation, the right of the owner of the bees still existed even if he could not retrieve them without being liable for trespass, and his inability to repossess them was not looked upon as abandonment of the animals to their former liberty.

In drawing analogies we find that cultured fish in the open sea may indeed be compared to bees in the limited sense of their freedom of movement and in the concept of being reclaimed wild creatures. However, bees roam out of the necessity to obtain food, while the hive serves as a fixed base for food

---

[1] Goff v. Kilts (Supreme Court of N.Y., 1836, 15 Wend. 550).

storage, reproductive activity, hibernation, and their highly structured social order. They do swarm when the hive becomes overcrowded, as was the case in the suit just reviewed, but a keeper normally reclaims the swarm and returns the migrating colony to a new hive within his keep. Cultured fish, on the other hand, might both feed and reproduce at the mariculture station, which may be somewhat less essential to species survival than the beehive, and roam by "choice" in the highly mobile schools that constitute their relatively unsophisticated social structure. Yet more significant, the economic value of the bees lies in their product, the honey, which is stored at their "station," whereas the intrinsic value of the fish lies in the creatures themselves. Therefore, the negotiable product of a fish-culture operation can be harvested more easily by nonowners than can the product of a bee-keeping operation.

If, as predicted [22], fish in the open sea can be domesticated to the point of being classified *animus revertendi*, their culture would be more analogous to a ranching operation since the strength of a qualified property right would depend on it being clearly evident to any fisherman, before he could be held liable for taking the fish, that a given school of fish was "owned" by someone else.[1]

The legal precedents we have briefly reviewed, then, are pertinent to the hypotheses upon which the feasibility of a mariculture operation involving fish in the open sea is based; that is,

1. Wild animals can be considered domesticated enough to become an absolute property if the issue involves an intrinsic value, as would be the case with a population of fish associated with a mariculture station.

2. Fish could bear indicia of ownership, either by special markings or morphological traits derived through selective breeding, or by the visible presence of the owner's agents; or the fish would remain in the immediate vicinity of a station.

3. Both casual and commercial fishermen could be kept informed of the area in which the cultured fish most probably would be roaming or "grazing," and made aware that the designated schools were the property of a legitimate industry.

4. The cultured species could be deemed "domestic fish stock" if they were part of an industry and considered taxable.

5. In regarding the mariculture station's fish herds as domesticated, a disposition of *animus revertendi* could be not only presumed (as was the fox's) but *de facto*.

---

[1] Commonwealth v. Chace (9 Pick Mass. 15). The defendant was indicted for stealing 14 tame doves owned, raised, and fed by Williams. The defendant was convicted but granted a new trial because no evidence was introduced to show the circumstances of the doves' deaths. The court said " . . . the doves, even though domesticated, were not the subject of larceny unless under the owner's care, for the reason that it is difficult to distinguish them from other fowls of the same species." Chace was acquitted.

Principle of Ratione Impotentiate. If we next look at Blackstone's second principle of qualified property rights, *ratione impotentiate,* we find significantly more consensus on legal applications pertinent to mariculture. Sessile forms have already been judged as being legitimate personal property, even though they are looked upon as animals *ferae naturae.* Oysters planted in public waters by a private citizen remain the property of the planter, if planted where oysters do not grow naturally, provided that the place is clearly designated by some form of markings. Such oysters can be objects of larceny [19]. In the case of State v. Taylor,[1] the court agreed that oysters are animals *ferae naturae* but do not come within the description of having to be dead, reclaimed, tame, or in actual possession of their owner. If at liberty, they have neither the inclination nor ability to escape. No abandonment claim against the owner could be made solely because they were placed in public waters, since they had been gathered elsewhere and specifically planted by the owner; in short, casting into the sea with the intention of reclamation is not considered to be abandonment.

The same rule can be applied to clams. In the case of People v. Morrison,[2] the opinion of the court stated

> ... Although in the nature of *ferae naturae,* to which qualified title may be acquired by possession, when reclaimed and transplanted they [clams] need not be confined, for as they cannot move about they cannot get away, even when placed in the water, as they must be in order to live. They and their produce thus cease to be common property and belong exclusively to the one who transplants them. ...

In light of the precedent set, no significant difficulties are foreseen in controlling and harvesting the sessile products of mariculture operations.

Principle of Propter Privilegium. Blackstone's third principle, *propter privilegium,* appears particularly germane to the issue of fishing rights in territorial waters, and relates to concepts embodied in the Geneva Conventions, the patrimonial sea of the Declaration of Santo Domingo, and the U.S. position stated to the Subcommittee on Peaceful Uses of the Seabed [15]. It may, however, provide a basis of precedent in the future should updated and expanded interpretation of the law be required to cover situations created through the leasing of ocean areas to mariculture stations. We shall be looking at the subject of fishery leases a little farther on.

Legal Bases for Liability

Having examined the question of property rights, we would be remiss if we did not also look at the reverse side of this issue, the parameters of liability

---

[1] State v. Taylor (27 N.J.L. 117, 72 Am. Dec. 347).
[2] People v. Morrison (194 N.Y. 175; 86 N.E. 1120; 128 Am. St. Rep. 552).

that may impose constraints on a culturing operation. To date the law has not had occasion to deal with this problem in a marine environment, so again we must turn to common law, applicable to animal husbandry in this instance, for precedent.

Since the body of law in the United States has its roots in English law, we look to Great Britain for the earliest examples pertinent to control of domestic animals. There we find that, with limited land area, an increasingly crowded environment led to common law requiring an owner of livestock to assume responsibility for keeping his animals fenced in; his neighbors were not expected to erect protective fences around their lands. The liability for failing to restrain livestock was placed solely upon the owner [23]. Liability was somewhat relaxed when an owner drove his herd over a public road or land; then the degree of liability for damage by a stray was tempered by the alacrity of the owner's efforts to regain control of the offending animal [24].

This common law was carried to America by the English settlers. However, as the population moved westward into the great plains, the vastness of the land area encouraged the development of huge cattle herds dependent upon grazing at large, giving rise to the custom of the open range. Midwestern and western settlers then rejected the fencing-in law of their eastern neighbors, and chose instead to regard range lands as "quasi-common." Herds of cattle were not trespassers; they had "implied license" of the landowners for their presence. The custom dictated that a landowner assume responsibility for fencing out cattle if he wished to exclude them from his land.

In time the population of the Western states grew and agriculture assumed increasing importance. As open land diminished, the "fence them out" states gradually adopted statutes requiring the livestock owner to fence them in once more[1] and making him liable for normal damages done by his animals.[2] Examples of such normal damages are

1. Communication of a disease to a landowner's stock.[3]
2. Misalliance of a scrub bull and a full-bred heifer.[4]
3. Personal injury or property damages sustained by a landowner while trying to route a trespassing animal[5] [23].

Ordinarily, a small animal such as a dog does little damage if it trespasses. However, if the animal is known to be vicious, a poultry killer, or the like, damage can be claimed. Other animals by their nature render an owner liable for absolute damages in case of trespass. Typically, these include swine, bulls, stallions, rams, boars, billygoats, and any other animal that is generally vicious or extremely destructive by its usual nature; the owner is held as an

---

[1] Williams v. Windham (3 La. App. 127).
[2] Goslar v. Reed (179 N.W. 621).
[3] Wood v. Wehr (6 p [2nd] 1105, 91, Mont. 280).
[4] Matthews v. Langhofer (202 P. 634, 110 Kan. 3).
[5] Walker v. Nickerson (Mass. 197, N.E. 451).

insurer. The owner or keeper of domestic animals is liable for injuries inflicted by them only when he has been negligent, the animals were wrongfully in the place where they inflicted injuries, or the injuries are the result of known vicious tendencies or properties [24].

The open sea would seem to be in little danger of soon developing the competition for space that plagues the relatively crowded settled land areas. Open sea mariculture stations presumably could be spaced far enough apart in the vast oceans to avoid interference with each other or with shipping lanes and other conflicting interests. But the practicality of stations on the high seas (analogous to "quasi-common" territory of the open range) is, at present, still to be demonstrated; culturing operations in coastal waters and in near-shore areas probably offer more immediate promise. These areas of more intense traffic and usage may give rise to situations requiring legal settlement of liability issues as well as territory rights. Nevertheless, if ocean areas should eventually become crowded, then similar liability rules could apply to ocean animals as to land animals; for example, the owner of predatory fish probably would be responsible for any normal damages his fish might do to a herbivorous-fish culturing operation unless the "open range" prevailed, whereupon the owner of the herbivorous fish would have to keep the predators out. This does not seem to be an immediate problem, but it is not too soon to begin considering the long-range aspects. If mariculture indeed becomes the way to conserve and increase our fish stocks, and at the same time supply the world with much-needed protein, some sea areas could become crowded, and we would have to turn to land-based precedent to settle the conflicts that would inevitably arise.

LEGAL AND POLITICAL CONSTRAINTS UPON OWNERSHIP, OPERATION, AND LOCATION OF MARICULTURE STATIONS

In the preceding sections of this chapter we have taken brief overviews of the broader legal issues dealing with sovereign rights over sea areas and the regulation of fishing rights, the control and the protection of property in the form of wild and cultured animals, and the bases for assigning liability for potential damages to property. Against this background we now narrow our view to more precisely defined, and often local, regulations that have specific applicability to the ownership and operation of mariculture stations. We shall examine these first in relation to the practical problems of jurisdiction and applicable lease laws. Then we shall consider the implications of current trends in environmental quality standards. Finally, we shall discuss the matters of pollution control and Food and Drug Administration (FDA) regulations as they pertain to the products.

Jurisdiction

The United States currently takes the position that a 12-nautical-mile exclusive zone, comprising a 3-mile territorial zone and a 9-mile contiguous zone, is authorized by customary international law.[1] In this context, mariculture will probably be considered a fishery for regulatory purposes. Whether the state or federal government has jurisdiction in the contiguous zone is in question and unresolved at this time.[2] Although the federal government has exercised its jurisdiction in this area, 3 to 12 miles offshore, in the granting of mineral leases,[3] there has not been, up to the present, any federal management of fishery resources in this zone other than the exclusion of foreign fishing vessels. We take no position on this question, as yet unresolved and in great contention; it is only necessary to point it out here. It is established, however, that the states have the authority within the territorial sea and internal waters to control their fisheries in the public interest, subject to the framework of their own constitutions.[4]

A recently completed study [25] considers many of the legal ramifications inherent in aquaculture. In summary the authors concluded that some of the hindrances to mariculture in the coastal region at this time are interests such as recreation, navigation, mineral claims, cable and pipeline laying, as well as more obviously polluting activities. In many cases these conflicts of interest have been tested in court. Relevant statutes of the concerned states are lengthy and detailed, but all share in common the same central issues and the same problems in resolving them.

We also see that " . . . in New England the body of law affecting private and public ownership of coastal water and the management of natural resources is complicated and has probably discouraged commercial enterprise in aquaculture" [26]. The authors maintain that the legal system can more readily respond if the economic potential and technological feasibility for specific ventures can be shown. One of their recommendations is that feasibility studies and pilot-scale experiments could be legally accommodated now; then when the specific needs of commercial aquaculture are established, the legal system should work out equitable, comprehensive aquaculture laws that offer greater assurance to commercial ventures. Some of the things requiring attention but not specifically considered in New England laws at present are horizontal and vertical zoning, leasing, licensing, building safety codes, easements, and the regulation of taxing power by variation in rates of taxation.

Laws and regulations vary greatly from state to state. Some states, in response to their special interests and economic needs, have encouraged aquaculture and provided positive, substantive laws. Others merely give

[1] Convention on the Territorial Sea and the Contiguous Zone, TIAS 5639, 15 & S.T. 1606, 1612, Art. 24 and Public Law 89-657, 1966.
[2] U.S.A. v. States of Maine, New Hampshire, Massachusetts, Rhode Island, New York, New Jersey, Delaware, Maryland, Virginia, North Carolina, South Carolina, Georgia, and Florida.
[3] U.S. v. California, 1947.
[4] State v. Stoutmire (131 Fla. 98, 179 So. 730, 1938).

priority to mineral interests. In California and Michigan strong priorities are given to sport fisheries whenever there is conflict with commercial fishery interests [27]. New laws presently are being considered in many states, and with the recent passage of the Coastal Zone Management Act by the U.S. Congress (October 27, 1972), all coastal states may be taking more comprehensive looks at their laws in this context. Consequently, we may see many changes in the next few years.

One major point of conflict is the distance to which municipal boundaries extend into tidal waters, with varying definitions depending upon the corporate charter of each locality [28]. Consequently, suitable aquaculture sites may or may not be within the jurisdiction of a municipality. In general, land under the low-water mark in tidal waters is not subject to private ownership, although states can and have sold such lands. Thus, private ownership usually stops at the high-water mark, with the state owning and holding the lands beyond it in a type of public trust. Exceptions include Maine, New Hampshire, Massachusetts, Pennsylvania, Delaware, and Virginia. Hawaii not only observes this general rule but extends it: the Hawaii Setback Law, designed to preserve the beauty of the shoreline by requiring a 300-foot building setback,[1] gives the counties added jurisdiction over their shorelines. In Hawaii, the Department of Land and Natural Resources is authorized to make lease agreements for the land lying between the high-tide line and the 3-mile limit; however, it is as yet unclear whether this authority extends to include the vertical column of water [29]. Florida was the first state with laws authorizing leases in the vertical water column.[2] The state of Connecticut has, in most cases, given nearshore area control to municipalities. Many states, too, hold that the riparian owner has certain rights beyond the low-water mark, even where that submerged land area is not subject to private ownership; these are common-law rights and are usually modified by statute. Generally speaking, they are the rights of ingress, egress, boating, bathing, and fishing. Other rights that "have been defined by law" are (1) the right to an unobstructed view of the waters (common law), and (2) the right to build a pier or wharf out to the point of navigability [25]. (The latter is not recognized in Florida and requires a permit in Hawaii owing to the Setback Law; however, it is recognized in many other states as a common-law right.)

Obviously, the structure of jurisdictional authority over coastal and open waters has many inconsistencies. If it were feasible to take a systems approach to the development of a logical regulatory framework, it might be possible to reach a practical settlement of the contiguous-zone dispute and to develop consistent laws to which all coastal states could subscribe—with sufficient flexibility to adjust to special needs or the pressures of evolving technology. However, political and legal traditions hold no encouragement

---

[1] Hawaii Revised Statutes, Chapter 205, Sec. 32-37.
[2] Florida Statutes Chapter 253, Section 68.

for a systematic approach in the foreseeable future. Consequently, any proposed mariculture operation will have to explore the laws of the specific state (and possibly municipalities) contiguous to the coastal waters in which it proposes to operate, and deal with these on an individual basis. Additional problems are, of course, posed by the several states that have not yet managed to clearly define their jurisdictional lines of authority. If mariculture operation becomes fairly common, it may generate sufficient demand for answers to force some resolution of these gray areas.

Lease Laws Applicable to Mariculture. Having just wrestled with the problem of who has jurisdiction over what marine areas, we may be mentally prepared now to consider the situation facing the seeker of a lease or other appropriate authorization for a mariculture station. He should be of a highly persistent nature, for he will once more find himself sorting out the applicable regulations for the specific area he hopes to lease.

Typical questions the would-be lessee will be seeking to answer are the following:

1. How large an area can be obtained (for a mariculture–aquaculture station)?

2. What type of area can be leased?

3. How is a particular parcel to be allocated (first application or competitive bid)?

4. How much total acreage may one individual hold and for how long?

5. Is the lease renewable and, if so, how often, and are there conditions under which the lessee can lose it (e.g., by not utilizing the area within a prescribed time)?

6. Can the area be dually shared by nonconflicting culturing systems such as oysters and salmon, or does one lessee acquire all fishery rights?

7. What protection does an aquaculturist lessee have from the competition of other marine activities (recreation, transportation, regular fishing)?

8. What state agency regulates and grants these leases? Does the state have provisions for aquaculture in its coastal-zone management plan—if it has a coastal-zone management plan?

Provisions covering these points vary widely from state to state [26]. For example, Maine has no area limit for oyster cultivation, but limits the area for cultivating Irish moss or other marine species to 1 square mile per parcel, no more than three parcels per person, and no more than 10 square miles in the territorial waters of the state. In Rhode Island, an oyster lessee's total acreage is not limited, but he may obtain only 1 acre at a time. Virginia and Maryland allow the first applicant to obtain a lease as long as other conditions are met. In Virginia the leased area may be up to 5,000 acres in Chesapeake Bay and only 3,000 acres in other waters. Maryland allows a maximum of 500 acres per person to be leased for oyster cultivation in Chesapeake Bay and a maximum of 30 acres in most other waters.

Connecticut and New York laws lease areas by competitive bid, and Connecticut statutes provide that residents and nonresidents are equally eligible for oyster cultivation. Maine, New York, New Jersey, Virginia, and California all have residency requirements.

In California the 1971 session of the state legislature enacted three laws dealing with aquaculture.[1] They concern mariculture, oyster cultivation, and the domestic anadromous (up-river spawning) fishery. The mariculture law recognizes mariculture per se and provides for protection against poaching, leasing of submerged land and water areas, and the right of access for the public to public beaches. It states a residency or domestic corporation requirement and the terms of lease length and renewal. The second law provides for oyster cultivation in areas where oysters were not native as of January 1971. Further regulations are similar to those for mariculture. The third law, concerning the domestic anadromous fishery, requires that operators of such a fishery be able to identify the fish they cultivate, and states that when such fish return to their wild state they become the property of the state and may be taken by anyone with a sport or commercial fishing license. It further provides for the examination of cultivated fish prior to their release to make sure they are free from any disease that might affect natural stock.

The examples of variance from state to state in regulatory statutes and laws which have been discussed so far seem to suggest that a very useful tool for prospective mariculturists would be a map of U.S. coastal waters, indicating the rules, regulations, and appropriate state (and possibly municipal) authorities pertaining to the marine areas adjacent to each state—in so far as this information can be ascertained. Such a graphic portrayal of the legal disarray could well be worth the expense in time and effort if it in any way speeded action by the states to develop a consistent structure of laws that would be conducive to widespread development of mariculture activity.

Environmental Quality Impact. In keeping with the relatively low interest in legislative circles and the resulting, virtually universal, lag in the planning and development of environmental controls, little policy has yet been developed that is applicable to mariculture. The relative novelty of the mariculture concept has doubtless been a factor also. Our research did uncover some developments in Washington state and Oregon that indicate the beginning of policy formulation in this area; a brief review shows us, however, that much remains to be done.

Although Washington became the first state to enact legislation establishing a comprehensive coastal-zone management program—the 1971 Shoreline Management Act (May 25, 1971)—there is no state aquaculture policy. However, the state's Department of Ecology *Final Guidelines* are directed toward aquaculture:

[1] California Revised Statues, Sec. 6480-6505.

Aquaculture (popularly known as fish farming) is the culture or farming of food fish, shell fish or other aquatic plants and animals. Potential locations for aquacultural enterprises are relatively restricted due to the specific requirements for water quality, temperature, flow, oxygen content, and, in marine waters, salinity. The technology associated with present day aquaculture is still in its formulative stages and experimental. Guidelines for aquaculture should therefore recognize the necessity of some latitude in the development of this emerging economic water use, as well as its potential impact on existing uses and natural systems.

Guidelines:

1. Aquaculture enterprises should be located in areas where the navigational access of upland owners and commercial traffic is not significantly restricted.
2. Recognition should be given to the possible detrimental impact aquacultural development might have on the visual access of upland owners and on the general aesthetic quality of the shoreline area.
3. As aquaculture technology expands with increased knowledge and experience, emphasis should be placed on underwater structures which do not interfere with the navigation or impair the aesthetic quality of Washington shorelines.

In Oregon, aspects of aquaculture relating to food fish are within the jurisdiction of the Fish Commission of Oregon and to some extent the State Game Commission. Oregon has specific legislation dealing with (1) the classification of suitable state-owned submerged land for oyster cultivation; (2) oyster cultivation fees and use taxes; and (3) chum salmon hatcheries, fees, and use taxes. The Oregon Coast Conservation and Development Commission was established[1] to prepare a comprehensive plan to aid in the conservation and development of the natural resources of the coastal zone and to provide a balance between possibly conflicting public and private interests in the coastal zone. The commission is required to present its recommendations and analysis to the governor and state legislature by January 1975. It is too early to tell in what direction the commission will move.

The main point to be made here is that little or nothing has yet been developed in the nature of environmental quality controls that suggests potential constraints upon mariculture activity. In fact, we shall see in the following paragraphs that it is that very lack of controls that poses potential, and often actual, threat.

Pollution Control and Food and Drug Administration Regulations. A problem common to all areas is the protection of a mariculture venture against pollution, which could be regarded as a two-edged sword. On the one

[1] Act 608-1971.

edge lies the threat to the survival and healthy growth of the cultured organisms that is inherent in noxious substances in their watery environment. On the other edge, assuming the organisms do thrive, lies the threat to their suitability for human consumption or for other nutritional uses caused by contaminants absorbed into the flesh of the organisms. Either situation could spell disaster for a culturing activity.

The problem has long been evident in the gradual disappearance from the marketplace of shellfish formerly harvested from natural beds in bays and more or less protected waters adjacent to now heavily populated coastal lands. Many of these had served as food sources since the days of the early American settlers, at which point in history their loss would have put a serious dent in the available food supply. With other sources of food to draw upon, the loss in the recent past has been reflected most visibly in the soaring prices for shellfish in the markets, and in a loss of livelihood for the former harvesters of these marine crops. The time is now at hand when the pressing nutritional needs of a burgeoning human population can ill afford the neglect of an important protein supply. Nor is the contamination problem confined to sessile organisms, although these often are the most helpless to evade it; contamination is becoming widely evident in the free-swimming fish population, even in the farthest reaches of the high seas.

Some indication of the sources of the problem can be seen in the following statement by the U.S. Department of Health, Education, and Welfare:

> Regarding detritus feeders (oysters, clams, and mussels), the National Shellfish Program, administered by the Food and Drug Administration, specifically prohibits the harvesting of oysters, clams, and mussels from waters subject to municipal wastes. This includes the discharge of raw sewage as well as effluents from both primary and secondary treatment plants. This program applies to all fresh and fresh frozen oysters, clams, and mussels whether for export or domestic consumption . . . There are no official tolerances for pesticides and heavy metals in fish and shellfish, but the FDA uses the following guidelines for enforcement of maximum permissible levels of some of these substances:

Pesticides:	DDT, DDE, and/or TDE	5.0 ppm
	Aldrin and/or Dieldrin	0.3 ppm
	Endrin	0.3 ppm
	Heptachlor and/or Heptachlor Epoxide	0.3 ppm
Heavy Metals:	Mercury	0.5 ppm

Each of these guidelines is based on the edible portion of raw seafood or on the entire amount of a processed seafood. The FDA has proposed a temporary tolerance of 5 ppm polychlorinated biphenyls (PCB) in fish. PCB's are environmental contaminants which are likely to be present in municipal wastes [*30*].

In looking at the protection offered by the law to the mariculturist facing this threat to his operations, we find an early case where an oyster cultivist was denied recovery of damages to his beds caused by sewage from the city's sewer system.[1]

On the other hand the record contains other cases where aquaculturists have been protected.[2] There is no doubt, however, that there are great variations in the positions of the individual states—from no protection to adequate protection. A mariculturist would have to look closely into the protection laws of the area he chooses to farm and keep as flexible a stance as possible in preparation for interpretations that cannot yet be foreseen.

CONCLUSIONS

Conclusions to be drawn from the substance of this chapter would appear to be as follows:

1. It is advisable that species to be cultured in any open sea operation be clearly and rather completely under the control of the culturist all or most of the time. This argues in favor of containment, sessile forms, or forms that can be caused to remain in a given area through employment of some combination of attraction devices.

2. Because of vagaries and wide differences in laws relating to territorial seas, and because of a relatively high level of usage conflicts in the nearshore zone, prospective mariculturists of the immediate future would be well advised to consider associating their operations with an enterprise holding an existing lease in territorial seas, or to look toward the high seas if work there proves feasible technically and economically.

3. Water pollution is perhaps one of the more significant dangers to any mariculture operation. Therefore, extreme caution should be exercised in siting mariculture operations anywhere near urban or industrial operations. Note that this is potentially in conflict with the shared-lease idea and an additional argument in favor of high-seas stations. At present, the FDA is also treating *discharge from* aquaculture operations as pollutants and is presently suspending judgment on whether they will be restricted, in greater or lesser degree. How this might apply to open sea mariculture is an interesting speculation.

[1] Darling v. City of Newport News (123 Va. 14, 96 S.E. 307, 3 A.L.R. 748 (1918)).
[2] Gibson v. City of Tampa (135 Fla. 637, 185 so. 319 (1938)). Huffsmire v. City of Brooklyn (162 N.Y. 584, 57 N.E. 176 (1900)). Grant v. United States (192 F20/482 (C.A. 4, 1957)).

4. State and federal governments might be well advised to watch aquaculture and mariculture developments closely and prepare themselves to respond appropriately with enabling and regulatory legislation as the needs may arise.

REFERENCES

1. Lord Esher. 1896. The gas float whitten (No. 2). In C. J. Colombos, ed., 1967, *International law of the sea,* 6th ed. Longman Group Ltd., Essex, England, p. 42.
2. Morrison, S., and G. W. Stumberg. 1954. *Cases and materials on the admiralty.* Foundation Press, Inc., Mineola, N.Y.
3. Ashburner. 1909. The Rhodian sea law: introduction. In C. J. Colombos, ed., 1967, *International law of the sea, 6th ed. Longman Group Ltd.,* Essex, England.
4. Sanborn, F. R. 1930. Origins of the early English maritime and commercial law. In C. J. Colombos, ed., 1967, *International law of the sea,* 6th ed. Longman Group Ltd., Essex, England.
5. Twiss. The black book of the admiralty (Introduction to Vol. II). In C. J. Colombos, ed., 1967, *International law of the sea,* 6th ed. Longman Group Ltd., Essex, England.
6. Colombos, C. J., ed. 1967. *International law of the sea,* 6th ed. Longman Group Ltd., Essex, England.
7. Reddie. 1841. An historical view of the law of maritime commerce. In C. J. Colombos, ed., 1967, *International law of the sea,* 6th ed. Longman Group Ltd., Essex, England.
8. Twiss, 1613. An abridgement of all sea laws. In C. J. Colombos, ed., 1967, *International law of the sea,* 6th ed. Longman Group Ltd., Essex, England.
9. Oppenheim, L. 1967. *International law, a treatise,* 8th ed., Vol. I. Constable & Company Ltd., London.
10. O'Connell, D. P. 1965. *International law.* Stevens & Sons Ltd., London.
11. McDougal, M. S., and W. T. Burke. 1962. *The public order of the oceans.* Yale University Press, New Haven, Conn.
12. Padelford, N. J. 1971. Public policy for the seas. *MIT-NSF Sea Grant Project GH-1,* rev. ed. The MIT Press, Cambridge, Mass.
13. National Academy of Sciences. 1972. *International marine science affairs.* Washington, D.C.
14. Jackson, H. M. 1972. The law of the sea crisis—an intensifying polarization. Staff rept. on U.N. Seabeds Comm., the Outer Continental Shelf and Marine Mineral Develop., Pt. 2. Govt. Printing Office., Washington, D.C.
15. McKernan, D. L. 1972. Alternate U.S. Representative to the Committee on the Peaceful Uses of the Seabed and the Ocean Floor Beyond the Limits of National Jurisdiction—Subcommittee II member. Statement of Aug. 4, 1972.

16. *Oceanol. Internl. Offshore Tech.*, Dec. 1971.,
17. Arnold, E. C. 1921. The law of possession governing the acquisition of animals *ferae naturae*, 55 Am. L. Rev. (now U.S.L. Rev.) In W. T. Fryer, ed., 1938, *Readings on personal property*, 3rd ed. West Publishing Co., St. Paul, Minn.
18. Salmond, J. W. 1947. *Jurisprudence*, 10th ed. G. L. William, ed. Sweet & Maxwell Ltd., London.
19. McKinney, W., and H. N. Greene. Annotated Cases 1917 B: Right of property in wild animals. Edward Thompson & Co., San Francisco, and Bancroft Whitney & Co., Northport, Long Island.
20. Bigelow, H. A., assisted by W. L. Eckhardt. 1942. *American casebook series*, 3rd ed. West Publishing Co., St. Paul, Minn.
21. Brown, R. A. 1936. *Personal property*. Callagan and Co., Chicago.
22. Craven, J. P. 1969. Item 9-10: *Res Nullius de Facto*—the limits of technology. Paper presented at Symp. International Regime of the Sea Bed, Rome.
23. Beuscher, J. H. 1960. *Law and the farmer*, 3rd ed. Springer Publishing Co., Inc., New York.
24. *Corpus Juris Secundum.*
25. Kane, T. 1970. Aquaculture and the law. *Sea Grant Tech. Bull. 2* (Univ. Miami).
26. Norton, D. 1971. In T. Gaucher, ed., *Aquaculture: a New England perspective*. Res. Inst. Gulf of Maine, New England Marine Resources Inform. Program.
27. Spangler, M. B. 1970. *New technology and marine resource development—a study in government—business cooperation*. Praeger Publishers, Inc., New York.
28. Henry, H. 1971. In T. Gaucher, ed., *Aquaculture: a New England perspective*. Res. Inst. Gulf of Maine, New England Marine Resources Inform. Program.
29. Trimble, G. 1972. *The legal and administrative aspects of an aquaculture policy for Hawaii: an assessment*. Center for Policy and Technology Assessment, Dept. Planning and Econ. Develop., State of Hawaii; and Resources Develop. Internship Program, Western Interstate Comm. for Higher Education.
30. Ratcliffe, S. D. 1972. Personal communication. Shellfish Sanitation Branch, Public Health Service, Dept. Health, Education, and Welfare, Washington, D.C., Nov. 1972.

Additional Readings

Gamble, J. K. 1973. *Index to marine treaties* (Sea Grant publication). University of Washington, Seattle, Wash.

McKernan, D. L. 1972. Alternate U.S. Representative to the Committee on the Peaceful Uses of the Seabed and the Ocean Floor Beyond the Limits of National Jurisdiction—member of Subcommittee II. Statement of Mar. 1972.

Schaefer, M. B. 1968. In J. F. Brahtz, ed., *Ocean engineering,* John Wiley & Sons, Inc., New York.

Stevenson, J. R. 1972 a. U.S. Representative to the Committee on the Peaceful Uses of the Seabed and Ocean Floor Beyond the Limits of National Jurisdiction—member of Plenary Committee. Statement of Aug. 1972.

Stevenson, J. R. 1972 b. U.S. Representative to the Committee on the Peaceful Uses of the Seabed and Ocean Floor Beyond the Limits of National Jurisdiction—member of Subcommittee II. Statement of Aug. 16, 1972.

PART **II**

THE OPEN SEA
MARICULTURE ENVIRONMENT

4

PHYSICAL OCEANOGRAPHY AND GEOLOGY

E. D. Stroup and S. V. Smith

"Open sea" has been defined in the mariculture context as unprotected waters anywhere offshore. Yet certain geographical constraints clearly would influence any nation's open sea mariculture activities. First, any nation contemplating open sea mariculture would probably find itself constrained to an absolute maximum range encompassed by its own territorial waters along with all international waters. But, in practice, no nation could expect to enjoy exclusive mariculture prerogative over international waters; some apportioning seems inevitable.

Any apportioning of traditionally international waters among nations, for any purpose at all, is one of the most controversial subjects in international relations, as we have seen in Chapter 3. Therefore, for the purposes of this study we have been partially arbitrary in proposing certain oceanic areas for consideration for U.S. open sea mariculture. We say "partially" arbitrary because certain criteria did influence our choice. First, waters very near the shores of foreign nations were not considered (excepting some overlap with Canada and Mexico). Second, we attempted to include the widest range of geographic latitude that appeared reasonable, simply because we wished to explore a wide range of mariculture possibilities. Third, we decided to include the Trust Territory of the Pacific, because it appears a likely candidate for an open sea mariculture industry for a plethora of reasons.

Our areas of consideration, as shown in Figure 4.1, include much of the North Pacific and North Atlantic oceans. Within the areas delineated in Figure 4.1 lies almost 20 percent of the world's oceans, and this portion includes most of the spectrum of worldwide environmental parameters. Generalities about this area tend to be generalities about the world's oceans in their entirety.

An adequate description of what amounts, then, to two major oceans covers a vast array of topics. A brief overview such as required here is probably a more difficult organization task than preparing a more lengthy and inclusive description. However, we can restrict the focus to factors pertinent to mariculture interests. Thus, it can be proposed that the oceans constitute the containers and, on the whole, the potential culture media for

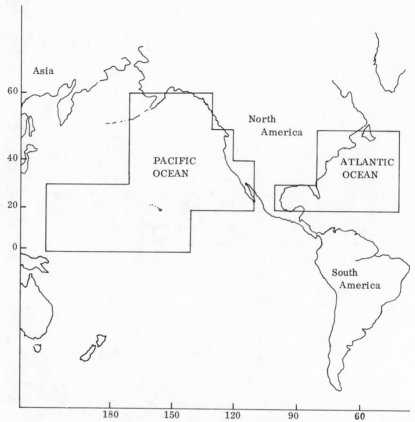

FIGURE 4.1 *Areas of interest for open sea mariculture.*

open sea mariculture, as well as containing potential sources of nutrients, organisms to be cultured, and possibly some of the culture stations themselves.

We begin with a description in this chapter of the containers (marine geology) and the physical characteristics of oceanic waters. In Chapter 5 we review the chemical properties of the proposed culture media. As we consider these factors, we attempt to discern what they imply for open sea mariculture. The questions of nutrient sources, culture organisms, and culture stations are discussed in Part Three.

GEOLOGY OF THE OCEAN BOTTOM

A basic aspect of the open sea is the nature of the ocean basin itself—the geology of the sea floor. Our discussion is limited primarily to the topography of the sea floor and the nature of its materials, for these two aspects seem to be of major potential importance to mariculture.

Topography

Depth Characteristics. Water depths of the open sea range from 0 to 11,000 meters (m), and depths throughout most of this range are found within the area of interest. For example, the Puerto Rico Trough is approximately 9,000 m deep. Open sea depths are neither randomly distributed about some mean nor evenly distributed throughout the range. As illustrated by Figure 4.2 [1], the majority of the world's ocean area is between 3,000 and 6,000 m deep; much of the Atlantic sea floor lies near the shallower portion of this mode, whereas the Pacific tends to be somewhat deeper. A minor secondary mode in the depth distribution (or hypsographic) curve occurs in water shallower than 200 m—the continental shelves. This latter mode is the seaward extension of the major dry-land mode between 0 and 1,000 m in altitude. The deep-water area, or abyssal sea floor, and the continental shelves together comprise over 80 percent of the world's ocean area.

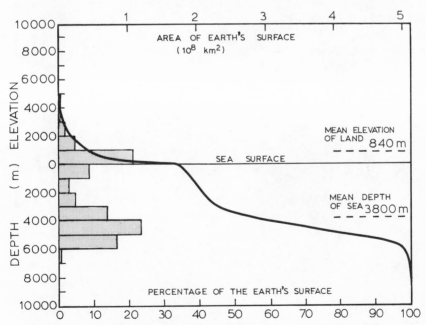

FIGURE 4.2 *Hypsographic curve showing the majority of the world's ocean area lying between 3,000 and 6,000 m deep. (Redrawn from Sverdrup et al., [1].)*

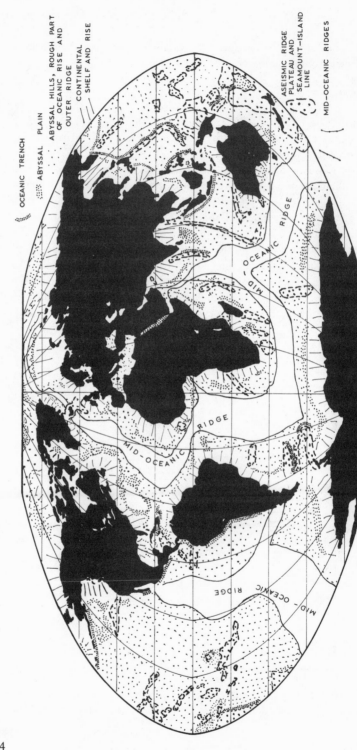

FIGURE 4.3 Distribution of some principal topographic features in the world ocean. (Adapted from Hill, Vol. 3, Chap. 14 [2].)

OCEANIC TRENCH

ABYSSAL PLAIN

ABYSSAL HILLS, ROUGH PART
OF OCEANIC RISE AND
OUTER RIDGE

CONTINENTAL
SHELF AND RISE

ASEISMIC RIDGE

PLATEAU AND
SEAMOUNT-ISLAND
LINE

MID-OCEANIC RIDGES

Geomorphology. Geomorphology is an additional topographic consideration. Figure 4.3 is a world map schematically illustrating the locations of
many features discussed here and a number of others as well [2]. The depths
of the North Atlantic and Pacific ocean floors differ from one another only
slightly, but the geomorphology of the two regions is distinct. Although it
contains extremely rugged areas, such as the mid-Atlantic ridge, much of the
North Atlantic ocean basin is smooth and flat; most of the North Pacific sea
floor is quite rugged. This difference is largely attributable to the widespread
occurrence of turbidity currents in the Atlantic and to their more restricted
occurrence in the Pacific; turbidity currents transport enormous volumes of
sediment, which tend with time to blanket topographic relief lying in their
path.

It appears, then, that a geographically significant but proportionately small
area of the oceans' water depths are shallow enough to allow bottom
mounting of mariculture stations. Therefore, although we are likely to begin
in shallow water, it seems reasonable to postulate that any grand-scale open
sea mariculture is likely to involve floating platforms.

Volcanic Topography. Much of the topographic relief on the Pacific sea
floor is volcanic in origin. Volcanic topography can vary from small lava flows
creating bumps a few decimeters high to volcanic islands rising over 10,000 m
from the adjacent sea floor. Volcanoes at or below the sea surface include
coral-rimmed *atolls*, essentially uneroded *seamounts*, and flat-topped *guyots*.
Seamounts that do not approach the surface are abundant throughout the
Pacific ocean; features reaching near or to the surface are particularly common in the western North Pacific. Cross sections in Figure 4.4 show the
essential differences among the three. Figure 4.5 shows the locations and
density distributions of atolls, low islands, and guyots [2]. Each may be
considered a potential mariculture site.

Tectonic Activity. Tectonic activity is a second major source of topographic relief on the abyssal sea floor (although it must be realized that
volcanic activity and tectonic activity are not by any means separable
processes). In particular, tectonism produces sinuous mid-ocean ridges (Figure
4.3) large enough to traverse entire oceans; such a ridge nearly bounds the
Atlantic Ocean area of mariculture interest on the east; another ridge enters
the Gulf of California, narrowly missing the Pacific Ocean area of interest.
The ridges are commonly areas of intense volcanic and associated earthquake
activity. Their magnitude is best appreciated by realizing that from flank to
flank the mid-Atlantic ridge spans approximately one third the width of the
North Atlantic Ocean. In some localities they may rise from abyssal depths to
the sea surface, as shown in Figure 4.6 [3].

Fracture zones commonly associated with the mid-ocean ridge systems are
usually long and tend to follow great circular paths more or less perpendicular
to the ridges, as may be seen in Figure 4.7 [4]. These zones are particularly

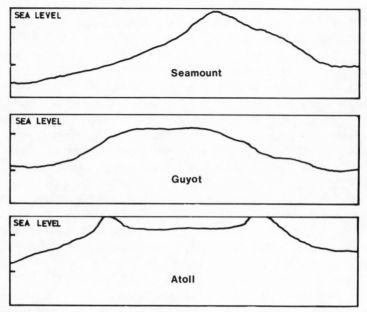

FIGURE 4.4 *Comparative cross section of seamount, guyot, and atoll.*

prominent in the Pacific Ocean, where they extend for up to half the ocean's width and may be associated with hundreds of kilometers of very rugged topography.

The topographic features just described rise above the plane of the surrounding sea floor. Arcuate grooves that extend several thousand meters below that plane are also an expression of tectonic activity. Known as *trenches,* these depressions are prominent around the margin of the Pacific Ocean, particularly adjacent to island arcs.

We have included this brief mention of tectonics for completeness and because tsunamis, which could pose threats to some open sea mariculture sites, result from tectonic activity; but at the moment any other implications for mariculture are unclear. Hill [2] is an excellent source on the state-of-the-art to about 1960 of geophysical information related to the topics we have covered, as well as for information on other aspects of marine topography. *Scientific American* [4] provides some more recent general discussions of the subject.

Composition

Materials present on the sea floor generally belong to one of two broad categories: sediments and volcanic materials.

Sediments. Sediments cover the bulk of the sea floor and originate from three sources: (1) terrestrial runoff, (2) chemical precipitation at the site of

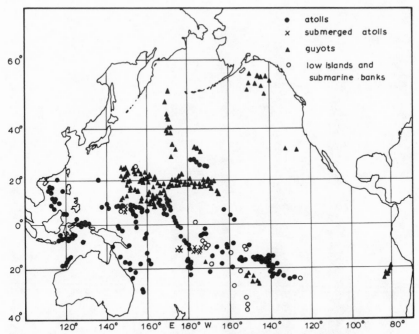

FIGURE 4.5a *Location and density distributions of volcanic features in the Pacific Ocean. (Adapted from Hill, Vol. 3, Chap. 15 [2].)*

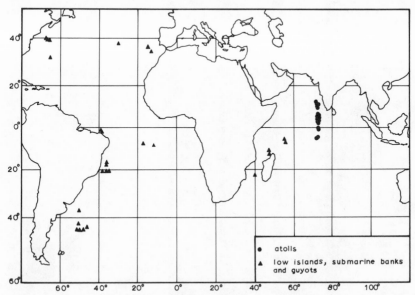

FIGURE 4.5b *Location and density distributions of volcanic features in the Atlantic Ocean. (Adapted from Hill, Vol. 3, Chap. 15 [2].)*

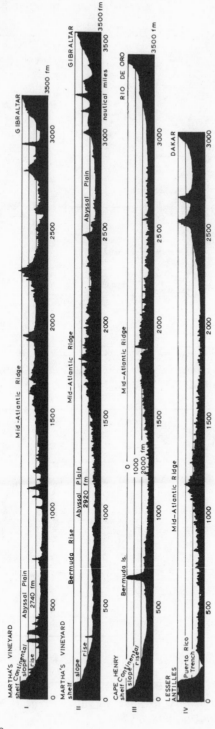

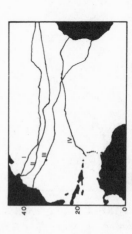

VERTICAL   EXAGGERATION   40:1

FIGURE 4.6   *Four trans-Atlantic topographic profiles illustrating the mid-Atlantic ridge. (Adapted from Heezen et al. [3].)*

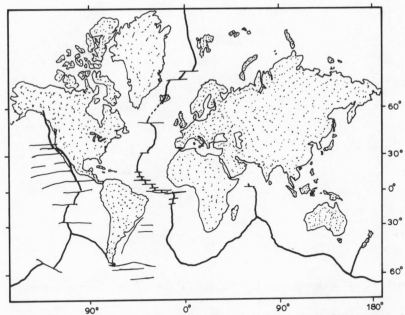

FIGURE 4.7 *Major areas of fracture zones along the mid-oceanic rift.*
(Adapted from Scientific American *[4].)*

deposite, and (3) precipitation elsewhere and subsequent transport to the site
of deposit. Transported precipitates may be of obvious biogenic nature (e.g.,
foraminifera tests) or of likely inorganic origin (e.g., silicate minerals precipitated interstitially). The mode of transportation may be wind, waves, or a
variety of ocean currents. Figure 4.8 illustrates the worldwide distribution
patterns of marine sediments [2].

Some generalities about these sediments and their origins are pertinent to
this investigation. The amount of calcareous materials in sediments is generally quite variable near land, and becomes more constant at a distance from the
coasts. When not influenced by proximity to land, the calcium carbonate
content of marine sediment generally decreases with increased water depth.
This calcareous material is largely biogenic (foraminifera and coccolitho-
phorids) and is frequently the major constituent of pelagic sediments from
water shallower than about 5,000 m. In deeper water, siliceous organisms
such as diatoms or radiolarians can be primary contributors to the sediment.
These siliceous components are prominent in three bands of Pacific Ocean
sediment: near the northernmost and southernmost extremes of the ocean,
and in an east–west strip near $5°$ north latitude. It is obvious from other
considerations in this chapter that the biogenic sediment distribution pattern
reflects three major processes: (1) biological productivity, (2) the degree of

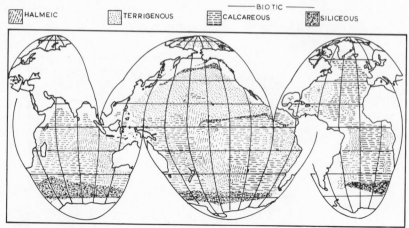

FIGURE 4.8 *Worldwide oceanic sediment distribution patterns. (Redrawn from Hill, Vol. 3, Chap. 25 [2].)*

masking by terrigeneous sediment, and (3) the degree of re-solution of calcareous or siliceous materials.

Land-derived sediments are prominent around the margins of the ocean basins, as would be expected. In the Atlantic, such "marginal" distribution (known as the *turbidity-current deposits*) includes much of the ocean basin, whereas the much larger Pacific basin is not so extensively affected.

Special mention should be made of clay-mineral deposition. These materials are of particular interest because of their potential contribution to the understanding of sediment sources, dispersal patterns, and chemical alteration. Two references that provide starting places on the subject are Griffin and Goldberg [5] and Biscaye [6]. These give extensive discourses on the Pacific and Atlantic Oceans.

The Pacific Ocean has large areas too deep for extensive calcium carbonate accumulation and too far from land for much terrigeneous influence. These regions characteristically have a high proportion of chemical precipitates (e.g., manganese nodules), sediment of volcanic origin, and windblown debris. These latter materials accumulate at extremely slow rates—less than 1 micro meter (μm)/year—compared with rates 10 or more times as fast for other sediments elsewhere in the world oceans.

Volcanic Materials. Volcanic materials include lava and other direct deposits, volcanic ash, and volcanically induced precipitates. There is an obvious overlap between sediments and volcanic deposits.

It seems then that knowledge of sedimentation can provide insight into the historical biproductivity of a prospective mariculture area as well as into possible problems related to terrigeneous materials. Again, Volume 3 of Hill's series is an excellent starting point for more complete information and additional references [2].

PHYSICAL CHARACTERISTICS OF OCEAN WATERS

Much of the behavior of the ocean is determined by the unique nature of water itself, thus warranting a general review of the basic physical properties of seawater. This includes discussions of thermal properties; density and stability; optical, acoustic, and colligative properties; and the effects of variations in salinity.

The main body of the open ocean is very different from the regions affected by the eastern and western boundaries; moreover, the eastern and western boundary regions are fundamentally very different from each other. Within the boundary regions, the conditions over the continental shelves—the *neritic zone*—can differ significantly from conditions farther offshore. Finally, the boundaries define various semiisolated bodies of water, which can be described separately. There is also a strong vertical density layering of the ocean, in which a thin, warm surface layer lies above the thick, cold, dense body of the deep sea. The *thermocline*, a highly stable transition zone between surface and deep water, effectively separates the motions and properties of the upper and lower layers of the ocean. We shall separate our descriptions into these two categories. Following oceanic water layers, we describe the offshore eastern and western boundary regions, with specific examples being the California Current and the Gulf Stream, respectively. The Gulf of Mexico, a large cul-de-sac with oceanic depths, is discussed separately, as are the neritic zones of the eastern and western United States.

Finally, we find some subjects that stand outside the above organization, but that must be covered because of their importance to possible mariculture projects. Windwaves are included in this category.

It is hardly necessary to emphasize the brief and sketchy nature of the coverage given here. Much more complete (although still general) discussions are available in any advanced general oceanography text, such as those by Dietrich [7] or Neumann and Pierson [8]. A short work by Pickard [9] is a valuable reference for descriptive physical information. Ecologically oriented discussions of a number of physical subjects are collected in a treatise edited by Hedgepeth [10]. The three-volume work edited by Hill [2] presents a collection of timely discussions by leading scientists on a wide range of topics, including a general description of eastern boundary currents by Wooster and Reid. Stommel [11] gives a comprehensive survey of the Gulf Stream system. Texts on waves include a comprehensive treatise by Kinsman [12], a still-useful descriptive work by Bigelow and Edmondson [13], and a detailed manual of sea and swell prediction by Pierson, Neumann, and James [14].

Thermal Properties

With respect to its thermal properties, pure water is a most remarkable substance, and these properties are only slightly modified by the presence of salts in the waters of the sea. The uniqueness of water goes largely unrecognized only because it is such a very common substance on the earth's surface.

For example, water has the second-highest specific heat of all known substances—approximately 1 calorie/gram/degree Celsius (cal/g/°C) at 18°C. (*Specific heat* is defined as the amount of heat energy in calories required to raise the temperature of 1 g of a given substance by 1°C under specific conditions. The specific heat of water is 1, because the definition of the calorie itself is based on the thermal properties of water.) In practical terms, the high specific heat of water means that water can absorb (or give off) very large amounts of heat energy with relatively small changes in its temperature. When this property is considered in the context of the huge amounts of water on the earth's surface, it can be seen that the oceans must act as climatic stabilizers for the entire globe.

As a second important example, water has the highest latent heat of evaporation of all liquids, approximately 580 cal/g at room temperature. (*Latent heat of evaporation* is defined as the heat energy absorbed in changing from a liquid to a gaseous state at the same temperature.) Thus, global-scale evaporation from the open sea involves the storage of enormous amounts of heat energy. The energy reappears as heat when the gas condenses to liquid, a conversion occurring primarily in the formation of clouds. Most of the input of solar heat energy to the atmosphere, the energy that is the basic cause of all atmospheric motions, goes through this route of ocean evaporation to cloud formation. Very little solar energy is absorbed directly by the atmosphere.

The phenomenon of thermal expansion illustrates one case in which the presence of dissolved material in the ocean causes behavior significantly different from that of pure water. It is common knowledge that pure water contracts as temperature falls down to 4°C, but that as temperature drops still lower the water expands. Pure water is thus most dense before freezing—at 4°C. Deep high-latitude lakes, for example, maintain this temperature all winter, except near the surface, since the colder surface water is less dense than the deep water, and thus cannot sink. The addition of salt, however, drops the temperature of maximum density below the freezing point for seawater. It is possible, therefore, to find very cold, dense water in the deep oceans.

The freezing point, −2°C, is thus a fixed lower limit to the range of temperatures in the ocean. The upper limit is climatologically fixed at approximately 30°C in the open sea. The mean temperature of the entire world ocean (after correction for the slight tendency for temperature to increase adiabatically downward with increasing pressure at the rate of about

0.1°C/1,000m) is 3.52°C. This low mean value reflects the relative thinness of the warm surface layer.

When one recognizes that the growth rates of marine organisms show strong tendencies to increase with temperature, up to 25 to 30°C, it becomes reasonably clear that open sea mariculture activities either will tend to be restricted to warmer surface waters or will require large heat energy inputs. For example, if the nutrients that occur naturally in waters below 1,000 m are to be employed for primary production, these colder waters probably must be warmed, as well as raised, in order to attain optimum bioproductivity.

Density

The density of seawater depends on its temperature, salinity, and pressure. When a body of water moves in the deep ocean, temperature and salinity remain constant unless there is mixing with different water bodies. The pressure, however, varies as the water moves to different depths. For many purposes it is most convenient to eliminate the effect of variable pressure by considering only the density a given water would have if it were at the sea surface. We thus consider only the effects of temperature and salinity on density.

The average density of seawater, if it were all at sea-level pressure, would be 1.025 grams/cubic centimeter (g/cm^3); the density range would extend approximately from 1.022 to 1.029 g/cm^3. These density variations seem small, but they strongly control the vertical structure of the ocean: the water is stably layered, with density increasing downward. This stability strongly suppresses vertical turbulence, since the motions are slow and the energy available to maintain turbulence is small. So, although horizontal motion and mixing in the ocean are common, vertical motion and mixing require special circumstances. Some of these circumstances, such as wind-driven upwelling or vertical mixing caused by surface cooling and instability in autumn, will be described later.

For many practical problems, the effect of temperature on density in the upper kilometer of the ocean so far outweighs the effect of salinity that the latter can be ignored, and the general pattern of density can be inferred directly from the temperature patterns. This is true because there is typically a rapid downward drop in temperature beneath the thin, warm surface layer. Within the same depth range, however, the salinity variations normally are small. Salinity variations become important in their effect on density only (1) where vertical temperature variations are small (i.e., in the deep layers of the sea or in polar regions); (2) in estuarine areas where salinity variations are relatively very large, or (3) when the precise value of the density must be known, rather than the general pattern.

The stability of vertical layering in the oceans is another indication that considerable energy will be required if deeper nutrient waters are to be raised in large quantities to support open sea mariculture. The relative importance of temperature over salinity in this process holds obvious implications for the utilization of heat energy.

Optical Properties

The absorption of radiation in the visible and near-visible spectrum in water is strongly determined by the wavelength, or color, of the light. In pure water, or the clearest filtered seawater, absorption is at a minimum at a wavelength of 0.47 μm (blue light), increases rapidly at shorter and longer wavelengths, and becomes very great for wavelengths outside the visible range. This is of particular importance in the infrared realm, since roughly half the incident solar energy is at these longer wavelengths.

The absorption of light shows striking differences in different oceanic water masses. As water becomes less clear, absorption is increased at all wavelengths, but it is increased relatively more at the short wavelengths. This seems to be caused primarily by dissolved organic matter, rather than by particulate substances. The result is a shift of the minimum absorption point from 0.47 μm (blue) to around 0.55 μm (yellow-green) for turbid coastal water. Only if the water is extremely turbid does the color of the suspended particles themselves affect the wavelength of the transmitted light.

The variations in absorption produce extreme variations in the penetration of daylight in different bodies of seawater. In the clearest ocean water, animals may be able to detect daylight down to about 1,000 m; the depth at which the light level equals 1 percent of the incident radiation is 100 to 150 m. This is approximately the minimum light at which sufficient photosynthesis can occur to allow growth, as opposed to mere survival, of plants in the sea. For coastal water, the respective figures more typically may be 200 m and 10 to 30 m; and, for very turbid harbor waters, only 20 m and 2 to 3 m.

From this we might draw two implications for mariculture. The first is the well-recognized fact that, at best, the photic zone (in which primary production can occur at significant rates) is only a little more than 100 m thick, and it becomes narrower with increasing turbidity. The second is that primary production can be self-limiting, since it increases dissolved organic matter and turbidity. Thus, without some sort of vertical mixing, we might conclude that intensive primary production could occur only in the first 10 or fewer meters below the surface.

Acoustic Properties

The dissolved salts in seawater cause it to be an electrically conductive medium, and thus nearly opaque to electromagnetic radiation except for the visible and near-visible light band. For this reason, sound must be depended upon for most of the information-carrying tasks for which we use radio waves above water. [Very low frequency (VLF) radio waves are in use for undersea communication also. Their effectiveness, however, is limited by the conductivity of seawater, as noted above, and by the very large transmitter power requirements and the large receiving antennae that are characteristic of such systems. Much of the VLF research and development information carries military security classifications.]

Since the compressibility of water is very much less than that of air, the speed of sound in seawater is typically about 1,500 meters/second (m/s), about five times faster than in air. The speed increases as temperature, salinity, and pressure (i.e., depth) increase. Under many conditions the effect of salinity can be ignored, since it is several times smaller than the other effects.

In the surface layer and in the deep water, the vertical temperature distribution either is uniform or changing very slowly. The increase in pressure with increasing depth causes a corresponding increase in sound speed in such layers. In the thermocline, however, a rapid decrease in temperature with increasing depth brings about a corresponding net *decrease* in sound speed, even though pressure continues to increase. This decrease in speed through the thermocline reverts back to an *increase* in the water beneath the thermocline, thereby creating a layer with minimum sound speed near the bottom of the thermocline, typically at around 1,000 m.

The rather sharp variations in sound speed with depth lead to refraction, or bending, of sound rays that are traveling laterally, and the refraction always bends the ray toward the layer in which sound speed is slower. The refraction patterns lead to complex zones of focusing and defocusing of acoustic energy. To predict these effects, the detailed thermal structure in the region of interest must be known.

Both the electrical and sound conductance properties of seawater may prove of importance in the design of attraction, repulsion, and harvesting subsystems of future open sea mariculture systems. Recent work has already shown the use of electricity in lobster and shrimp harvesting and the employment of sound in repulsing certain predators (see Chapters 9 and 10).

Colligative Properties

Colligative properties are those which depend only on the concentration of solute in a solution. Freezing-point depression is one of these properties, and it has already been mentioned that the freezing point of seawater of average salinity is near $-2°C$. Another colligative property of particular biological importance is osmotic pressure.

The phenomenon of osmotic pressure comes about when two solutions with different concentrations are separated by a semipermeable membrane that will allow passage of the small molecules of the solvent, but not of the large molecules of the dissolved substances, such as salt. There will be a net flux of solvent molecules through such a membrane toward the side with higher concentrations of the solute. In a closed system, this will increase the pressure on the "saltier" side. If the pressure becomes high enough, the flux across the membrane will stop; the resulting osmotic pressure corresponds to the difference in the concentrations present. For instance, if the liquids are fresh water and normal seawater at about 25°C, the osmotic pressure is approximately 26 kilograms/square centimeter (kg/cm^2).

Many biological membranes are semipermeable; water will flow into or out of a marine organism depending on the relative saltiness of its body fluids versus that of the surrounding water. The organisms have mechanisms for adjusting the fluid balance, but usually these mechanisms are restricted in their capacity. Large salinity variations may present an impenetrable physiological barrier. Most pelagic organisms can, however, handle any salinity variations over the normal oceanic range; only in inshore or estuarine waters may the variations be large enough to act as a barrier.

Other colligative properties include elevation of the boiling point and lowering of vapor pressure, but these are of little interest in the present context and can be omitted from this discussion.

To this point we have been examining characteristics common to all seawater. We have noted, too, the existence of three discrete layers of ocean water that are distinguishable not only by the depth ranges at which they occur, but also by the unique qualities each exhibits. For the purposes of mariculture, the dissimilarities in ocean waters located at different depths, as well as their behavioral patterns at varying latitudes and longitudes, may be as significant as the similarities.

Surface Water Layer

The temperature and thickness of the ocean surface layer are determined by the oceanic heat budget and the pattern of surface currents. Surface salinity is determined by the difference between evaporation and precipitation plus runoff from adjacent land, modified again by the pattern of currents. It is possible to characterize the surface waters in our area of interest by broad zonal bands within which similar climatological characteristics prevail. These zonal bands, or general categories, are (1) equatorial, (2) tropical, and (3) transitional.

Equatorial Surface Water. Warm Equatorial Surface Water lies equatorward of about 20° north. Its temperatures are over 25°C. Temperatures increase toward the west, with maximum values approaching 30°C. Average annual fluctuations are less than 3°C. Salinities over the whole zone are somewhat lowered by the high rainfall of the doldrum belt, where the salinity reaches a minimum between 34.0 and 34.5 parts per thousand (ppt) in the Pacific and near 35 ppt in the saltier Atlantic. High temperatures and low salinities make the Equatorial Surface Water the least dense in the open oceans.

The thickness of the equatorial surface layer is at a minimum along approximately 10° north, which defines the southern edge of the North Equatorial Current. There is a general overall increase in thickness to the west, whereas to the east the lower boundary drops from less than 50 m at 10° north to approximately 150 m at 20° north. The comparable south-to-north slope in the western Pacific is 150 to 250 m.

The light Equatorial Surface Water lies above a shallow but very intense permanent thermocline that strongly suppresses vertical flow or mixing. Nutrient concentrations, in consequence, are permanently low—generally between 0.25 and 0.5 microgram (μg) atom/liter of phosphate—phosphorus.

Tropical Surface Water. Tropical Surface Water lies approximately in the zone 20 to 30° north. Temperature here is more variable in space and time than in the equatorial zone. Along 20° north latitude it ranges from about 18°C in winter to a high of about 25°C in summer; the annual fluctuation increases northward to a low of around 8°C along 30° north latitude. This category of surface water is particularly characterized by a zonal band of high salinity, with maximum values about 35.5 ppt in the Pacific and above 37.0 ppt in the Atlantic, as a result of relatively high evaporation and low rainfall in the horse latitudes.

The band of Tropical Surface Water is relatively thick, extending down to 150 m in the east and to 250 m in the west. Although the thermocline beneath is less intense than that in the equatorial zone, it lies deeper. Consequently, relatively nutrient rich water, which can be found at a depth of 100 to 150 m in the central ocean near 10° north, is confined below 600 m at a latitude of 25° north. This factor, together with a tendency toward convergence and sinking in this zone between the major wind belts, very strongly reduces the possibility of upward transport of nutrient-rich water. Surface nutrient concentrations are typically minimal in the Tropical Surface Water, with values under 0.25 μg atom/liter of phosphate. Because of the resulting low level of biological productivity, and the concomitant low production of dissolved and particulate organic material, these waters have the greatest clarity of any in the open ocean.

Transitional Surface Water. Transitional Surface Water is highly variable within the band extending from 30° north latitude to the region of more uniform, and very cold, Polar Surface Water. (The latter lies outside our area of interest.) Both temperature and salinity vary widely within the transition zone: progressing poleward from its southern to northern boundaries, the temperature sinks from approximately 18 to about 5°C. Salinity decreases over the same distance from 37 down to 35 ppt in the Atlantic, and from 35 down to 32 ppt in the Pacific. These northward gradients are sharper toward the west, and spread out toward the east.

The transition zone is strongly affected by seasonal climatic fluctuations, with an annual surface temperature range of more than 10°C in some areas. This leads to a characteristic annual variation in surface-layer thickness as follows:

- *Winter:* Severe cooling, aided by storminess, causes a deep convective overturn of the surface layers, leading to a thick but cold surface layer above a deep, fairly weak, permanent thermocline. The isothermal surface layer may be 200 to 300 m deep, with relatively high nutrient levels but low productivity because of low light intensity.
- *Spring:* Increasing solar input in the spring leads to two results. First, biological productivity increases rapidly. Second, the upper layers begin to warm, developing a shallower seasonal thermocline above the deep permanent thermocline. The increased productivity begins to lower surface nutrient concentrations, and the increasing stability of the water column inhibits turbulent transfer of nutrients from below.
- *Summer:* The heating process begun in the spring reaches its maximum. A thin, 20- to 30-m surface layer is warmed as much as 8 to 12°C above the winter temperatures in these latitudes, and the shallow seasonal thermocline (or *summer thermocline*) is most intensely developed. Between this and the deep permanent thermocline there may be an isothermal layer (or *thermostad*) still at the winter temperature. Surface nutrients have been so reduced that biological productivity is at low levels in spite of the long hours of daylight.
- *Fall:* Surface cooling leads to convective overturn (aided by storms of increasing severity), which progressively deepens the surface layer and mixes away the seasonal thermocline. The deepening level of mixing increases surface nutrients, leading to somewhat increased biological productivity until fall becomes so far advanced that light again becomes limiting. Cooling and mixing continue until the deep isothermal layer of winter is once again established.

The higher Atlantic salinities lead to an increased density of the surface water in winter, and thus to increased depth of overturn. This reaches its greatest extent in the Norwegian and Irminger Seas (north of our area of interest) where convection to the bottom occurs in winter. No such deep convection occurs in the North Pacific. Figures 4.9, 4.10, and 4.11 show surface temperatures and salinity distributions in the area of interest [1].

It would appear that strong seasonal variations, along with limitations in light availability toward more northern regions, are likely to make Transitional Surface Water a less attractive medium for year-round, primary-production-based culture than Tropical Surface Water might be. These variations quite possibly play a role in the migratory behavior of some nekton and we know very little of the potential results of containing

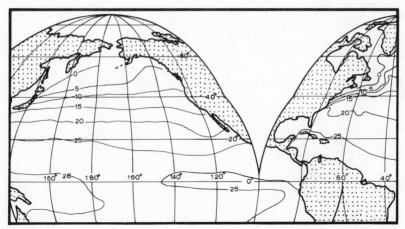

FIGURE 4.9 *Surface temperature distribution (°C) in February for areas of interest. (Adapted from Sverdrup et al. [1].)*

migratory species in northern latitudes when they would normally migrate. This does not imply, however, that Transitional Surface Water may not be attractive for nonmigratory forms or for culture systems that are independent of primary production.

Large-Scale Surface Currents. Large-scale surface currents tend to bound the areas of interest since they essentially coincide with the major northern hemisphere gyre in each ocean. As Figure 4.12 shows, the westerly flow of the North Equatorial Current, at speeds of 25 to 50 cm/s, lies in the region of

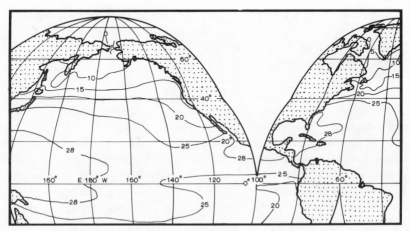

FIGURE 4.10 *Surface temperature distribution (°C) in August for areas of interest. (Adapted from Sverdrup et al. [1].)*

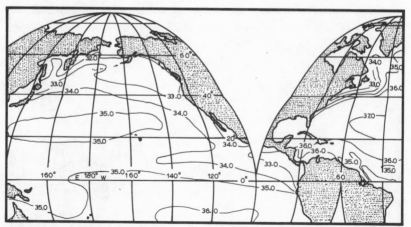

FIGURE 4.11 *Surface salinity distribution (ppt) in areas of interest. (Adapted from Sverdrup et al. [1].)*

the trade winds, roughly 10 to 20° north, and the easterly North Pacific and North Atlantic currents, flowing at speeds of 25 cm/s, are found in the region of prevailing westerly winds, 30 to 50° north. These bands of westward and eastward flows are connected by the northward and southward flow of the boundary currents (which will be described separately) to form the quasi-closed circulation of the gyres.

The easterly flowing Equatorial Countercurrent lies at about 5 to 10° north, just to the south of the North Equatorial Current, in both oceans. It varies seasonally, flowing broadest and strongest at 25 to 50 cm/s in summer, at which time it extends from the Philippines to the Gulf of Panama in the Pacific, and from about 50° west to the Gulf of Guinea in the Atlantic. In winter the countercurrent is narrower, weaker, and more variable. In the Pacific it is discontinuous eastward of 120° west; in the Atlantic the easterly flow is confined to the Gulf of Guinea.

The belts between the North Equatorial currents and the North Pacific or Atlantic currents are characterized by weak, meandering, highly variable currents. These are the horse-latitude zones of Tropical Surface Water, with high salinity and very low surface nutrient concentrations.

Very little is known about the variability of the major currents. We do know that one variation is periodic: the tidal effect tends to superimpose a rotary current on whatever unidirectional flow is present. In some areas, such as Hawaii, the tidal current may be in fact the main component of the flow. However, until very recently all tide measurements were made along coasts, so data on deep-sea tides are lacking.

There are also many aperiodic variations. In general, the major transfer of momentum from wind to ocean occurs in the trade-wind zones, so one would

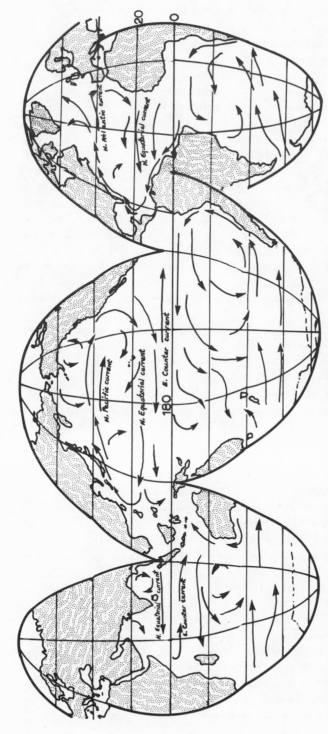

FIGURE 4.12 Major surface currents in the Atlantic and Pacific oceans. (Adapted from Sverdrup et al. [1].)

expect the equatorial currents to be relatively steady in velocity. The more variable winds of the westerlies impart a large component of transient motions to the flow there. The details will be better understood after the results of moored-buoy observational programs now being developed have been analyzed. But it can be said that these regions of mean easterly flow are characterized by eddies, shifting bands of strong and weak flow, and transient periodic and aperiodic oscillations. Accurate prediction of a specific current at a particular time and place is not possible.

Even the trade-wind-driven equatorial currents may prove much more complex than heretofore believed. For example, a Russian buoy array deplyed in the central Atlantic about 15° north, in early 1972, revealed a very high degree of variability, with the surface current often running east against the trade wind. Much more information will be needed before generalizations can be made, except perhaps to emphasize that there is no place in the open ocean in which the currents can be depended upon to be constant in direction and speed.

Our present inability to understand and predict variability in surface currents holds implications for open sea mariculture. If a station were to depend on currents for either power or water circulation, or if it were to be so designed that unexpected strong currents might damage its operations, the dearth of present knowledge would leave such a station particularly vulnerable. This implies that current-sensitive stations should exist only in those areas where local currents are well understood, and, conversely, stations in general should be designed to be as nearly insensitive to currents as possible.

Intermediate Water

Intermediate Water comprises the layers of water in and just beneath the main thermocline, roughly from just beneath the surface layer to a depth of 1,500 to 2,000 m. These layers are intermediate in both depth and density— between the surface layer and the cold, dense, relatively uniform Deep and Bottom waters. The increase in density with depth, especially rapid in the upper Intermediate Water, makes these layers the most stably stratified in the ocean. Lateral mixing and flow (along the density layers) can occur easily, but vertical motions are strongly suppressed.

The density layers of the Intermediate Water rise toward the surface as they approach the higher latitudes (as the temperature of the surface water drops and the density thereby increases), with deeper-lying layers reaching the surface farther to the north in our area of interest. Many of the characteristic features of the Intermediate Water can be explained this way: the surface-water properties typical of a certain latitude are carried by mixing and/or advection down, and equatorward, along the density layer which comes to the sea surface in that latitude. These characteristics change slowly with distance because of the suppression of vertical mixing. Typically, then, the *vertical* pattern of characteristics in the intermediate layers of the inter-tropical regions bears a close relationship to the *horizontal* sea-surface pattern of these characteristics in a poleward direction.

Other characteristic features of the Intermediate Water are caused by processes occurring at these depths, rather than by mixing or sinking of water from the surface. Light intensities are too low for biological production by plants to occur beneath the surface layer, except perhaps rarely and briefly in the very uppermost thermocline. The Intermediate Water is thus a region of consumption by animals, with all life depending on the productivity of the surface layer just above. The proximity of the surface layer supports a sizable animal population in these mid-depths, however, and the deep penetration of detectable daylight triggers a marked daily vertical migration of many of the components of this population. The vertical migrators rise toward the surface at night to feed, and return to the deep layers during the day.

This migration pattern—the upward swimming of hungry animals at dusk, and the downward swimming of fed animals at dawn—represents a significant net downward transport of organic matter, and thus of nutrients. The sinking of any nonliving particulate organic matter from the surface layer represents an additional downward transport of nutrients across the barrier of the stable thermocline. Oxidation of the nonliving organic matter and the vertical migrator's metabolism at depth during the day have two effects on the characteristics of the Intermediate Water: (1) oxygen concentrations are reduced, and (2) inorganic nutrient concentrations are increased. The resulting oxygen minima and nutrient maxima characteristic of the Intermediate Water are discussed in Chapter 5. These features are especially strongly pronounced in the eastern intertropical oceans because of high surface productivity and relatively slow currents, but the effects extend over very wide regions.

The motions of the Intermediate Water are very complex. The major wind-driven circulation, discussed earlier, affects the upper Intermediate Water with an intensity that decreases with depth. At the same time there is a tendency for sinking and equatorward flow along density layers associated with surface convergence zones. The patterns of flow can best be determined by examining plots of property distributions along layers of constant density (*isentropic analysis* of property distributions). The speed of average flow varies downward from the typical surface layer rate of 25 cm/s to very slow motions of perhaps 1 cm/s in the deeper Intermediate Water. The actual instantaneous current may be much stronger at all depths, with transient motions of 50 cm/s or more with variable direction.

Intermediate Water temperature varies over a wide range. At the upper boundary, the temperature matches that of the sea surface, varying from over 25°C in equatorial regions to about 5°C at the poleward limit of the transition zone. (Intermediate Water, in the sense used here, does not exist in polar regions.) Temperatures at the depth of the lower boundary are very uniform with latitude, but the boundary itself is somewhat arbitrarily defined. We can choose approximately 2°C as the lower boundary in the Pacific at a depth of about 2,000 m, and a temperature of 4°C in the Atlantic at about 1,500 m (see Figure 4.13).

A pronounced layered structure of salinity maxima and minima is another characteristic of the Intermediate Water. As shown in Figure 4.13, the surface high-salinity zones in mid-latitudes (Tropical Surface Water) produce layers of high salinity that extend downward and toward the equator in the uppermost Intermediate Water. These shallow maxima are prevalent in most of the central and western ocean between 10 and 25° north. The maxima do not extend into the eastern ocean because of the pattern of wind-driven circulation, which moves west in these latitudes.

The deeper Intermediate Water of the North Pacific is characterized by a broad layer of low salinity, which receives its characteristics from the sea surface in the far-northern Pacific. In the Atlantic, the higher surface salinities

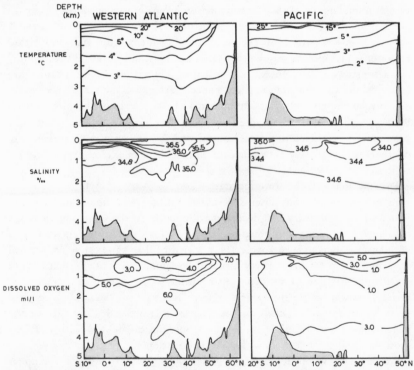

FIGURE 4.13 *Vertical sections showing distribution of temperature, salinity, and oxygen in the western Atlantic and Pacific oceans. (Adapted from Sverdrup et al. [1].)*

in high-northern latitudes lead to winter production of dense Deep and Bottom Water, rather than Intermediate Water. A relatively small low-salinity layer is found in the North Atlantic only in the northwest corner; this is formed in the region of convergence between the Gulf Stream and the colder, fresher water of the Labrador Current.

North Atlantic Intermediate Water is also influenced by high-salinity outflow from the Mediterranean Sea. This outflow crosses the shallow Straits of Gibraltar and sinks, with mixing, to a depth of about 1,200 m, at which level it spreads into the ocean. Near its source this water is visible as a salinity maximum layer; farther removed, salinity is reduced by mixing, but remains noticeably higher near the 1,200-m depth (Figure 4.13).

The rate of renewal of Intermediate Waters by circulation and mixing is much slower in the North Pacific than in the North Atlantic. The longer subsurface residence times of this water in the Pacific have allowed the biomechanical depletion of oxygen and the enrichment of nutrient concentrations to become far more advanced than in the Atlantic.

In the Pacific, very wide areas based along the eastern boundary have an intermediate oxygen minimum with concentrations less than 0.5 milliliter (ml)/liter. The minimum extends westward across the ocean, with the lowest values found in broad, vertically thin tongues. One prominent tongue is centered in the North Equatorial Current at about 3,000 m; concentrations of less than 0.5 m/liter extend as far west as 170° east. The pattern of this minimum can be related closely to the current direction.

A minimum tongue extending westward near 1,000 m, in 30 to 40° north, must be related more to oxidation occurring within the area itself, since the weak flow at this depth is actually directed toward the east and southeast. Water with an oxygen concentration below 0.5 ml/liter is found to about 170° west in this tongue.

In the North Atlantic, a minimum layer extends west from the coast of Africa in latitudes 10 to 20° north at depths of about 500 m. Oxygen concentrations are 1 to 1.5 m/liter in the east, increasing to 3 ml/liter in the west. This shallow minimum extends into the western North Atlantic, with concentrations approaching 4 ml/liter. There is no Atlantic counterpart to the mid-latitude minimum of the North Pacific.

Nutrient maxima tend to occur slightly deeper than the oxygen minima in the Intermediate Water. In the Pacific, nutrients have a broad maximum centered near 1,000 m in mid-latitudes, and at about 700 m equatorward of 20° north. This maximum is especially well developed along the eastern and northern borders of the ocean. Phosphate concentrations of over 3 μg atoms/liter are found over most of the eastern and central North Pacific within this maximum. In the Atlantic there is only a weak nutrient maximum, also centered near 1,000 m, that extends northward from high southern latitudes. This brings concentrations of 1 μg atom/liter of phosphate no farther than to mid-latitudes in the North Atlantic. (Nutrient distribution is discussed in more specific detail in Chapter 5.)

The importance of Intermediate Water to open sea mariculture stems primarily from the fact that this water contains a vast reserve of inorganic nutrients and that these nutrients offer degrees of accessibility that vary directly with the proximity of Intermediate Water to the surface. From the information summarized here, we suggest two tentative generalities: (1) the Pacific generally seems more attractive for primary-production-based culture systems than the Atlantic; and (2) because northern latitudes (where Intermediate Water is nearer the surface) pose seasonal variation problems, the area between the equator and about $20°$ north generally appears slightly more attractive for primary-production-based culture systems than areas farther north. This is because of the shallower depths at which nutrients may be expected, the greater annual light flux, and higher ambient temperatures. Clearly, however, these broad generalities will not apply to all areas; thorough studies at a proposed site should precede the decision to locate there.

Deep and Bottom Waters

There are only two regions in which large volumes of water become dense enough, during winter, to sink to abyssal depths. These regions are the sources for the Deep and Bottom waters of the entire world ocean. Both regions are associated with the Atlantic: the Weddell Sea on the coast of Antarctica is at the far southern end of this ocean, the Norwegian Sea, between Greenland and Scandinavia, is at the northern end.

In the Weddell Sea, the dense water formed on the continental shelf sinks down along the continental slope until, after some mixing, it arrives at the abyssal floor as Antarctic Bottom Water (AABW) with a temperature slightly below $0°C$. From here it spreads northward up the Atlantic and eastward toward the Indian and Pacific oceans.

In the Norwegian Sea, winter convection forms dense water that is blocked from the Atlantic proper by a submarine ridge running from Greenland to Iceland to Scotland. The dense, cold water intermittently cascades over this ridge and mixes with water in the Atlantic to form the North Atlantic Deep Water (NADW), with a temperature about $3°C$. The NADW then spreads south through the Atlantic, and turns east, south of Africa, toward the Indian and Pacific oceans.

In the Atlantic the AABW and NADW are distinct; the first travels north along the bottom, and the second moves south near depths of 3,000 m. They also differ in salinity. The AABW is at about 34.65 ppt near its source, whereas the NADW is near 34.9 ppt in the North Atlantic.

By the time the water from these two sources has reached the Pacific, mixing has blended them into a single mass. The entire deep Pacific is filled with a nearly uniform body of water with temperatures of 1 to $2°C$ and salinities of 34.6 to 34.8 ppt. This single mass represents nearly one third of all the water on earth; it has been given the name *Common Water* [15].

The mean rates of flow are very low at these great depths, perhaps less than 0.5 cm/s in general, with some regions having mean flows as high as 5 cm/s where flow is concentrated by bottom topography. The actual motions at given times include erratic motions of speeds to several decimeters per second. Little is known about the causes or time–space scales of these random motions.

The proximity of the sources implies that the "age" of Deep Water in the Atlantic (the time since the water was last in contact with the sea surface) must be much less than that in the Pacific. It is estimated that the overall residence time of Atlantic Deep Water is 300 years; in the northern North Atlantic it must be significantly less. In the North Pacific, the residence time is estimated to be roughly 2,000 years. These differences in residence time are reflected in the distributions of oxygen and nutrient concentrations, as was also the case for the Intermediate Water. The biochemical alteration of properties is much slower, however, at great depths because of the limited supply of organic matter from above.

The deep North Atlantic is essentially filled with NADW, near 3°C and with 34.9-ppt salinity; temperature and salinity decrease slightly toward the bottom. Oxygen concentration is high, near 5 ml/liter. Nutrient concentrations are low: phosphate is near 1 μg atom/liter in mid- to high latitudes.

The deep North Pacific is filled with the exceptionally uniform Common Water, near 1.5°C and with 34.7-ppt salinity. Oxygen concentration increases downward from 2 ml/liter to near 4 ml/liter at the bottom. Nutrient concentrations are high: phosphate is near 2.5 μg atom/liter, with a slight decrease with depth. The vertical distribution pattern of oxygen concentration for both oceans is shown in Figure 4.13; and vertical distribution of phosphate is discussed and illustrated (Figure 5.2) in Chapter 5.

This brief review of Deep Water serves to emphasize that, even if the technology to exploit abyssal depths existed, which it does not, this layer of the ocean appears to offer a rather unattractive opportunity for mariculture. Not only is it essentially inaccessible, but its temperatures and chemical properties are largely undesirable. Moreover, its natural fauna are not common dietary items, as we shall see in Part Three.

Western Boundary Currents

The strong, narrow ocean currents found only near the western boundaries of the oceans are, as a class, termed *western boundary currents*. The phenomenon has a single, fundamental physical cause: the increase in the deflecting effect of the earth's eastward rotation (the Coriolis effect) with increasing latitude. When water is forced to move north or south by the presence of boundaries, the Coriolis effect will cause intensification of the currents along the western boundary. Such intense currents are dynamically prohibited from occurring along eastern boundaries; and we note that the intensification would be in the west even if the flow were in the reverse direction. Within our area of interest, the Gulf Stream system represents this class of currents. The Gulf Stream, or at least an equivalent current, would thus exist even if the Caribbean Sea, Gulf of Mexico, and Florida Straits were absent.

The western boundary currents are typically of the order of 150 kilometers (km) wide, with maximum speeds of 150 to 250 cm/s. Highest speeds are in the surface layer, but appreciable motion exists deeper than 1,000 m. There is evidence that these currents occasionally may involve water all the way to the bottom.

The currents first form close to the physical boundary of the ocean. As they flow poleward, they reach a point at which they separate from the boundary, turn gradually eastward, and flow out into the ocean basin. Here the narrow current meanders in a continually shifting pattern that is unpredictable in detail. Some meanders have been observed to develop into loops, which close and pinch off from the main current, becoming large independent eddies that move alowly away. They may maintain their identity for many weeks. Farther downstream, the boundary currents gradually spread, break up, and slow down in a complex manner, developing into the broad, slow variable easterly flow of the west-wind zones.

In structure, the boundary currents are not rivers of warm, low-latitude water flowing poleward through a cold sea. Instead, they are currents that flow along the very edge of the boundary between a mass of cold water on the poleward side and a mass of warm water on the other. The band of high velocity actually acts as a dynamic barrier, preventing the warm water from spreading poleward over the cold. The surface-layer temperatures in the intense currents are not significantly different from the surface temperatures outside the flow on the warm-water side; but there is a sudden, large temperature drop on the poleward side. This abrupt change from warm to cold waters often takes the form of a visible line, or *front;* but the visible front does not necessarily coincide with the edge of the main current.

Beneath the surface, the most striking feature of the boundary currents is the steep slope of the thermocline across the flow. It is shallow on the cold side and deep on the warm side of the current. Such thermocline slopes are necessarily associated with ocean currents because of the dynamics of flow on a rotating earth. The narrowness and speed of the boundary currents make the slopes associated with them the most strongly developed in the oceans.

Gulf Stream. The Gulf Stream is the prime western boundary current in our area of interest. It derives much of its flow from water that has entered the Caribbean Sea with the North Equatorial Current and Guiana Current (an offshoot of the South Equatorial Current that crosses the equator). This water moves west and then northward through the Yucatan Channel into the

Gulf of Mexico with little change in its properties. At this point, the flow is still relatively broad and slow.

Most of the water entering the Gulf of Mexico turns directly east and enters the Straits of Florida, the long, right-angled channel formed by Florida on one side and Cuba and the Bahamas on the other. Off Miami, the straits constrict to a minimum of about 90 km wide and 800 m deep. The initial acceleration and narrowing of the current is clearly related directly to this physical constriction; but we have seen that a narrow, intense current would develop off the east coast of North America even if the Straits of Florida were nonexistent.

Within the straits, the flow is referred to as the Florida Current, a name sometimes extended to include the region as far as Cape Hatteras. The current has an average maximum speed at the surface and near the center of about 150 cm/s. Its speed decreases with depth. Flow is greatest in early summer and at a minimum in late fall, with a seasonal fluctuation of 10 to 20 percent. Random fluctuations of similar magnitude are superimposed on the seasonal trends.

The subsurface temperature distribution in the Florida Straits is dominated by the thermocline slope: the thermocline deepens and thickens from west to east across the flow. For example, although the surface layer only changes from somewhat less than 100 m thick off Miami to somewhat more than 100 m thick off Bimini, the 15°C isotherm near the center of the thermocline descends from 100 m to about 500 m over the same distance.

The Florida Current contains a maximum salinity layer in the upper thermocline (derived from the central North Atlantic by way of the North Equatorial Current) and a layer of minimum salinity near the bottom (derived from the far South Atlantic by way of the Guiana Current). Nutrient and oxygen distributions are typical of intertropical waters offshore.

The Florida Current follows the continental slope as far as Cape Hatteras, where the coast turns northward but the current continues northeast. It is separated from the shore by a narrow zone of shallow, slow-moving coastal water of very different characteristics. The current may decrease slightly in mean maximum speed, but it increases greatly in the amount of water transported, since it is widened by the addition of warm offshore water and deepened by the addition of water colder than 8°C as the bottom depth increases. The current is taking on its true western-boundary character in this region.

Between Cape Hatteras and the tail of the Grand Banks, at about 45° west, the current, now properly called the Gulf Stream, forms a sinuous, shifting, narrow boundary between the thick, warm surface layer of the Sargasso Sea and the thin, cold, low-salinity *Slope Water* to the northwest. Mean maximum speed still approaches 150 cm/s, but the actual flow at any given location is widely variable because of the shifting pattern of the current, with eddies often present on both sides. This shifting pattern is poorly understood, largely because of practical difficulties in performing large-scale, rapid surveys. The volume transport of the flow has more than doubled over that of the Florida Current.

In oxygen and nutrient distributions, the Gulf Stream water is not significantly different from the mid-latitude Sargasso Sea water offshore. However, the slow-moving Slope Water, shoreward of the Gulf Stream, is higher in nutrients at the surface.

The detailed behavior of the Gulf Stream system beyond the tail of the Grand Banks, where the flow should properly be called the North Atlantic Current, is not well known. There is even some question whether any true Gulf Stream water moves northeast toward Scandinavia, or whether there is a separate large gyre north of the Gulf Stream system. The complexity and variability of the flow patterns make even such major questions hard to answer unambiguously.

The comparative constancy and predictability of the Florida Current and Gulf Stream in low to mid-latitudes may make it possible to tap the kinetic energy they contain to induce artificial upwelling and to provide power for mariculture operations. But frequent hurricanes in this area indicate that any offshore stations would have to be designed for high survivability.

Eastern Boundary Currents

Eastern boundary currents result from broad easterly flows of water that extend across the oceans in mid- to high latitudes. When they reach the eastern boundaries they divide, with most of the water moving equatorward. This equatorward flow subsequently turns west to supply the westerly transport of the North and South Equatorial currents in low latitudes. The eastern boundary currents exhibit similar characteristics in the various oceans.

In distinct contrast to the poleward western boundary currents, the eastern boundary currents are slow, broad, and shallow, with relatively small transport. Numerous eddies and irregularities are characteristic of these weak flows, so that the current at any location is predictable only in the average. Mean speeds are 25 cm/s or less, with most of the flow occurring above 500 m.

The eastern boundary currents have no definite outer boundary, but the bulk of the flow seems to occur within perhaps 1,000 km of the coast. The region within a few tens of kilometers of the coast is more complex than that farther offshore, because of various boundary effects. Weak but persistent surface and/or subsurface poleward coastal countercurrents are common, and wind-induced coastal upwelling, with its resulting physical and biological effects, is one of the most characteristic features of eastern boundary currents. The vertical motion in this upwelling is slow: indirect estimates indicate speeds of perhaps 50 m/month.

The surface temperature in the eastern boundary currents is lower than that at the same latitudes in the central oceans for two reasons: (1) the currents are transporting cooler high-latitude water equatorward; (2) the coastal up-welling is introducing cooler water from below. Beneath the surface, the thermocline is relatively shallow near the coast, sloping gradually down toward the west. At the coast, the upper isotherms bend up to intersect the surface if there is upwelling; the lower isotherms bend down as they approach the coast if there is a subsurface countercurrent.

Similarly, surface salinity is generally lower in the boundary current than offshore at the same latitudes, since these currents are carrying water from high latitudes into the mid-latitude zone of maximum surface salinity. Be-neath the surface the salinity patterns are complex, reflecting the sources of the different water layers in each ocean.

Subsurface oxygen minima are most strongly developed in the eastern boundary current regions, with the thickest and most extreme minima found along the coasts themselves. This fact reflects the low rate of flow, which restricts replenishment of oxygen by horizontal currents, and the high surface biological productivity, which supplies relatively large amounts of oxidizable organic matter to subsurface layers. In certain areas, intense coastal upwelling can raise sufficient low-oxygen water to make the surface layer significantly undersaturated with oxygen.

The oxygen minima are associated with strong nutrient-maximum layers centered in the lower thermocline in the eastern boundary regions. Even though coastal upwelling only involves water from the upper thermocline, it can raise significant amounts of this relatively rich water to the surface. Nutrient levels near shore can be tens of times higher than the very low levels typical of mid-latitude offshore areas. The outstanding high biological pro-ductivity of the ocean along such coasts is attributable to this enrichment.

Coastal upwelling is most frequently a result of current‑wind interaction. Consequently, there are seasonal variations that impart similar variations to the enrichment. Occasionally, a large and persistent anomaly in the wind pattern causes an extensive breakdown in the upwelling pattern, leading to catastrophic readjustments in the dependent biological communities.

California Current. The California Current is an entirely normal representa-tive of the class of eastern boundary currents. It is broad, highly variable in detail, and slow, with temperature, salinity, oxygen, and nutrient patterns following the general descriptions already outlined. The upwelling is strongly seasonal, being present in spring and summer, with the strongest upwelling shifting northward from Baja to northern California during this period. In fall and winter, upwelling is basent, with a northward countercurrent (the David-son Current) running close to the coast.

The seasonal pattern is reflected in surface enrichment and biological productivity, which are high in spring and summer and low in fall and winter. The annual range in surface temperature is reduced, since the summer addition of cold upwelling water counteracts the warming effects of solar radiation. This range of $3°C$ or less along much of the California coast, in contrast to an annual range of over $10°C$ at similar latitudes in the western North Pacific Ocean. Nonseasonal fluctuations in wind strength and direction induce unpredictable variations in the patterns of upwelling and surface temperature. The California upwelling is apparently never intense enough to bring low-oxygen water into the surface layer.

A subsurface northward countercurrent is present throughout the year, extending along the coast from Baja California to Oregon. The flow is weak, but its relatively high salinity water is a striking subthermocline feature along the northern coast.

Alaska Current. A relatively small component of the North Pacific Current turns northward at the eastern boundary, feeding the poleward Alaska Current rather than the equatorward California Current. The division occurs at about $48°$ north (the latitude of the United States–Canada boundary, at Juan de Fuca Strait). Water in the Alaska Current, moving poleward, tends to be slightly warmer than in mid-ocean at the same latitudes. Motion is slow but relatively deep, perhaps involving water as deep as 2,000 m. The prevailing wind drift is convergent with the coast, so no upwelling is seen in these high latitudes. The salinity of the surface layer is very low owing to high precipitation and river runoff. In summer a shallow, weak thermocline forms, to be eliminated by cooling in fall and winter. The low surface salinities keep the water column stable even in winter, however, and prevent the deep convective overturn typical of the northern Atlantic. This stability restricts the seasonal enrichment of the surface layer that would accompany such overturn. Figures 4.14 and 4.15 give cross-sectional profiles of the California and Alaska currents, respectively.

It appears that the California Current in our area of interest provides a stabilizing and nutrifying effect along the western U.S. coast, which might serve to allow extension of some mariculture activities more to the north than may be possible farther out at sea where seasonal variations are wider.

Gulf of Mexico

The Gulf of Mexico is a unique, semienclosed arm of the ocean. It is typical of no other general situation, but it should be described here because of its possible importance to U.S. mariculture efforts.

The gulf was formed by the remarkable sinking of a block of the earth's crust some 1,400 km long by 1,000 km wide. Present depths are oceanic in character: the maximum is near 4 km, and about one third of the basin is deeper than 3 km. Continental shelves are broad around much of the gulf. Drowned limestone plateaus some 200 km wide are found off Florida in the east and the Yucatan Peninsula in the south. An alluvial shelf of roughly the same width is found off the Texas–Louisiana coasts in the north. Only along the Mexican coast, in the west, is the shelf narrow.

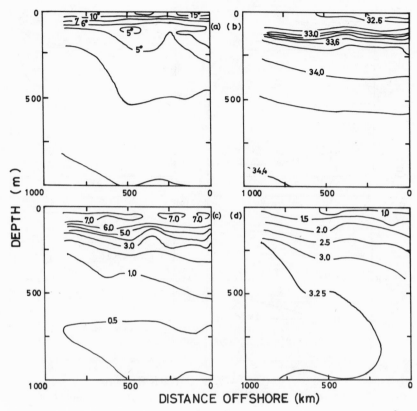

FIGURE 4.14 *Cross-sectional profile of the California Current at 32.5°N: (a) temperature, °C (9/55); (b) salinity, 0/00 (9/55); (c) dissolved oxygen, ml/liter (9/55); (d) inorganic phosphorus, μg atoms/liter (8-9/55). (Redrawn from Hill, Vol. 2, Chap. 11 [2].)*

As described in the discussion of the Gulf Stream, the water that has been pushed westward through the Caribbean by the trade winds flows north into the Gulf of Mexico, through the Yucatan Channel. Since most of this water turns to the right and immediately leaves the gulf through the Florida Straits, the main body of the gulf is not directly involved in this powerful current system. Indirectly, however, the effects of this flow are important. There is sufficient horizontal mixing and exchange between this current and the gulf waters to maintain the nearly oceanic properties of the latter, in spite of the almost landlocked character of the basin. This mixing process accounts for the variable and sometimes strong currents observed in the eastern half of the gulf; in the western half, farther from the energy source, the currents are weak and variable.

Surface-water temperature is more variable with season than it is in the

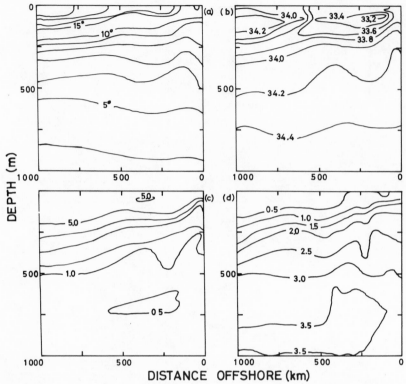

FIGURE 4.15 *Cross-sectional profile of the Alaska Current at 53.5°N (7-8/57): (a) temperature, °C; (b) salinity, 0/00; (c) dissolved oxygen, ml/liter; (d) inorganic phosphorus, μg atoms/liter. (Redrawn from Hill, Vol. 2, Chap. 11 [2].)*

open ocean at similar latitudes as a result of the influence of the surrounding land masses on air temperature. The range is 29°C and above in August, and 20 to 25°C in February. Under the surface layer the temperature decreases in a normal thermocline, but below 2 km, which is the silldepth of the Yucatan Channel, it remains very constant, near 3.95°C.

The surface salinity is about 36 ppt, except in the area near the mouth of the Mississippi; a salinity of only 26 ppt can be found as far as 90 km from the mouth. Beneath the surface a trace of the North Atlantic salinity maximum and of the deeper South Atlantic minimum remain, as mentioned in the Gulf Stream discussion. The Deep Water is uniform at 34.98 ppt.

Oxygen concentration would be expected to reflect any tendency for the deep gulf water to stagnate. No such tendency is seen: deep-water concentrations of around 5 ml/liter indicate relatively rapid renewal of the water, even though the entrance sill is at 2 km and the exit sill is at only 800 m.

Tides in the gulf are small, only 0.3 to 0.6 m in general, but water-level rise during hurricanes can be 5 m or more.

This oceanic arm with its broad continental shelf and, except for hurricane conditions, comparatively calm waters probably constitutes one of the more attractive areas for open sea mariculture. As we see in Chapter 12, oil concerns have already extended the coastal zones of the Gulf Coast states many kilometers seaward with bottom-mounted, surface-piercing platforms. These platforms demonstrate that hurricanes can be resisted with current technology, and offer ready-made bases for pilot-scale open sea mariculture.

Neritic Zone

The neritic zone extends from the shore out to a depth of 200 m, basically comprising the water lying over the continental shelves. It is very difficult to make any valid generalizations about this zone, except for one: its properties are so variable in space and time that generalizations are dangerous.

Variability here is caused by the climatic influence of the nearby landmass, by river runoff, by tidal currents, and by the effects of the shoreline on circulation patterns. The impact of these influences is increased by the shallowness of the water. Both seasonal and random fluctuations with time tend to be much larger than at similar latitudes offshore. Variations with space also can be very rapid; this is most obvious near a point of significant river discharge, but can be true in more subtle ways elsewhere.

Nutrient concentrations tend in general to be higher in the neritic zone than offshore. This is caused partly by runoff from land; but probably more important is the presence of the shallow bottom, which eliminates the downward loss of nutrients that is characteristic of the deep sea. Where wind-driven coastal upwelling occurs, of course, nutrient concentrations in the neritic zone are relatively very high. The advantages of increased biological productivity associated with the richer water of the neritic zone are at least partially offset, however, by the increased likelihood of environmental fluctuations large enough to be harmful to some of the organisms present.

The great neritic variability also makes for extreme observational difficulties. In the open sea, in many cases, data taken as long ago as the 1920s may be combined with modern data as though they were observed simultaneously, and data taken at one location may safely be assumed to characterize a broad surrounding area. But in shallow water, observations separated only by a few hours in time, or by a few kilometers distance, may be impossible to compare meaningfully. In short, compared to deep-water surveys, shallow-water surveys must be both more detailed and more rapid. Successful compromises between these antithetical requirements cannot always be made.

Keeping in mind the unreliability of generalizations, we can briefly summarize the characteristics of the neritic zones off the continental United States as follows:

- *East and Gulf coasts* (general): Because of the general west-to-east movement of weather systems, the continental climatic influence is very much stronger off eastern coasts than off western. Seasonal changes consequently are pronounced in the neritic zone off the Gulf and Atlantic coasts. In addition, winter cyclonic storms occur in considerably lower latitudes on the eastern sides of continents than on the western, so eastern coastal waters tend to be relatively stormy in that season. And hurricanes are very much commoner on eastern coasts, as described later. On the other hand, the frequency of large swells from distant storms is much lower on eastern coasts, so that nonstormy conditions tend in general to be calmer than in the west.

- *Gulf of Maine* (north of Cape Cod): The glaciated continental shelf is wide (250 to 400 km), rocky, and irregular. Winter storms are severe, and summer hurricanes relatively infrequent. Tidal and variable currents predominate. The water is derived from the cold Labrador Current (mean temperature is $8°C$; mean salinity is 30 ppt).

- *Middle eastern states* (Cape Cod to Cape Hatteras): There is a relatively broad (90 to 150 km), smooth sedimentary shelf. Winter storms are severe, and summer hurricane frequency moderate. Tidal and variable currents predominate. The water is derived from the cold Labrador Current (mean temperature is $12°C$; mean salinity is 31 to 32 ppt).

- *Southeastern states* (Cape Hatteras to Miami): A narrower (50 to 100 km), smooth sedimentary shelf decreases in width along the Florida peninsula, becoming extremely narrow within the Straits of Florida south of Palm Beach. Winter-storm severity and frequency decrease southward through this zone; summer hurricane frequency increases. Tidal and variable currents predominate; the strong flow of the Gulf Stream (properly, called the Florida Current here) remains off the shelf. Only within the Straits of Florida does the Florida Current induce occasional strong motions in the narrow neritic zone. The neritic water is normally much warmer than it is north of Cape Hatteras, since the Labrador Current water usually does not extend south of the cape (mean temperature is $24°C$; mean salinity is 35.5 ppt).

- *Gulf of Mexico:* The continental shelves are broad (100 to 240 km) and smooth. Active sedimentation is occurring in that region influenced by the Mississippi River. Winter storms are only moderately frequent and severe, but summer hurricane frequency is relatively high. Currents are predominantly variable, stronger off Florida and weaker along the Alabama–Texas coasts. The water is derived from the Caribbean, but strong continental influence is reflected in the annual temperature variation (mean temperature is $23°C$; mean salinity is 32 ppt), which is much higher at these latitudes than in the open ocean. The mean neritic salinity is low largely because of the strong effect of the Mississippi in the central northern region.

- *Pacific Coast* (general): In contrast to the continental climate of the east coast, the west coast has a maritime climate: milder, less variable, and less stormy. Wind-driven coastal upwelling is present in spring and summer, with the resultant cooling partially offsetting increased solar radiation during this period and thereby keeping the annual range of the surface temperature extremely small.
- *Pacific Southwest* (Mexico to Cape Mendocino): The continental shelf is narrow (15 km), irregular, and rugged, with numerous submarine canyons reaching across the shelf nearly to the beach. Severe storms are infrequent, but swell from distant storms is common. Currents are slow, generally southerly in spring and summer and northerly in fall and winter, but with many irregularities in the flow. Upwelling in spring and summer brings about strong enrichment of the neritic zone (mean temperature is $15°C$; mean salinity is 34 ppt).
- *Pacific Northwest* (Cape Mendocino to Canada): The shelf is narrow (15 km) and irregular off Oregon, but it broadens to 60 km and becomes smooth off the Washington coast. Winter-storm frequency increases northward; hurricanes are unknown; swell frequency is fairly high. Currents are slow and variable, with northerly flow prevalent in fall and winter and southerly flow in spring and summer. Upwelling is present off Oregon in spring and summer. Abundant river discharge off Washington produces a light, low-salinity surface layer that masks any tendency for upwelling, except occasionally (mean temperature is $10°C$; mean salinity is 31 ppt).
- *Oceanic Islands:* Oceanic islands with coral reefs almost universally drop steeply to great depths, with the neritic zone outside the reef so narrow as to be nonexistent. Where a deep lagoon exists within the reef, this may be considered a sort of neritic environment. On low islands, water within these lagoons has essentially oceanic, surface-layer characteristics; in particular, nutrient concentrations are low. On high islands, where the lagoon is formed by the gap between the island and an offshore barrier reef, conditions are variable, depending on the amount of runoff and the freedom of connection with the ocean.

ENERGY SOURCES

In this chapter we have touched on the heat exchange functions of the oceans and on their role as climatic stabilizers for the entire globe. We have also reviewed the horizontal and vertical movements of the huge volumes of water. Obviously, there is a constant exertion and expenditure of both heat and kinetic energy. There must also be a constant input of energy to keep the system going. Some of it, in the form of kinetic, or mechanical, energy, derives from the stresses created in part by the earth's rotation and in part by the gravitational pull of the moon and sun.

Another major, and more variable, input of energy to the ocean occurs through its interface with the atmosphere, the relatively shallow layer of Surface Water. Energy is received by this surface layer in two important forms, radiant solar energy and the mechanical energy from the drag of the wind on the surface. Energy from these sources is the prime moving force of the ocean system and hence of vital interest to mariculture.

Radiant Solar Energy

Radiant solar energy is the only effective heat source for the ocean. Outside the atmosphere, this radiation amounts to 2 cal/cm^2/minute (min) on a surface perpendicular to the radiation. When averaged over the whole sphere of the earth, this is equivalent to 0.5 cal/cm^2/min. Only about half this amount, on the average, reaches the surface. The rest is either reflected back to space or absorbed by atmosphere and clouds.

This average surface-incident value of approximately 0.25 cal/cm^2/min is, of course, far from equally distributed over the earth. In high latitudes the sun is always at a low angle. This not only spreads the radiant energy out over a greater surface area than would be the case with the sun vertical, but also causes the radiation to travel a longer slant path through the atmosphere before reaching the surface, thus causing increased attenuation through increased atmospheric absorption and scattering. In summer these effects, which reduce the input of energy per unit surface area, are partially offset by the increased length of day in higher latitudes. As a result, in the northern-hemisphere summer the radiant input maximum, about 0.30 cal/cm^2 min, is found at about 35° north latitude; even at the north pole the input averages about 0.15 cal/cm^2 min. In the northern-hemisphere winter, the maximum shifts south of the equator, although seasonal variation in input is not very great below 20 to 30° north. In polar regions winter radiation input drops to zero for periods of time that increase with latitude, reaching 6 months at the actual poles.

In the ocean, long-wave solar energy is absorbed very rapidly; as a result, even in the clearest seawater 27 percent of the total energy is absorbed within 1 cm, and 62 percent within 1 m. Thus the heating of the ocean is confined to a very thin skin. Only about 0.5 percent of the energy remains at 100 m in clear water; this is blue light with a wavelength near 0.45 μm. Most of the energy absorbed is converted directly to increased heat energy of the surface layer, but a very small amount of the visible light is intercepted by plants and changed to chemical energy. This is largely converted again after some delay, appearing later as waste heat associated with biological processes.

To maintain long-term thermal stability, the ocean must lose, on the average, just as much heat as it gains. In contrast to the single source of solar energy, three processes account for this loss: (1) direct conduction from warmer sea to cooler air accounts for only about 10 percent of the total loss; (2) long-wave heat radiation from the oceans accounts for about 40 percent; (3) evaporation accounts for about 50 percent.

Outward radiation and conduction from the sea vary relatively little with latitude. Evaporation has a greater variation, being highest in the broad trade-wind zone, centered between 10 and 20° north latitude in the areas of interest. Moreover, evaporation is at a peak in winter, being especially intense off north eastern coasts, where cold continental air is blowing over warmer water moving northward from low latitudes. In the Gulf Stream region, this winter evaporation rate can amount to over 1 cm of water/day.

South of 20° north latitude, heat gain always exceeds loss to the atmosphere. This region must therefore disperse heat by ocean circulation. In the remainder of the principal northern oceans there is an overall net gain in summer and loss in winter, resulting in the observed seasonal cycle of heating and cooling. On the annual average there is a net heat loss in high ocean latitudes, so that over the year those latitudes must gain heat from lower latitudes by the transport of ocean currents.

High evaporation rates resulting from warm water in a cool atmosphere could possibly pose minor problems for some mariculture system designs. But the problems seem solvable. Probably the most significant aspect of solar energy input to the oceanic surface is the possibility of employing future technology to tap it, either in its original radiant form or in one or more of its secondary forms (winds, currents, and thermal gradients). This possibility is explored further in Chapter 13.

Wind and Wave Energy

Winds are the other major source of solar-derived energy—wind energy being transferred mechanically to the oceans. The details of this process are little known, since the energy transfer depends in a complex way on the roughness of the sea surface, which itself depends in turn on the energy transfer. We do know that the energy is imparted to the water in two ways. First, one part of the energy goes into surface wave motion; that is, the wind pushes against the surface, making the sea rough or wavy. Wave energy moves rapidly across the sea surface, often traveling great distances from the area of wind impact before being dissipated in the form of surf on some coast. Transmission of wave energy in deep water, however, produces a vertical rotation rather than a significant forward motion of the water itself. The input of energy into waves is a rapid effect, and a single storm can generate waves that will affect very large areas. For wave prediction, then, one must have detailed day-to-day wind information over a broad area.

The second part of the wind's energy imparts bodily movement to the waters of the oceans. Water motions due to an individual temporary wind condition are complex and transitory, affecting a relatively small area. But if there is an average stability to the wind pattern, the imparted movement, modified by the earth's rotation, leads to a pattern of horizontal circulation roughly in the direction of the average winds. So, for insight into the general surface circulation of the ocean, one may look at the mean wind patterns over the ocean.

For an in-depth study of wind characteristics and movement, general patterns are available in atlases and detailed information on averages is available in the series of Pilot Charts published by the U.S. Navy Oceanographic Office. In the broadest terms, the wind distribution in the area of interest includes two major wind belts: the northeast trades, in the zone from 10 to 25° north; and the westerlies, from 30 to 50° north. Figures 4.16 and 4.17 show general summer and winter wind patterns in the area of interest [16].

Trade winds. The trade winds are the most steady of all winds on earth. They blow almost without pause, with strengths usually between 5 and 10 m/s. On islands in their path the effects of these nearly unidirectional winds are seen in vegetation patterns, reef development, erosional sculpting of mountains, and the culture of the inhabitants. For the sea, the trade-wind zone is the most important region for transfer of momentum from air to oceanic circulation. The trade-wind zone shifts seasonally, following the sun; for instance, in the winter, Hawaii, extending from 19 to 23° north, occasionally finds itself north of the trades.

Hurricanes. Although hurricanes do not significantly affect the general oceanic circulation, they are a factor to be considered for mariculture proposed in areas where they occur. The mild and predictable nature of the

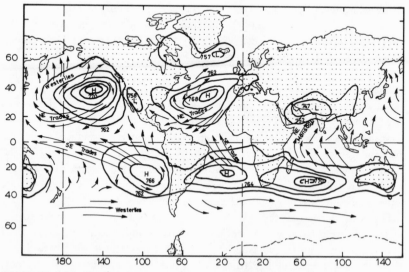

FIGURE 4.16 Predominant wind pattern: July. Pressure data (mm): L, low; H, high. (Adapted from Bartholomew [16].)

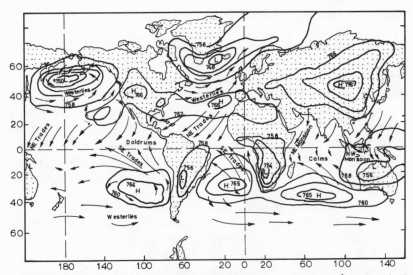

FIGURE 4.17 *Predominant wind pattern: January. Pressure data (mm): L, low; H, high. (Adapted from Bartholomew [16].)*

trades give no hint of the overwhelming destructiveness of wind and mountainous seas in these most powerful of all general storms. Hurricanes tend to form most frequently in late summer at low latitudes in central and western ocean areas (although not within about 5° of the equator itself). The storms initially move westward with the trades in which they are imbedded. As they near the western sides of the oceans their paths curve poleward and, if they reach mid-latitudes, gradually around toward the east. The track of a hurricane that remains over the ocean thus tends to form a huge parabola.

The storms are roughly circular, with winds rotating around a central, calm, low-pressure "eye." The eye has an average diameter of about 25 km, but this may vary between 8 and 80 km. Rotation is counterclockwise in the northern hemisphere and clockwise in the southern. Wind speed increases as the eye is approached. Hurricane-force winds are defined to be 110 km/hour (h) or more; in a major storm the speeds near the eye may be over 300 km/h. The path of great wind damage in such a hurricane may be 80 to 160 km wide.

The huge waves generated by these storms can do great damage even to large vessels at sea. This is true even in the calm eye, where phenomenal, chaotic cross seas make evasive maneuvering impossible. When the storms move onshore, the tremendous surf they generate is made much more destructive by an overall rise in sea level, caused by two factors: the direct shoreward push of the wind on the water, and the lowering of atmospheric pressure on the sea surface. The effect of the wind on the sea level is most pronounced on a coast with a broad, shallow area offshore; here sea level can be raised by tens of feet, allowing the surf to beat directly on structures and natural features in low-lying areas.

After hurricanes come ashore they weaken rapidly because of increased friction at the earth's surface and a sharply reduced supply of water vapor.

Most North Pacific hurricanes, locally known as typhoons, originate between the Marshall Islands (170° east) and the Philippines. This area has the greatest hurricane frequency on earth. The storms move westward, turning northward to the east of the Philippines, or else crossing the islands to turn northward in the South China Sea. If they do not dissipate over land, they finally turn northeastward near the coast of Japan. The months of greatest frequency are July through October; February through April are hurricane-free.

Relatively infrequent, short-lived hurricanes also form off the west coast of Central America. They move generally northwest, occasionally affecting the coast as far as California; only rarely do they move westward to reach the vicinity of Hawaii, by which time they have usually weakened below hurricane strength. September and October are months of highest frequency; December through May are hurricane-free.

In the North Atlantic, many hurricanes form near the Cape Verde Islands off the African coast. They initially move west across the ocean; some continue westward through the Caribbean to strike Central America, but most curve northward in or near the West Indies. Many of these affect the eastern coastline of the United States. A second region of formation exists in the western Caribbean itself. Storms originating here mostly move northward to strike the Atlantic and Gulf coasts of the United States. On these coasts the average hurricane frequency is about two per year, with severely damaging storms about once in 3 years. The months of highest frequency are August through October, with December through April being hurricane-free.

Northern Hemisphere Westerlies. The northern hemisphere westerlies are much more variable than the trades. This is especially true in winter when these higher latitudes are swept by a succession of easterly moving cyclonic storms. The winter-storm frequency increases rapidly above 30° north, reaching its maximum at 40 to 50° north in both oceans. The storms generally take a track that is north of due east, so they are observed in lower latitudes on the western sides of the oceans than on the eastern sides. The average westerly winds shown in these latitudes on winter maps thus actually represent a great variability in wind speed and direction, with gales alternating with light and variable winds.

Summer storms are weaker and less frequent, and their tracks are shifted somewhat to the north. Wind direction in the region of the westerlies is steadier, but wind speed is typically low.

The zone between the trades and the westerlies is a region of calms and variables with fine weather, especially in summer. South of the northeast trades is another belt of calms, the doldrums, which represents the zone of convergence between the southeast and northeast trades. In the central oceans this zone is between 5 and 10° north, but in the western Pacific the zone can be much broader and may shift south of the equator in winter. The doldrum belt is characterized by high average cloudiness and rainfall.

Wind and wave energy fall into two categories of interest to open sea mariculture: energy that might be tapped to provide or subsidize power for mariculture activities; and energy that might exert destructive influences. Here we discern an unfortunate pattern, for the oceanic areas that tend to be the most attractive from the viewpoint of wind- and wave-energy utilization also have more frequent and intense storms. As noted, however, there are exceptions to this general pattern, such as the Hawaiian Archipelago. The prime implication for open sea mariculture appears to lie in the balancing of site selection, free-energy utilization, and storm-damage resistance in the context of optimum cost—benefit ratios.

CONCLUSIONS

In this chapter we have delineated broadly the oceanic areas considered most appropriate for consideration for U.S. open sea mariculture on a geopolitical basis. These areas are sufficiently large that generalizations about them tend to be generalizations about the entire northern Pacific and Atlantic oceans. These generalizations have been drawn somewhat sketchily in an attempt to illuminate physical oceanographic characteristics of potential concern to open sea mariculture. The most important points that seem to have emerged in this review are as follows:

1. A geographically significant, but still proportionately small, amount of the total oceanic area under consideration is shallow enough to allow the use of bottom-supported platforms; therefore, open sea mariculture should expect to eventually employ floating platforms.

2. The dynamics of seawater and marine biological processes result in a huge nutrient pool residing in Intermediate Water, which are colder and more dense than Surface Water in tropical and subtropical latitudes. If Intermediate Water is to be employed to stimulate bioproductivity, it will require heating, both for optimum biological temperature and for sufficiently long-term residence in the photic zone, where uncontained.

3. Because of the light absorption qualities of seawater itself, and of seawater containing dissolved organic materials, intensive photosynthesis in waters that are not vertically mixed (to bring phytoplankton repeatedly into the narrow photic zone) can occur only from the surface to a depth of 10 or so meters at most.

4. Equatorial and Tropical Surface waters are low in inorganic nutrients; but Equatorial Surface Water may be within 100 to 150 m of nutrient-rich

Intermediate Water. Tropical Surface Water, on the other hand, may be as much as 600 m thick, overlying the Intermediate Water.

5. In the delineated areas of interest, waters north of about 30° north latitude generally appear less attractive for artificial upwellings than waters southward; not only do the nutrient-rich waters lie deeper to the north, but the annual vertical stability of the water column is less.

6. There is no firm evidence that currents of constant velocity and direction can be expected anywhere in the oceans.

7. Nevertheless, the western boundary current on the east U.S. coast is adequately strong and directional that, theoretically, energy could be extracted from it.

8. Eastern boundary currents, such as the California Current, tend to be relatively high in nutrients and conducive to natural upwellings.

9. The combination of natural and man-made conditions in many parts of the Gulf of Mexico appears attractive for open sea mariculture purposes.

10. Free energy from winds is most reliable and constant between 10 and 25° north latitude in the area of interest.

After a mariculture-oriented review of chemical oceanography topics in Chapter 5, we shall explore oceanographic implications for open sea mariculture in Chapter 6.

REFERENCES

1. Sverdrup, H. U., M. W. Johnson, and R. H. Fleming. 1942. *The oceans: their physics, chemistry and general biology.* Prentice-Hall, Inc., Englewood Cliffs, N.J. 1087 pp.

2. Hill, M. N., ed. 1963. *The sea.* Vol. 1, *Physical oceanography,* 864 pp.; Vol. 2, *Composition of sea water; comparative and descriptive oceanography,* 554 pp.; Vol. 3, *The earth beneath the sea; history,* 963 pp. Wiley-Interscience, New York.

3. Heezen, B. C., M. Tharp, and M. Ewing. 1959. The floor of the oceans, I. The North Atlantic. *Geol. Soc. Amer. Spec. Paper 65.* 122 pp.

4. *Scientific American 221*(3).

5. Griffin, J. J., and E. D. Goldberg. 1963. Clay-mineral distributions in the Pacific Ocean. In M. N. Hill, ed., *The sea,* Vol. 3, pp. 728–741.

6. Biscaye, P. E. 1965. Mineralogy and sedimentation of recent deep-sea clay in the Atlantic Ocean and adjacent seas and oceans. *Geol. Soc. America Bull.* 76:803–831.

7. Dietrich, G. 1957. *General oceanography.* Translated by F. Ostapoff. Wiley-Interscience, New York. 588 pp.

8. Neumann, G., and W. J. Pierson, Jr. 1966. *Principles of physical oceanography.* Prentice-Hall, Inc., Englewood Cliffs, N.J. 545 pp.

9. Pickard, G. L. 1964. *Descriptive physical oceanography.* Pergamon Press, Inc., Elmsford, N.Y. 199 pp.

10. Hedgepeth, U. W., ed. 1957. Treatise on marine ecology and paleoecology, Vol. I. *Geol. Soc. America Mem. 67.* 1296 pp.
11. Stommel, H. 1960. *The Gulf Stream.* University of California Press, Berkeley, Calif. 202 pp.
12. Kinsman, B. 1965. *Wind waves.* Prentice-Hall, Inc., Englewood Cliffs, N.J. 676 pp.
13. Bigelow, H. B., and W. T. Edmonson. 1947. Wind waves at sea, breakers and surf. U.S. Navy Hydrographic Office, Washington, D.C. *H.O. Pub. 602.* 177 pp.
14. Pierson, W. J., Jr., G. Neumann, and R. W. James. 1955. Observing and forecasting ocean waves. U.S. Navy Hydrographic Office, Washington, D.C. *H.O. Pub. 603.* 284 pp.
15. Montgomery, R. B. 1958. Water characteristics of Atlantic Ocean and of world ocean. *Deep Sea Res. 5:*134–148.
16. Bartholomew, J. 1950. *The advanced atlas of modern geography.* McGraw-Hill Book Company, New York.

Additional Readings

Cox, R. A. 1965. The physical properties of sea water. J. P. Riley and G. Skirrow, eds., *Chemical oceanography,* Vol. 1, pp. 73–120.

Fisher, R. L., and H. H. Hess. 1963. Trenches. In M. N. Hill, ed., *The sea,* Vol. 3, pp. 411–436.

Fuglister and Worthington. 1951. *Tellus,* Vol. 3. pp. 1–14.

Heezen, B. C., and M. Ewing. 1963. The mid-oceanic ridge. In M. N. Hill, ed., *The sea,* Vol. 3, pp. 388–410.

Heezen, B. C., and A. S. Laughton. 1963. Abyssal plains. In M. N. Hill, ed., *The sea,* Vol. 3, pp. 312–364.

Hsueh, Y., and J. O'Brien. 1971. Steady coastal upwelling induced by an along-shore current. *J. Phys. Ocean. 1*(3):180–186.

Johnson, M. W., and E. Brinton. 1963. Biological species, water-masses and currents. In M. N. Hill, ed., *The sea,* Vol. 2, pp. 381–414.

Menard, H. W. 1964. *Marine geology of the Pacific.* McGraw-Hill, Inc., New York. 271 pp.

Menard, H. W., and H. S. Ladd. 1963. Oceanic islands, seamounts, guyots and atolls. In M. N. Hill, ed., *The sea,* Vol. 3, pp. 365–387.

Oceanography. 1971. *Readings from Scientific American.* W. H. Freeman and Company, Publishers, San Francisco.

Pinchot, G. B. 1970. Marine farming. *Scientific American 223* (6):14–21.

Staff, National Weather Record Center. 1969. Selective Guide to Climatic Data Sources. *Key to Meteorological Records Documentation No. 4.11.* Govt. Printing Office, Washington, D.C.

User's Guide to NODC's Data Services. 1973. *Key to Oceanographic Records Documentation No. 1.* Govt. Printing Office, Washington, D.C.

5

CHEMICAL OCEANOGRAPHY

Oceanic Foundation Staff with S.V. Smith

As pointed out in the introduction to Chapter 4, the oceans comprise both the containers and the culture media for open sea mariculture. Having described the containers and reviewed the physical characteristics of ocean waters, it is pertinent now to examine the purely chemical properties of the culture media. These include the salinity (the types and amounts of major organic salts present in seawater) as well as the presence of gaseous elements and other substances potentially supportive of, or prejudicial to, the healthy growth of cultured organisms. The topic of nutrients is touched upon only briefly and in terms of the distribution of chemical elements basic to life and growth, since this subject as a whole is treated separately in Chapter 7.

Marine chemistry is a diverse field that for years has been dominated by physical oceanographers, biologists, and geologists. Chemists, per se, have entered the field in numbers only rather recently. The historical development of marine chemistry at the hands of oceanographers, however, may merely reflect the role of chemistry as a basic discipline serving the applied disciplines. A clear example of this relationship is the importance of seawater salinity in the physical oceanographer's work and his consequent development of techniques for measuring this quality. Density is perhaps the single most important physical characteristic of seawater, and this physical property is highly sensitive to salinity. Because of the importance of density, and of salinity in determining density, oceanographers have developed an ability to measure salinity to five or six significant figures—far better than can be done for most other routine chemical analyses.

SALINITY OF THE WORLD'S OCEANS

Salinity, defined as grams of dissolved solids per kilogram of seawater, is a well-known property of seawater and remarkably constant throughout the world's oceans. If one neglects the strictly local effects of coastal runoff or of evaporation in situations of restricted water flow, virtually the entire ocean has a salinity of 35 ± 3 parts per thousand (ppt). In short, with the stated local exceptions, the range of variation in salinity worldwide is less than 10 percent of the mean value. From the standpoint of most purely chemical

TABLE 5.1 *Major Constituents of Seawater*

Constituent	g/kg of water at salinity 35 ppt
Chloride	19.353
Sodium	10.76
Sulfate	2.712
Magnesium	1.294
Calcium	0.413
Potassium	0.387
Bicarbonate	0.142
Bromide	0.067
Strontium	0.008

interpretations, this salinity variation can be neglected; the range and distribution are important principally to physical oceanography, as we have seen.

Not only is the total salt content of seawater relatively constant, but so also are the ratios of the major constituents relative to each other. In Table 5.1 we list the concentrations of major constituents at the "standard salinity" of 35 ppt, commonly used for reporting normalized values [1]. The principal positively charged ions are sodium, magnesium, calcium, potassium, and strontium; the principal negative ions are chloride, sulfate, bicarbonate, and bromide. These nine ions comprise over 99 percent of the total dissolved constituents of seawater. The deviation of most of these normalized numbers from the values reported in the table is near the analytical precision with which each constitutent can be measured. Numerous authors have been intrigued by this remarkable constancy, and many models of seawater origin have been erected to explain it—and its minor exceptions.

The minor components of seawater are not nearly so consistent with total salinity as are the nine principals in Table 5.1. Consequently, they spur more scientific curiosity than the major constituents. Furthermore, in any mariculture venture these other elements may be more likely than the major constituents to cause unpredictable trouble or to offer unusual benefits.

NUTRIENTS

Of special interest to mariculture are the nutrients, in particular, phosphate, nitrate, nitrite, ammonia, and silicate. Of these, phosphate, nitrate, and silicate have been most intensively studied. In fact, their oceanic distribution patterns are almost as well known as the distribution of salinity. The various nutrients show relatively similar distribution patterns, although the similarity is by no means as striking as it is among the nine major elements discussed earlier. *Chemical Oceanography*, edited by Riley and Skirrow, deals with the various nutrients and contains particularly relevant articles on phosphorus (by Armstrong), inorganic nitrogen (by Vaccaro), and silicon (by Armstrong).

Figure 5.1 shows typical vertical distribution patterns for the three major nutrients in seawater [2]. Figure 5.2 shows longitudinal sections of phosphate distribution through the Atlantic and Pacific oceans [2]. The other nutrients show generally similar patterns, but different absolute values. Several generalities emerge from these figures. First, the Pacific Ocean is consistently richer in these three nutrients than is the Atlantic. Second, all oceans show low nutrient values in surface waters, increasing with depth toward maxima, which occur near 1,000 m in most areas. Below the maxima there is little or no tendency for nutrient concentrations to decrease.

As noted in Chapter 4, Pacific Deep Water is generally considered to be older than Atlantic Deep Water. The richer nutrient levels in the Pacific Ocean, and the variations in vertical enrichment patterns of the two oceans, both can be explained as a regeneration of nutrients, formerly bound into particulate matter in surface waters, by oxidation and release of the nutrients back to the dissolved state.

A final characteristic of the three sets of curves in Figure 5.1 is the maximum value toward which each nutrient tends. It can be seen that silicate, reaching concentrations of about 170 μg atoms/liter in the Pacific Ocean is by far the most abundant of the three nutrients. It has been suggested by numerous authors that either nitrate or phosphate is always limiting to the growth of organisms in the sea. However, we consider that such a generality

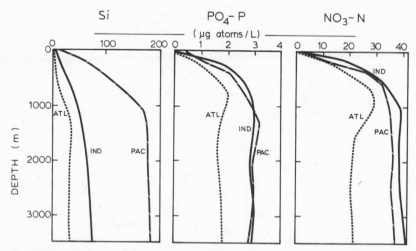

FIGURE 5.1 *Vertical distribution patterns of silicate, phosphate, and nitrate in the North Pacific, South Atlantic, and Indian oceans. (Adapted from Sverdrup et al., [2].)*

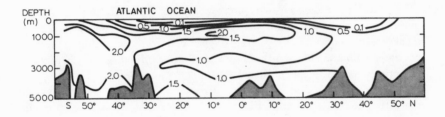

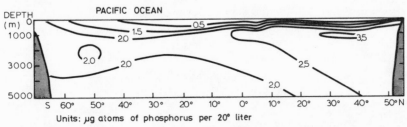

Units: μg atoms of phosphorus per 20° liter

FIGURE 5.2 *Longitudinal section of phosphate distribution through the central Atlantic and Pacific oceans. (Adapted from Sverdrup et al., [2].)*

does not describe the entirety of the world's oceans. Rather, the limiting nutrient must vary among localities and among populations. For example, Menzel and Ryther demonstrated that iron is the limiting factor in the Sargasso Sea [3], and unpublished data by Raymond indicate similar factors may be limiting in some Hawaiian waters [4].

The subject of vertical and horizontal nutrient distributions has also been discussed in Chapter 4. The implications for mariculture are drawn in Chapter 6.

DISSOLVED GASES

Three dissolved gases, oxygen, nitrogen, and carbon dioxide, are present in seawater in abundant amounts. Carbon dioxide hydrates to form carbonic acid, which then dissociates to form bicarbonate. As bicarbonate it is commonly counted as one of the major constituents of seawater. Because the totality of its forms is so abundant and because reaction rates are high, carbon dioxide probably is of no real concern as a limiter to biological activity.

To a first approximation, free nitrogen (N_2) is not affected by marine reactions, and its abundance in seawater is fixed primarily by the solubility of atmospheric nitrogen. This is somewhat of an oversimplification, since nitrogen fixation and denitrification are known to affect the gaseous nitrogen pool. But in the open seas these reactions appear likely to be a minor factor in affecting the total nitrogen content of seawater.

The distribution of oxygen is more complex. Although surface oxygen values are largely controlled by the solubility of this gas in seawater, biological activity is important in regulating oxygen concentration within the rest of the water column. Photosynthesis and respiration (discussed in Chapter 7), respectively, produce and consume large amounts of oxygen.

As might be expected from the intimate relationship between oxygen production and nutrient utilization, there is generally a negative relationship between oxygen abundance and nutrient abundance in the marine environment. Where oxygen is high, nutrients tend to be low, and vice versa. As shown in Figure 5.3, there is commonly an oxygen minimum, which is even more pronounced than the phosphate maximum; but the oxygen minimum

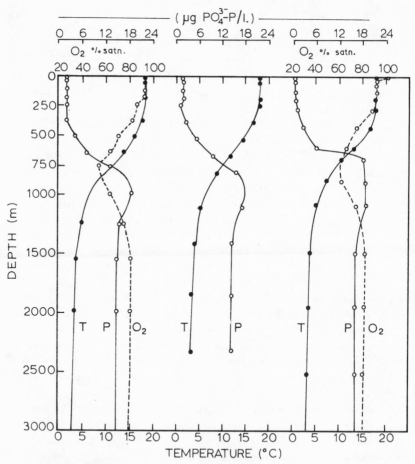

FIGURE 5.3 *Vertical profiles of temperature, oxygen, and phosphate from three different areas of the Sargasso Sea. (Adapted from Sverdrup et al., [2].)*

does not coincide perfectly with the phosphate maximum [2]. The dynamic relationships are complex, and Wyrtki [5] and others have built theoretical models to explain the dynamics of oxygen distribution in terms of a combination of mixing, diffusion, and oxidation.

Each of the three major gases in seawater behaves quite differently. Nitrogen is little affected by most marine reactions, and carbon dioxide seems largely related to salinity, which is relatively constant. Therefore, we conclude that oxygen is the major gas in seawater of concern to the mariculturist.

TRACE ELEMENTS

Thus far, less than a dozen elements have been discussed as constituents of seawater. Yet it is widely accepted that most, if not all, elements found on earth occur in seawater in at least trace concentrations. Since most trace elements are present only in minute quantities, the techniques available for their detection and measurement are frequently inadequate. Moreover, available data suggest that concentrations of most trace elements are extremely variable. In light of these two facts it is difficult to generalize further on the subject of trace-element distributions in seawater. Table 5.2 shows elements that occur consistently in significant concentrations and those elements that have been detected in trace concentrations only.

Whether or not their distribution patterns are known at this time, trace elements may be of extreme biological importance. In fact, the success or failure of a particular organism or population in a given marine ecological niche may depend in large part upon the availability or the absence of certain trace elements. For example, there is good evidence that a number of trace metals function either as cofactors for enzymes or as structural components of enzymes. Conversely, some organisms may take up naturally or artificially occurring trace constituents and concentrate them in certain tissues or organs to the extent that the constituents become lethal, either to the concentrating organism or to some higher-trophic-level organism; and man may be the highest-trophic organism in the chain.

It is reasonable, then, that elements occurring in seawater in trace concentrations be divided on the basis of biological properties into those which are active in and necessary for normal metabolism, and those which are not. The second category consists almost entirely of the heavy metals, and its members have the capacity to become toxic even when present in very minute concentrations; that is, rather than being useful in and necessary to normal metabolism, their presence in an organism generally acts to inhibit or disrupt metabolism. But in the first category, some elements, such as copper, although necessary in low concentrations, become toxic as concentrations increase. Others, such as molybdenum, are used only in small quantities, but are not toxic except in relatively high concentrations.

TABLE 5.2 *Composition of Seawater*

Element	Seawater concentration (μg/liter)	Principal dissolved species
H	1.1×10^8	H_2O
He	7×10^{-3}	He (gas)
Li	1.7×10^2	Li^+
Be	6×10^{-4}	—
B	4.5×10^3	$B(OH)_3^-$, $B(OH)_4^-$
C	2.8×10^4	HCO_3^-, CO_3^{2-}
C (organic)	2×10^2	—
N	1.5×10^4	N_2 (gas)
N	6.7×10^2	NO_3
O	8.8×10^8	H_2O
O	6×10^3	O_2
O	1.8×10^6	SO_4^{2-}
F	1.3×10^3	F^-
Ne	0.12	Ne (gas)
Na	1.1×10^7	Na^+
Mg	1.3×10^6	Mg^+
Al	1	—
Si	3×10^3	$Si(OH)_4$, $SiO(OH)_3^-$
P	90	HPO_4^{2-}, $H_2PO_4^-$, PO_4^{3-}
S	9.0×10^5	SO_4^{2-}
Cl	1.9×10^7	Cl^-
Ar	4.5×10^2	Ar (gas)
K	3.9×10^5	K^+
Ca	4.1×10^5	Ca^+
Sc	$<4 \times 10^{-3}$	$Sc(OH)_3$
Ti	1	$Ti(OH)_4$
V	2	$VO_2(OH)_3^{2-}$
Cr	0.5	CrO_4, Cr^{3+}
Mn	2	Mn^{2+}
Fe	3	—
Co	0.4	Co^{2+}
Ni	7	Ni^{2+}
Cu	3	Cu^{2+}
Zn	10	Zn^{2+}
Ga	3×10^{-2}	—
Ge	7×10^{-2}	$Ge(OH)_4$
As	2.6	$HAsO_4^{2-}$, $H_2AsO_4^-$
Se	9×10^{-2}	SeO_4^{2-}
Br	6.7×10^4	Br^-
Kr	0.2	Kr (gas)
Rb	1.2×10^2	Rb^+
Sr	8×10^3	Sr^{2+}

TABLE 5.2 *continued*

Element	Seawater concentration (μg/liter)	Principal dissolved species
Y	1×10^{-3}	$Y(OH)_3$
Zr	3×10^{-2}	—
Nb	0.01	—
Mo	10	MoO_4^{2-}
Rh	7 to 11×10^{-3}	—
Ag	0.3	$AgCl_2^-$
Cd	0.1	Cd^{2+}
In	4×10^{-3}	—
Sn	0.8	—
Sb	0.3	—
I	60	IO_3^-, I
Xe	5×10^{-2}	Xe (gas)
Cs	0.3	Cs^+
Ba	20	Ba^{2+}
La	3×10^{-3}	$La(OH)_3$
Ce	1×10^{-3}	$Ce(OH)_3$
Pr	0.6×10^{-3}	$Pr(OH)_3$
Nd	3×10^{-3}	$Nd(OH)_3$
Sm	0.5×10^{-3}	$Sm(OH)_3$
Eu	0.1×10^{-3}	$Eu(OH)_3$
Gd	0.7×10^{-3}	$Gd(OH)_3$
Tb	1.4×10^{-3}	$Tb(OH)_3$
Dy	0.9×10^{-3}	$Dy(OH)_3$
Ho	0.2×10^{-3}	$Ho(OH)_3$
Er	0.9×10^{-3}	$Er(OH)_3$
Tm	0.2×10^{-3}	$Tm(OH)_3$
Yb	0.8×10^{-3}	$Yb(OH)_3$
Lu	0.1×10^{-3}	$Lu(OH)_3$
Hf	$<8 \times 10^{-3}$	—
Ta	$<3 \times 10^{-3}$	—
W	0.1	WO_4^{2-}
Re	8×10^{-3}	—
Au	1×10^{-2}	$AuCl_2^-$
Hg	0.2	$HgCl_4^{2-}$, $HgCl_2$
Tl	1×10^{-2}	Tl^+
Pb	3×10^{-2}	$PbCl_3^-$, $PbCl^+$, Pb^{2+}
Bi	2×10^{-2}	—
Rn	6×10^{-13}	Rn (gas)
Ra	1×10^{-7}	Ra^{2+}
Th	$<5 \times 10^{-4}$	$Th(OH)_4$
Pa	2.0×10^{-6}	—
U	3	$UO_2(CO_3)_3^{4-}$

With possible exceptions, it is probable that most trace elements are not generally limiting to biological productivity in nature where population densities typically remain rather low. However, one cannot yet conclude that some trace elements will not become limiting if population densities are drastically increased through culturing.

An obvious implication for open sea mariculture is that concentrations of both essential and toxic trace elements must receive attention in the design and siting of open sea mariculture systems.

TOXINS

A variety of toxins may be encountered in significant concentrations in the open sea. They include compounds generated by human activities, such as industrial and domestic wastes, poisons, and discharges, as well as antibiotic agents from natural sources. In any given area where mariculture activities are contemplated, this hazard must be taken into account so that the presence of any such toxins is identified and their effects are anticipated. A brief review of their characteristics, by category of toxin, follows.

Ectocrines

Biologically active excretory products of marine organisms that affect the functions of other associated organisms are called ectocrines [6, 7]. Among pelagic organisms, the dinoflagellates and blue-green algae have been most commonly associated with ectocrine productions, with the dinoflagellates *Gymnodinium breve* and *Gonyaulax cantenella* receiving the most attention. The compounds produced by these organisms are most probably heat-stable alkaloids having effects similar to strychnine, muscarine, and aconitine. In the open sea environment such productions may be only of academic interest, posing no threat to fish. It is more probable that ectocrines are beneficial to fish life under normal circumstances, owing to their antibacterial and antiviral activities. But in an intensive mariculture system, more serious ectocrine effects might occur.

As pelagic open sea organisms are generally low in number but high in diversity, the concentration of ectocrines remains quite low. However, waters enriched with nitrogen, phosphorus, and organics favor the growth of ectocrine-producing organisms; when encountering such enriched water these organisms may suddenly and dramatically increase in numbers. Associated with this, of course, will be an increase in the concentration of the ectocrines. At high concentrations these compounds can become toxic to marine life, and may even produce respiratory, dermatitic, or gastrointestinal ailments in humans who come in contact with the material.

Control of toxic phytoplankton growth has been a problem plaguing aquaculturists in Israel for several years. Their difficulties have demonstrated the problems of reducing contamination to a minimum. In the open sea, however, the high natural diversity of toxin-producing species poses an even higher potential contamination rate, and extra care must be taken. Control through adjustment of water chemistry (nutrient composition and ratios) may be feasible, but only after we have developed a considerably greater understanding of chemical–organism relationships. For the present, large-volume turnovers may sufficiently dilute contaminant growth so that toxin production levels can be kept below critical values.

Wastes from Human Activities

The large majority of compounds entering the oceans as a result of human activities have localized effects. For example, essential trace elements in concentrations that exceed tolerable limits will be destructive to aquatic life until sufficiently diluted, but it is unlikely that sufficient quantities of essential trace elements will ever be discharged to produce large fish kills in the offshore environment. Furthermore, these elements are not known to accumulate through the food chain, making it improbable that threats from this source will endanger open sea mariculture activities. However, a notable number of exceptions occur as a result of accumulating human wastes.

Heavy Metals. Heavy metals such as cadmium, mercury, and lead accumulate in tissues in a fashion that results in progressive concentration from one trophic level to the next. As a result, even when heavy-metal concentrations in water are relatively low at the outset, by the time they have been concentrated by the food chain and are available to a terminal predator, such as man, concentrations may be very high in the tissues. It is this fact that poses a problem to aquaculture activities.

The majority of heavy metals enter the oceans directly through industrial outfalls and produce serious, but comparatively localized, effects. The greatest threat would thus appear to be in coastal waters near industrial centers. A significant portion of discharged heavy metals, however, reaches water surfaces in the form of airborne particulates. Atmospheric particulates collect in a thin layer at the sea–air interface and may be many times more concentrated there than in deeper waters. Being airborne, their atmospheric distributions will follow wind patterns closely. Thus, lower initial concentrations occur in areas receiving mild winds from areas of minimal industrialization. Subsequently, currents and natural dispersion rates will tend to bring distributions toward homogeneity. It would be wise to consider this more fully during site evaluation of open sea systems.

Pesticides. Pesticides applied in the form of dusts or sprays also tend to be distributed on the basis of wind patterns. Such great quantities of these materials have been applied that no area of the world is known to be entirely free of pesticide effects.

Of the 200 or so pesticides in current use, the majority are synthetic organic compounds such as DDT and dieldrin. The remainder are either inorganic (mercurous chloride) or naturally occurring organics (pyrethrum). For the purpose of this report, it is perhaps safe to describe a generalized pesticide as "a nonspecific synthetic compound having highly toxic and persistent properties (accumulative in tissues and through the food chain) and which can affect wide geographical areas." The sensitivity of various species to individual pesticides differs. Thus, widespread applications of these compounds will probably have important effects on species composition.

Many pesticides are only slightly soluble in water, but have a strong affinity for fats and oils. Thus, they tend to accumulate in cells containing large quantities of fat or oil food reserves. Among the pelagic marine plants, diatoms and dinoflagellates store the majority of their excess photosynthetic products in the form of oils and may thus be considered primary accumulators of pesticides in the marine food chain. Since similar phenomena occur with higher-trophic organisms, many pesticides concentrate as they are passed up the trophic chain and can reach dangerous concentrations at higher-trophic levels, even when overall concentrations are moderate.

Radioactive Materials. It is well known that low-level radioactivity (background radiation) is natural through the globe. Some of this is induced by radiation sources outside the earth's ecosystem, and the rest is produced by radioactive elements that occur naturally on earth. Organisms that have evolved in the earth ecosystem are, of course, able to tolerate natural levels of radiation without significant ill effects. But with the development of nuclear energy, man began introducing increased levels of radioactivity into the earth's ecosystem. As these accumulate in the environment, and especially in organisms, the organisms are exposed to increased amounts of radiation. A radiobiological rule of thumb states that all high-energy radiation is biologically damaging, and that the amount of damage is an integral function of the dosage rate and total dose. This is not meant to imply that there are no biologically "good" uses to which well-controlled nuclides can be put, e.g., malignancy treatment and biological tracing.

The oceans receive radioactive elements over and above background levels from aboveground nuclear weapons testing, both directly and from atmospheric fallout; from nuclear reactor discharges; and from radioactive wastes that enter the oceans. It should be emphasized here that aboveground nuclear weapons testing appears to have declined in intensity since the mid-1960s, and that nuclear-reactor and nuclear-waste standards appear to be holding present levels of discharge well within known human tolerances, and very likely well within marine ecosystem stability limits [8]. However, with the increasing utilization of nuclear energy, it must be realized that *potential* radiation dangers to the oceans and their ecosystem will increase with time. Concurrently, with reasonable expectations that man will continually expand his understanding and control of radioactivity and its consequences, the effect of increasingly intelligent controls and standards could serve to reduce *actual* radiation hazards. But accurate prediction of future oceanic radio-activity levels and their distributions as increasing usage and ever more effective controls counterbalance one another is, of course, not yet possible.

Radioactivity's adverse biological effects fall into two categories: (1) genetic effects, and (2) direct somatic effects. Genetic effects consist of radiation-induced mutations, which more frequently than not are adverse to succeeding generations. They can occur at dosage levels that produce little direct somatic effect on the irradiated individual or population. Direct somatic effects occur at higher dosage levels and, of course, would usually be accompanied by genetic effects. Somatic effects may be manifested in decreased vitality and longevity, inhibition of metabolic processes, and in near-immediate death if dosage levels are very high.

The exposure of marine organisms to high-energy radioactivity may occur from (1) proximity to radioactive materials; (2) ingestion, and in some cases selective concentration, of radionuclides during respiration and feeding; and (3) adsorption on external body surfaces. Radionuclides in organisms are, of course, passed on up the trophic levels—in some cases decreasing, and in others increasing, in concentration as they do so.

Because of large body surface-to-volume ratios and their location in surface waters, plankton probably tend to collect more radionuclides per unit weight than larger organisms. Thus, organisms feeding on plankton may ingest more radionuclides per gram of food than organisms that feed on larger organisms. In the potential case of man employing plankton as food, there is another factor that would further intensify this effect. Man tends to ingest the whole body of small organisms, whereas in the case of the larger organisms he usually eats only the striated muscle tissue. Thus, with the smaller organism man tends to ingest adsorbed radionuclides as well as those contained in the gut; with the larger organisms he ingests only those relatively few radio-nuclides that have migrated to striated muscle tissue.

The implications of radioactivity in the marine environment for mariculture are probably many, varied, and complex. The most significant seem to be as follows:

- Owing to food-chain concentration of radionuclides, the culture of organisms at each succeedingly higher trophic level is likely to produce radionuclide carriers of increasing significance; if radioactivity levels are at all high in a polyculture environment, higher-trophic products can exceed acceptable radioactivity contamination levels.
- Radioactivity, either in general or in specific locations in the marine environment, probably threatens mariculture interests in direct proportion to its intensity.
- Selection of mariculture sites and designs for systems should minimize radiation exposure.

Although these conclusions are no surprise, they suggest an unhappy picture for those of us who envision primary-production-based polyculture systems associated with the heated effluents from future offshore nuclear power plants.

ORGANIC CARBON COMPOUNDS

The roles played by organic carbon compounds in oceanic bioproductivity are obscure. Although their levels are high in the oceans and there is evidence that they are metabolized, no hypothesis so far advanced is without a legion of opponents. Suffice it to state here that, although nonliving organic carbon occurs in both dissolved and particulate forms in the oceans in concentrations ranging from a few micrograms to 1.0 mg/liter, the oceanic biomass is based upon autotrophic primary production, which depends on inorganic rather than organic carbon compounds. And, as explained elsewhere in this volume, the availability of inorganic carbon in the oceans is such that it is impossible to perceive it as a limiter of autotrophic production.

CONCLUSIONS

Probably the most important chemical oceanography concern for any mariculture venture that is based upon primary production is the availability of inorganic nutrients; the extent to which these can be upwelled and controlled readily will certainly be a major determinant of the feasibility of such a venture. Superficially at least, this indicates that tropical and the more southerly subtropical waters as well as boundary-current regions generally will be most attractive for artificially induced upwellings (see Chapter 4).

There are also a number of other points of some importance:

1. The constancy of salinity throughout the world's oceans is sufficiently great that salinity is unlikely to be a matter for open sea mariculture concern.

2. Although organic carbon may or may not be biologically important, knowledge in this realm is presently so limited that it is difficult to draw any defensible conclusions. Probably, a primary-production-based open sea mariculture venture would be safest if naturally occurring levels of both dissolved and particulate organic carbon were maintained.

3. Of the three major dissolved gases in seawater, oxygen deserves the highest level of attention when assessing potential mariculture viability: the dynamics of this dissolved gas are such that it could under some circumstances limit intense productivity, and its dynamics therefore should be considered in siting and designing open sea mariculture systems.

4. The availability of iron and beneficial trace elements should be considered in siting involving autotrophic productivity.

5. It seems entirely possible that ectocrines could cause toxicity problems in intense productivity systems. Their occurrence and concentrations might

someday be controlled by controlling autotrophic species compositions through control of trace elements and nutrient compositions, if the right formulas can be found. This approach, however, could be expensive.

6. Toxic societal wastes in the form of heavy metals, pesticides, and radionuclides gradually spread through the oceans, and tend to concentrate through food chains. At present, they probably remain at dangerous levels only near the sites of their original introduction into the oceans, and elsewhere in pelagic higher-trophic organisms. Obviously, designs for open sea mariculture should attempt to avoid these sources of contamination.

REFERENCES

1. Culkin, F. 1965. The major constituents of sea water. In J. P. Riley and G. Skirrow, eds., *Chemical oceanography,* Vol. 1, Academic Press, Inc., New York, pp. 121–162.
2. Sverdrup, H. U., M. W. Johnson, and R. H. Fleming. 1942. *The oceans: their physics, chemistry and general biology.* Prentice-Hall, Inc. Englewood Cliffs, N.J. 1087 pp.
3. Menzel, D. W., and J. H. Ryther. 1961. Nutrients limiting the production of phytoplankton in the Sargasso Sea, with special reference to iron. *Deep Sea Res.* 7:276–281.
4. Raymond, L. P. 1973. Oceanic Institute, Hawaii. Unpublished data.
5. Wyrtki, K. 1962. The oxygen minimum in relation to ocean circulation. *Deep Sea Res.* 9:11–23.
6. Rounsefell, G. A., and W. R. Nelson. 1966. Red Tide research summarized to 1964 (Including an annotated bibliography). *U.S. Dept. Interior SSR-F 535.* 85 pp.
7. Syker, J. E. 1965. B.C.F. Symposium on Red Tide. *U.S. Dept. Interior SSR-F 521.* 11 pp.
8. Nat. Acad. Sci. 1971. *Radioactivity in the marine environment.* 265 pp.

Additional Readings

Anonymous. 1970. *Man's impact on the global environment: assessment and recommendations for action.* The MIT Press, Cambridge, Mass. 319 pp.
Arrhenius, G. 1963. Pelagic sediments. In M. N. Hill, ed., *The sea,* Vol. 3, pp. 655–727.
Barber, R. T., A. Vijayakumar, and F. A. Cross. 1972. Mercury concentrations in recent and ninety-year-old benthopelagic fish. *Science* 178:636–639.
Bramlette, M. N. 1961. Pelagic sediments. In M. Sears, ed., *Oceanography.* Amer. Assoc. Adv. Sci. Publ. 67, Washington, D.C., pp. 345–366.
Broecker, W. 1963. Radioisotopes and large scale ocean mixing. In M. N. Hill, ed., *The sea,* Vol. 2, pp. 88–108.
Duursma, E. K. 1965. The dissolved organic constituents of sea water. In Riley and Skirrow, eds., *Chemical oceanography,* Vol. 1, pp. 433–477.

Garrels, R. M., and F. T. Mackenzie. 1971. *Evolution of sedimentary rocks.* W. W. Norton & Company, Inc., New York. 397 pp.

Goldberg, E. D. 1957. Biogeochemistry of trace elements. *Geol. Soc. Amer. Mem. 67,* Vol. 1, pp. 345–358.

Goldberg, E. D. 1965. Minor elements in sea water. In Riley and Skirrow, eds., *Chemical oceanography,* Vol. 1, pp. 163–196.

Gordon, D. C., Jr. 1971. Distribution of particulate organic carbon and nitrogen at an oceanic station in the central Pacific. *Deep Sea Res.* 18:1127–1134.

Harvey, H. W. 1955. The chemistry and fertility of seawaters. Cambridge University Press, New York. 224 pp.

Hill, M. N., ed. 1963. *The sea.* Vol. 1, *Physical oceanography,* 864 pp.; Vol. 2, *Composition of sea water; comparative and descriptive oceanography,* 554 pp.; Vol. 3, *The earth beneath the sea,* 963 pp. Wiley-Interscience, New York.

Horne, R. A. 1969. *Marine chemistry.* Wiley-Interscience, New York. 568 pp.

Martin, D. F. 1970. *Marine chemistry* Vol 2, *Theory and applications.* Marcel Dekker, Inc., New York. 451 pp.

Menzel, D. W., and J. H. Ryther. 1968. Organic carbon and the oxygen minimum in the South Atlantic Ocean. *Deep Sea Res.* 15:327–338.

Raymont, J. E. G. 1963. Plankton and productivity in the oceans. Pergamon Press, Inc., Elmsford, N.Y. 660 pp.

Riley, J. P., and G. Skirrow, eds. 1965. *Chemical oceanography,* Vols. 1 and 2. Academic Press, Inc., New York. 712 pp.

Skirrow, G. 1965. The dissolved gases—carbon dioxide. In Riley and Skirrow, eds., *Chemical oceanography,* Vol. 1, pp. 227–322.

Sillen, L. G. 1961. The physical chemistry of sea water. In M. Sears, ed., *Oceanography.* Amer. Assoc. Adv. Sci. Publ. 67, Washington, D.C., pp. 549–582.

Smith, S. V., J. A. Dygas, and K. E. Chave. 1968. Distribution of calcium carbonate in pelagic sediments. *Marine Geol.* 6:391–400.

Sorokin, J. I. 1971. On the role of bacteria in the productivity of tropical oceanic waters. *Intl. Revue der Gesellschaft Hydrobiologie.* 56:1–48.

Strickland, J. D. H. 1965. Production of organic matter in the primary stages of the marine food chain. In Riley and Skirrow, eds., *Chemical oceanography,* Vol. 1, pp. 433–477.

Vaccaro, R. F. 1965. Inorganic nitrogen in sea water. In Riley and Skirrow, eds., *Chemical oceanography,* Vol. 1, pp. 365–408.

Zenkevitch, L. A. 1961. Certain quantitative characteristics of the pelagic realm. In M. Sears, ed., *Oceanography.* Amer. Assoc. Adv. Sci. Publ. 67. Washington, D.C., pp. 323–335.

6

OCEANOGRAPHIC IMPLICATIONS FOR OPEN SEA MARICULTURE

J. A. Hanson

With the information on the open sea environment summarized here in Part Two, we can attempt to draw general conclusions concerning the more attractive geographic areas for open sea mariculture that lie within our previously selected overall area of interest. But to achieve this, it is necessary to establish criteria for site selection. We suggest that the following attributes are desirable for any open sea mariculture site:

1. Actual or potentially available energy sources: preferably these should be conservative energy sources, but nonconservative energy availability should not be ignored.

2. Either the availability of natural platforms upon which to base mariculture stations, or ease of emplacement of man-made platforms.

3. Freedom from destructive storms and waves as well as overly strong currents.

4. Natural availability of nutrients suitable to artificial food-chain establishment or enhancement, or to the direct nutrition of given desirable species.

5. General suitability to the growth of desirable species, as manifested in physical and chemical characteristics, a lack of detrimental environmental factors, inclusion in the natural ranges of desirable species, and the like.

6. Freedom from civilization-derived contaminants and from an obvious threat thereof.

7. Minimization of transportation requirements between mariculture stations and major consumer centers.

If these are indeed the attributes for which we should search, then how have the preceding chapters helped us, and where have they failed us? First, they seem to have provided a fair understanding of which areas have the higher conservative energy potentials. For example, the equatorial- and boundary-current areas appear to contain the higher wind, wave, and current energy potentials. Generally, the lower latitudes have higher daily and annual solar radiation levels, not only incident to the atmosphere but also at the sea surface. Areas of volcanic activity have been indicated, and these should provide a rough index to geothermal energy potentials. And then, of course, simply general proximity to either major transportation routes or major fossil-fuel reserves implies economic availability of this conventional, nonconservative energy source.

Second, the topography and bathymetry in the areas of interest suggest that Micronesia, the Hawaiian Archipelago, the Line Islands, the waters off the North American continent's east and west coasts, the waters of the Florida Keys and U.S. Caribbean, and the Gulf of Mexico offer either mariculture platforms themselves (high islands and atolls) or the more suitable sites for emplacement of man-made platforms.

Third, large swells appear to occur in the open sea without a degree of geographical discrimination that is significant for our purposes. Large wind waves are, of course, associated with severe storms in both oceans. Strong steady currents can be predicted geographically with some overall statistical confidence, even though precise predictions for given areas are not yet possible. Current along the southeast U.S. coastline appears to be the most predictable at present; it is also the area of strongest current and most susceptible to hurricanes.

Fourth, surface nutrient levels and organic carbon production vary widely, reaching their maximum values near major landmasses or in natural upwellings. Deeper nutrients, however, appear to have higher values in the Pacific than the Atlantic; the Pacific may be more attractive for artificial upwelling. Artificial upwellings might be possible anywhere with an adequate material and energy investment, although usually most easily obtained in low latitudes. This might attract us to areas in which conservative energy sources can be tapped to produce artificial upwellings, the main cost for their creation thereby being reduced to capital expenditures and maintenance.

Fifth, except for localized terrigeneous effects there seems to be, at the level of generality we must adhere to here, little to choose from one area to another in terms of major chemical features. Although regular geographic patterns in salinity, dissolved oxygen, and carbon dioxide appear to be of some biological significance, too little is known about trace-element variations to allow any general conclusions. Temperature, on the other hand, varies widely with geography and may be both an upper and lower limiter of productivity and reproduction, varying in effect with the organism of interest. Thus, we might not be surprised to find upper and lower temperature ranges as well as light ranges imposing significant constraints on our choice of species for given mariculture locations.

Sixth, the localized effects of civilization-derived contaminants (concentrated trace elements, pesticides, and radionuclides) suggest that the mariculturist should approach areas subject to the direct effects of industrialization with caution; unless (or until) levels of global-scale contaminants begin to threaten most life, he can hope for relative immunity in all other areas.

Seventh, the closer we can locate open sea mariculture to major centers of consumption, the lower will be the portion of overall costs consumed by transportation; but such a location would be more likely to cause contamination problems. Locating mariculture near global shipping routes would also reduce transportation costs. However, it should be kept in mind that transportation is only one of several major subsystems of a total mariculture system. Therefore, any decision to lengthen or shorten transportation links must be made within the context of a variety of other cost–benefit assessments. Although always desirable, the shortening of transportation links is not necessarily an overriding consideration in selecting potential open sea mariculture sites.

PRELIMINARY EVALUATION OF POTENTIAL OPEN SEA MARICULTURE ENVIRONMENTS

With the foregoing open sea environmental considerations in mind, and without necessarily asserting that mariculture operations of any sort will be economically feasible there, we suggest that six regions within the overall area of interest are the most attractive in terms of their potential for open sea mariculture. These are:

- Micronesia
- East coast of the continental United States
- West coast of the continental United States
- East Pacific North Equatorial Current
- Hawaiian Archipelago
- Gulf of Mexico

In the following paragraphs (which *do not* suggest an order of priority) we summarize the major attractions and limitations of each area.

Micronesia

Micronesia offers an abundance of mariculture "platforms" in lower latitudes that are essentially free from point-source contamination. A prospective labor force is either already in residence on many "platforms" or potentially available in the immediate area. Moreover, the United States seeks actively to assist the island groups of the Trust Territory of the Pacific Islands in developing economically. The potential availability of conservative energy seems good. But tropical storms and typhoons are not uncommon. Proximity to major transportation routes is not very attractive presently, but feeder routes are in existence and could no doubt expand with demand. With sufficient investment there seems a good possibility that many commercial species could be cultured in Micronesia on commercial scales. Cultural and political problems are many, though, and large capital investments may involve considerable risk.

East Coast of the Continental United States

The main attractions of the East Coast of the United States are the broad and rather shallow continental shelf, the persistent Gulf Stream, and the proximity to major population centers. But seasonal variability is high, as is the potential for contamination problems. Hurricanes occur about every year, and conservative energy potential, except from wind and currents, tends to be comparatively low.

West Coast of the Continental United States

The West Coast of the United States offers about the same major advantages as does the East Coast, except that the shelf is much narrower. The area is much less subject to destructive storms or seasonal variations, but has more frequent and severe swell and surf, and currents are slower and seemingly less predictable with present knowledge. The occurrence of natural upwellings in its Eastern Boundary Current may imply that artificial upwelling could be induced with less energy than might be required in other areas.

East Pacific North Equatorial Current

Nutrient-rich Intermediate Water is somewhat nearer the surface in the East Pacific North Equatorial Current region and most forms of conservative energy are abundant. Freedom from contamination is good, seasonal variations are small, and destructive storms are infrequent. Conversely, most of the region is far from shore, off major shipping lanes, and in very deep water, except for a very few islands and seamounts at its far eastern end. Commercial-scale mariculture in this region would demand self-sufficient, floating platforms.

Hawaiian Archipelago

This region between 20 and 30° north latitude contains a number of islands and atolls that are reasonably near a major shipping route. Work-force proximity is fair, weather factors are favorable, conservative energy availability is good, and point-source contamination problems are essentially nonexistent. But waters near islands and atolls are very deep so that, unless atolls were employed, mariculture operations would require floating platforms, although these would not necessarily have to be of a size of self-sufficiency comparable to those which would be required in the East Pacific North Equatorial Current. However, the atolls northwest of the major Hawaiian Islands are designated conservation areas at present, and the legality of mariculture activities in such areas is in doubt.

Gulf of Mexico

This region features a broad continental shelf zone with "open sea" platforms already in place off the Gulf states. Its attractiveness is degraded by frequent hurricanes and, in some areas, by major river mouths that constitute point sources of contaminants. Seasonal variations are moderate for the most part. Because of the existence of a large number of offshore platforms and the proximity to urban areas, transportation, and research centers, the Gulf seems to offer a most attractive area for pilot-scale open sea mariculture activities in spite of the stated drawbacks.

RECOMMENDATIONS FOR FUTURE RESEARCH

To state comprehensively all the important avenues for future oceanographic research would require a book in itself. Moreover, the staff involved in this study is not up to the task. Consequently, this short section is devoted to a brief mention of those rather broad avenues that appear most critical to open sea mariculture.

Upwelling Dynamics

As our examination progresses, the apparent desirability of artificial upwelling of Intermediate Water increases. How can this nutrient-rich water be brought to and kept in the surface regime in volumes sufficient to support large-scale bioproductivity? At what rate should it flow through the culture area to stimulate high bioproductivity while minimizing undesirable side effects? These are only two of many important facets of the artificial upwelling question.

The questions of artificial upwelling and primary productivity dynamics offers an important and intriguing subject for intensive investigation. We believe a research program in this area would best be a closely managed combination of empiricism and dynamic modeling. This subject is considered further in Chapters 7 and 15.

Organic Carbon Dynamics

In Chapter 5 we mentioned the current state of our ignorance with respect to the role or roles of organic carbon compounds in the oceanic ecosystem in general and in bioproductivity in particular. Although most authorities believe these compounds do play an important role or roles, the simple fact is that there is much speculation and little agreement on what the roles may be. That this piece of the puzzle is missing is adequate justification for concerted research in marine organic carbon dynamics in itself. That these substances may be important to bioproductivity in open sea mariculture adds further urgency.

Fate of Contaminants

Toxic metals, radionuclides, pesticides, ectocrines, and other potentially damaging substances enter and diffuse through the oceanic ecosystem in quantity, continuously and repeatedly. For the most part, rough characterizations of the dynamic interactions between these substances and the marine environment have been developed. In some instances these characterizations have not been well verified empirically, and in essentially all cases details of the physical and chemical dynamics are incomplete, as is knowledge of their ultimate fates. Since these compounds and/or their derivatives can be damaging to bioproductivity directly, and since they most likely could eventually affect the equilibrium of the oceanic ecosystem as a whole, it appears imperative that our knowledge in this realm be advanced greatly and soon.

Research, in the form of a large number of individual projects, is underway now. But a concerted and closely managed program with well-defined goals is indicated.

Conservative-Energy Localities

Conservative-energy levels (winds, waves, currents, thermal gradients, and direct solar radiation) undoubtedly are greater in some oceanic regions than in others. A worthwhile objective, therefore, would be to pinpoint those regions in which either a given source is very high or combinations of several sources represent a particularly high total flux. In Chapter 13 we examine the energy question in more depth and show why this question may be important to world energy problems as a whole, as well as to the future of open sea mariculture.

PART III

SCIENCE AND TECHNOLOGY
IMPORTANT TO
OPEN SEA MARICULTURE

7

NUTRITIONAL CONSIDERATIONS OF OPEN SEA MARICULTURE

L. P. Raymond, P.K. Bienfang, and J. A. Hanson

The trophic-level concept and its relation to the bioenergetics of the sea have been explored in Chapter 1 in the context of the bioeconomic perspective on open sea mariculture. The role of nutrition as one factor in determining the economic feasibility of large-scale commercial mariculture was outlined there. In this chapter, devoted wholly to the question of nutrition, we begin with a review of the kinetcis of energy transfer between the trophic levels, starting with the storage and transfer of solar energy at the point of primary production, and moving on through primary, secondary, and tertiary consumers–producers. Then the chemistry of nutrition is reviewed, with a look at the nutrient elements required by the various classifications of organisms and some of the possible food forms and sources. The third part of this chapter deals with potential mechanisms for feeding cultured organisms, from both the design and economic standpoints, and explores the implications of the nutrition issue for open sea mariculture. A discussion of the role of nutritional requirements in the selection of species for culture follows in Chapter 8.

KINETICS OF NUTRITION

An ecosystem is dynamic in character. Its viability and integrity are maintained by the persistent flow of energy through the various components of the system. In keeping with the trophic-level concept, the organisms within an ecosystem may, from an energy viewpoint, be grouped into a series of reasonably discrete trophic levels as producers, primary consumers (herbivores), secondary consumers, and so on. Each level is dependent on the preceding level as a source of energy, with initial production relying on radiant energy. Figure 7.1 gives an energy-flow view of the trophic chain. Profitable mariculture will depend upon the effective application of energy principles to establish criteria such as optimal crop density, fertilization, and yield.

A comprehensive overview of nutrition in marine food chains includes the following elements: (1) the nature of the various components; (2) the essential substrates and the kinetics of their assimilation; (3) the biomass and

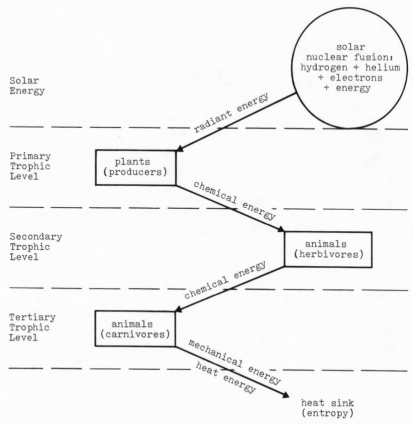

FIGURE 7.1 *Trophic energy flow. This schematic energy-flow diagram depicts the flow of energy, originating with the sun, through the trophic levels. Note that this simplified presentation neglects the feedbacks of recycling activities. [Modified from Russell-Hunter, 1971 (Additional Readings).]*

production rates of the components; and (4) the kinetics of energy transfer between trophic levels.

Plankton comprise most of the oceanic biomass. Their components are the phytoplankton (plants), which are the primary producers in the food chain, and the zooplankton (animals), which include the majority of the primary consumers.

Primary Production

Primary producers have the capacity to convert radiant energy into chemical energy via photosynthesis. Owing to the fluid character of the marine environment, most primary production results from the growth of suspended microscopic plants called *phytoplankton*. When provided with adequate substrates, these single-celled plants have the ability to synthesize all the complex organic substances required for growth and reproduction. The production and distribution of phytoplankton is controlled in a complex way by physical factors (light, temperature, dissolved gases), chemical factors (nutrients), and biological factors (intrinsic growth rate and grazing).

Taxonomically, the phytoplankton contain relatively few phyla: the Chrysophyta, Pyrophyta, Cyanophyta, and Cryptophyta. However, there is considerable diversity within phyla to complement the nutritional and habitat diversity of the marine environment. Members of the Chrysophyta nearly always dominate a phytoplankton population; principally, these include the diatoms (Bacillariophyceae) and the chrysophytes (Chrysophyceae). The Pyrophyta is most commonly represented by the class Dinophyceae, or dinoflagellates, which are frequently responsible for "red tides." The remaining three phyla are generally minor components of the population.

On the basis of size, the phytoplankton are classified as either ultraplankton (0.5 to 10 μm), nannoplankton (10 to 50 μm), or microplankton (50 to 200 μm). These three classifications account for 99 percent of all phytoplankton. There exists a good deal of evidence that nannoplankton constitute the largest plant biomass of the oceans. Open-ocean phytoplankton samples are dominated by diatoms, but commonly contain representatives of all phyla. Neritic samples demonstrate greater heterogeneity, and species dominance varies greatly with location. Phytoplankton biomass is commonly determined by measurements of chlorophyll *a* concentrations, cell density, or dry weight. The rate of production of organic matter dictates the relative contribution, or impact, of a group of organisms on an environment. Phytoplankton productivity determinations measure the rate of photosynthesis,

$$\longrightarrow n CO_2 + n H_2 O \xrightarrow{\text{light}} n(CH_2 O) + n O_2$$

that occurs per volume of water. The most popular and reliable productivity measure, the radiocarbon technique, measures the rate at which inorganic carbon dioxide is transformed into organic carbon. Simply stated, this procedure involves adding a known amount of distinguishable carbon atoms (i.e., radioactive) to a plankton-containing water of known alkalinity. After appropriate incubation, the phytoplankton are isolated by filtration. Subsequent determination of the amount of radioactive carbon atoms in the phytoplankton permits calculation of the total amount of organic carbon produced.

Light and Photosynthesis in the Sea. The amount of solar energy impinging on a given volume of water is an important variable influencing primary production. The literature contains some excellent reviews on the relationship of phytoplankton production and the flux of radiant energy [1–4].

Light limitation is of great importance to oceanic productivity. Subsurface light intensities define the base of the euphotic zone, worldwide. The relationship between photosynthesis and light intensity is illustrated in Figure 7.2. At low light intensities, the relationship is essentially linear; this corresponds to the *light-limitation* zone shown in the figure. The direct proportionality diminishes with increasing light intensity until a saturation point is attained (the *saturation zone*). In this region, photosynthesis is not controlled by the supply of light but by some other limiting factor. At higher intensities, photosynthesis is actually depressed owing to photooxidation, corresponding to the *inhibition zone* [2, 5].

The influence of light is expressed seasonally, geographically, and vertically. Phytoplankton productivity shows most seasonal variation in temperate regions where climatological variations are most extreme and occurs to a lesser degree in the tropical and subtropical regions. Radiant energy variation on a geographic scale becomes important in high latitudes where light limits primary production. The extent of light penetration into the sea defines the comparatively thin region where photosynthesis can occur. In a vertical section, productivity commonly shows a relatively low surface value and an intermediate maximum, which declines exponentially with depth. This typical pattern (Figure 7.3) corresponds to the region of photosynthetic inhibition, the saturation zone where maximum production occurs, and the region of light limitation. The intersection of the photosynthesis and respiration curves defines the depth at which net production is zero. Generally taken

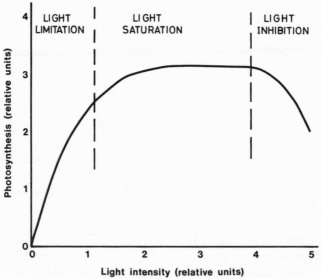

FIGURE 7.2 *Photosynthesis as a function of light intensity.*

to be at 1 percent surface light intensity, this arbitrary point delimits the depth of the euphotic zone.

Essential Nutrients for Phytoplankton Growth

Phytoplankton have requirements for nitrogen, phosphorus and, in the case of diatoms, silica. The demand for these nutrients frequently exceeds the supply. In such cases, the nutrient in shortest supply relative to need becomes the substrate that limits primary production. The overwhelming influence of nutrients on oceanic productivity is shown in Figure 7.4. The highest productivities occur in areas proximate to land where nutrients originating from land and/or upwellings become available.

The cellular metabolism of phytoplankton also requires elements such as sulfur, potassium, and sodium, which are abundant in seawater, and the less

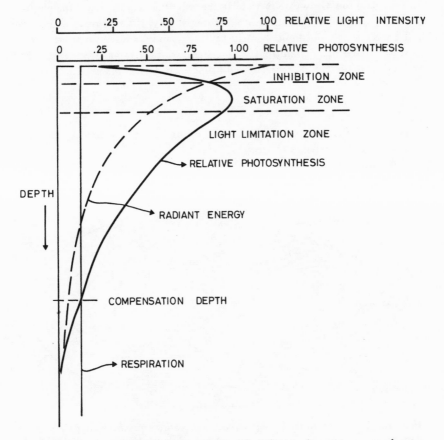

FIGURE 7.3 *Vertical relationships of light and primary production. (Adapted from Yentsch, and Ryther.)*

abundant trace metals, iron, calcium, manganese, magnesium, zinc, and boron. The presence of these metals is usually adequate and they are seldom limiting. Table 7.1 describes a typical media for phytoplankton culture *in vitro*. This particular recipe also illustrates the constituents of seawater, thus providing a realistic picture of the requirements of marine plants.

A variety of organic substances have been shown to be assimilated by laboratory cultures of phytoplankton, but the utilization and nutritional importance of such compounds to natural populations is poorly understood. The vitamin B group (B_{12}, thiamine, and biotin) is an exception to this rule. At least one of these vitamins is required by some species of every phyla. Most marine vitamin production is by phytoplankton, particularly where the phytoplankton biomass is substantial. Vitamins are thought to be a determining factor in the composition and/or distribution of phytoplankton populations [6, 7].

Phytoplankton Growth Kinetics. If the essential requirements for phytoplankton growth are present in nonlimiting supply, the rate of algal growth will be maximum and limited solely by the organism's intrinsic capacity for increase. Under these conditions, each cell grows at the same rate as its parent and the total number, N, increases with time according to the equation

$$N_t = N_0 e^{kt}$$

where N_0 is the original number, N_t the number at any time t, and k a constant representative of the species.

Both closed (batch culture) and open (continuous culture) systems are used

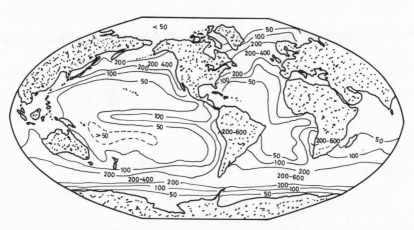

FIGURE 7.4 *Generalized map of organic carbon production in grams of carbon per square meter per year. [Adapted from Hedgpeth, 1957 (Additional Readings).]*

TABLE 7.1 *Media for Laboratory Culture of Phytoplankton*

Artificial seawater medium	Amount per liter of distilled water
Solution 5	
KNO_3	101.1 g
KH_2PO_4	6.8 g
$NaHCO_3$	4.2 g
Trace elements X	
$ZnCl_2$	4.0 mg
$(NH_4)Mo_7O_{24} \cdot 4H_2O$	36.8 mg
$CuCl_2 \cdot 2H_2O$	3.4 mg
$MnCl_2 \cdot 4H_2O$	40.0 mg
H_3BO_3	60.0 mg
Trace elements Y	
$FeCl_3 \cdot 6H_2O$	240.0 mg
0.05 M EDTA = 1.86 g EDTA/ml	100.0 ml
Mix 8	
Thiamine–HCl = 10 mg/50 ml	200.0 mg
Biotin = 2.5 mg/50 ml	50.0 mg
B_{12} = 0.25 mg/50 ml	5.0 mg
Solution	
NaCl	32.1 g
$MgSO_4 \cdot 7H_2O$	6.66 g
$MgCl_2 \cdot 6H_2O$	5.49 g
$CaCl_2$	1.11 g
Solution 5	10.0 ml
Tris buffer[a]	20.0 ml
Trace elements X	1.0 ml
Trace elements Y	1.0 ml
Mix 8	2.0 ml

Seawater enrichment medium (Medium "F")	Amount per liter of distilled water
Seawater	1 liter
$NaNO_2$	150 mg (1.765 μM)
$NaH_2PO_4 H_2O$	10 mg (72.5 μM)
$Na_2SiO_3 \cdot 9H_2O$	30 mg ($>$10 μM)
Ferric sequestrene[b]	10 mg (1.3 mg Fe)
$CuSo_4 \cdot 5H_2O$	0.0196 mg (0.005 mg Cu)
$ZnSO_4 \cdot 7H_2O$	0.044 mg (0.01 mg Zn)

continued

TABLE 7.1 *continued*

Seawater enrichment medium (Medium"F")	Amount per liter of distilled water
$CoCl_2 \cdot 6H_2O$	0.02 mg (0.005 mg Co)
$MnCl_2 \cdot 4H_2O$	0.360 mg (0.1 mg Mn)
$Na_2MoO_4 \cdot 2H_2O$	0.0126 mg (0.005 mg Mo)
Thiamine–HCl	0.2 mg
Biotin	1.0 μg
B_{12}	1.0 μg

[a]Tris buffer solution: 243 g of Sigma 7–9 Tris, 102 ml of concentrated HCl, 800 ml of distilled water to pH 7.6.

[b]Ferric sequestrene (Geigy) is the sodium iron salt of ethylenediaminetetraacetic acid, 13 percent iron.

Source: Jones, L. 1960. From letter to F. T. Haxo, March 1960.

to describe phytoplankton growth kinetics. In both systems all essential growth requirements *except one* (the limiting nutrient) are provided in excess.

Closed System. Consider a closed system (e.g., a flask) that contains sufficient quantities of all requirements, except phosphorus, which is present in a limited supply. The resulting phytoplankton and limiting-nutrient kinetics are described by Figure 7.5. Initially, there is very little growth. During this *lag phase* (A), the cells assimilate essential nutrients and organize, physiochemically, for growth and division. The length of the lag phase is primarily a function of the physiological condition of the innoculum.

The population's maximum rate of increase is expressed during the *logarithmic growth phase* (B). Note that the ambient limiting nutrient concentration is almost exhausted, which indicates that the organism is utilizing intracellular reserves, rather than environmental nutrients, for production. The cessation of exponential growth, caused either by the exhaustion of nutrients or by *autoinhibition,* or both leads into the *stationary phase* (C). During the stationary phase, the cell density does not vary appreciably. The decrease in growth reflects the depletion of internal nutrients and a concomitant accumulation of toxic metabolites. The *decay phase* (D) describes a senescent culture in which the mortality rate exceeds the production rate. The mortality rate is nearly exponential, mainly in response to the drastic increase of lysogenic metabolites, and the production rate approaches zero as more and more nutrients become locked in either dead or dying cells.

Open System. The open-system technique for the culture of phytoplank-

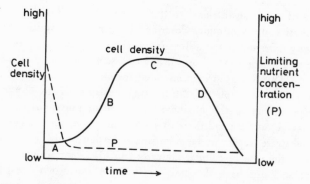

FIGURE 7.5 *Phytoplankton growth and nutrient kinetics in a nutrient-limited batch culture.*

ton provides the essential feature that growth occurs at a constant rate in a constant environment. Those environmental factors responsible for the cessation of exponential growth in batch culture are within the control of the experimentor in a continuous system. The growth rate of the population is controlled by the supply rate of an essential requirement (typically a dissolve nutrient).

Media are added to a constant-volume growth chamber at a constant rate (f) and emerge at an equal rate. The ratio of the flow rate (f) to the growth-chamber volume (v) determines the dilution rate (w) of the chamber:

$$w = \frac{f}{v}$$

Phytoplankton growth in the chamber is expressed by μN, where μ is the specific growth rate, $1/N \, dN/dt$, and N is the concentration of cells. The net change in phytoplankton concentration is given by

$$\frac{dN}{dt} = \mu N - wN$$

where wN represents the wash-out, or dilution, rate. Thus, if μ is greater than w, the concentration will increase; if μ is less than w, it will decline. When the concentration does not change significantly with time, that is, $dN/dt = 0$, the growth rate must equal the wash-out rate. This situation of constant density and continuous turnover of organisms is called a steady state. Although the operation of such continuous-culture systems is elaborate and rigorous, it does provide a mimicry of natural conditions fundamental to the study of nutritional kinetics.

Nutrient Assimilation. The marine environment is seldom constant and/or nonlimiting. Phytoplankton productivity (growth) is usually controlled by

the supply of essential nutrients to the euphotic zone. For this reason, algal growth kinetics are commonly described in terms of a limiting "nutrient," although *any substrate* essential for growth can be a limiting factor.

The limiting factor and the organism response are described by a hyperbola (Figure 7.6). This has been demonstrated for numerous organisms at various trophic levels, and is applicable for limiting substrates of any physical character. Perhaps reflecting the biochemical nature of nutrition, the hyperbola implies that (1) under conditions of severe limitation, small increases of the limiting substrate result in large organism responses; (2) there is a finite capacity for the rate of cellular increase; and (3) substrate increases beyond this point will not have detectable effects on the growth response. The relationship between limiting-nutrient concentration and phytoplankton growth rate is obscured owing to the ability of plankton to assimilate and store nutrients intracellularly prior to their use for growth. The two-stage process predicates that nutrient assimilation and growth be considered separately.

The rate of nutrient uptake by phytoplankton plotted as a function of that nutrient concentration in the surrounding media is defined by the equation

$$V = V_{max} \frac{S}{K_s + S}$$

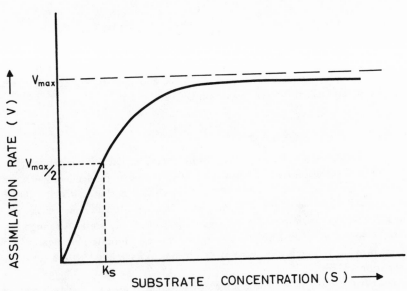

FIGURE 7.6 *Relationship of limiting-substrate concentration and assimilation rate.*

where V is the uptake rate, V_{max} the maximum uptake rate, S the substrate concentration, and the half-saturation constant K_s represents the substrate concentration at which $V = V_{max}/2$ [8–10].

The values of K_s and V_{max} vary among different species of phytoplankton and for different limiting nutrients. These variations are throught to reflect evolutionary adaptations to existence in various types of environments, and may, in part, determine species dominance for a given set of nutritional conditions.

The storage of assimilated nutrients in an internal reservoir suggests a stability factor, in that the phytoplankton do not merely "live for today" by immediate responses to nutrient increases, with subsequent growth increases. The amount of limiting nutrient contained in the reservoir is related to the growth rate. Through this *yield coefficient* we have a measure of the nutritional state of the unit of population, the phytoplankton cell.

Growth. The relationship of the specific growth rate, μ, and the environmental limiting nutrient concentrations follows the hyperbolic pattern and is given by

$$\mu = \frac{\mu_{max} S}{K_s + S}$$

Note that a relation of strict equality is not valid for the growth–substrate relationship. The reasons for this departure lie in the variable character of the yield coefficient and the maximum uptake rate.

Although it is not within the scope of this presentation to go into any greater detail, it should become clear that the relationship of phytoplankton growth to even a single nutrient is complicated. This complexity arises from the necessity of relating the ambient nutrient concentration to the internal nutrient concentration, which in turn is related to the growth rate. Thus, the nutritional state of the phytoplankton population (reservoir content) has a profound effect on the resulting growth rate.

Secondary Production

The *zooplankton* represent a heterogeneous group of organisms having great diversity of shape, size, life history, taxonomic composition, and nutritional habits. Within zooplankton are found animals that either express their entire life cycle while drifting, (*holoplankton*) or are planktonic for only a portion of their life cycle (*meroplankton*). On the whole, holoplankton dominate the zooplankton; the concentration of meroplankton in the zooplankton depends on geographic location and proximity to land.

Taxonomically, zooplankton consists of protozoa, coelenterates, ctenophores, mollusks (e.g., gastropods and cephalopods), chaetognaths, chordates (e.g., tunicates, salps, fish larvae), and arthropods. Persistently a major component, arthropods include copepods and Malacostraca (euphausids,

mysids, isopods, amphipods, and decapods). More detailed discussions of the character and composition of zooplankton are presented in works by Hardy [11], Friedrich [12], and Wimpenny [13].

Classified by trophic position, the zooplankton consist of herbivores, carnivores, and omnivores. The herbivores, which graze upon phytoplankton, are in turn grazed by larger carnivorous zooplankton. Omnivores can satisfy their nutritional needs by consumption of either plant or animal biomass, and the type of food eaten is greatly influenced by its relative availability. Furthermore, certain zooplankton may feed herbivorously during early life stages and carnivorously during adult stages. Because of zooplankton's overwhelming contribution in the conversion of plant energy to animal energy, zooplankton grazing is often taken to be synonymous for herbivorous feeding in the ocean. Marine energy-transfer dynamics between the first and second trophic levels are best understood as they pertain to zooplankton.

Herbivorous Feeding and Growth. In a stable system, the biomass produced at each successive trophic level must be less than that produced at the preceding level. A portion of the ingested "energy" is oxidized to support basal metabolism, and the remainder is directed into additional biomass production. Approximately 80 percent of ingested energy is dissipated as mechanical energy used for food capture, excretions, respiration, molting, locomotion, and the like (Figure 7.7). Of the respiratory expenditures, the mechanical energy involved in the food-gathering process is the most expensive. Such energy expenditures have been reported to be twice that of the entire remaining basal metabolism. Thus, the maintenance of high phytoplankton densities results in greater returns per unit feeding effort, and the

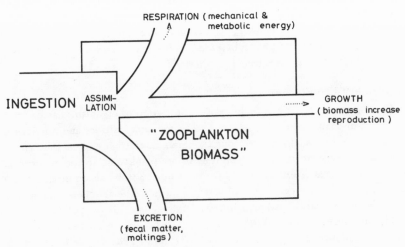

FIGURE 7.7 *Model of a zooplankter.*

zooplankton are able to channel greater proportions of energy toward growth.

Certain physical characteristics of marine herbivores are dictated by the microscopic nature of the primary trophic level: they must (1) be of relatively small size, and (2) possess feeding mechanisms that permit effective utilization of such a substrate. Filter feeding, a response to feeding in a food-limited environment, is the most efficient way to graze the 5- to 50 -μm phytoplankton. Energy is conserved because the thoracic limbs are multipurpose: their beating serves for propulsion, respiration, and the creation of currents for filter feeding. A complete description of the mechanism and design of filter feeding may be found in Barnes [14] and Lasker [15].

Food intake is dependent upon food size. There exists an optimal size range upon which herbivores will preferentially feed, and a maximum and minimum size, predicated by the physical constraints of the ingestion mechanism and efficiency considerations.

The relationship of uptake, or grazing rate, to the phytoplankton concentration describes a hyperbola (Figure 7.8). The lower portion of the curve is essentially linear and rises proportionate to increases in phytoplankton concentration. The asymptotic portion describes the intuitive notion that persistent increases in phytoplankton concentration cannot enhance the grazing rate indefinitely. Obviously, the efficiency of the organisms' food collection apparatus and the subsequent production rate have finite limits. At phytoplankton concentrations less than A, the metabolic maintenance costs are greater than the energy derived from feeding, and the herbivore does not graze. This relationship has been shown to apply to many isolated species of zooplankton, although the relative values for A, B, and C will vary with the species, size, and age of the zooplankton and with the nutritional state of the phytoplankton [16-20].

The maintenance of short, discrete food chains maximizes the production efficiency of the desired culture organism. Because energy transfer through biological cycles involves energy losses at each trophic level (gross ecological efficiency is equal to 20 percent), a given amount of phytoplankton normally will produce no more than roughly one fifth as much carnivore biomass as it will herbivore biomass. Furthermore, competition with ubiquitous natural carnivores, such as ctenophores and chaetognaths, would reduce the production efficiency of the desired organism from the given 20 percent value. Thus, it is not coincidental that mariculture ventures have a tendency to concentrate on culturing herbivorous products. Unless the market value of a carnivore is high, the lower production efficiencies of higher-trophic-level organisms argue against their selection for aquaculture. Therefore, this path of inquiry terminates here. Direct feeding of carnivores is, however, treated in later sections of this chapter.

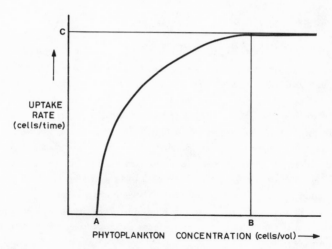

FIGURE 7.8 *Zooplankton uptake (grazing) rate compared to phytoplankton concentration.*

CHEMISTRY OF NUTRITION

Nutritional Requirements of Zooplankton

Many of the dietary requirements for zooplankton growth and reproduction have been discovered through laboratory experimentations. Frequently, attempts have been made to culture zooplankton *in vitro,* using unialgal foods, and to evaluate the suitability of the food from the growth rates and survival data. Table 7.2 shows phytoplankton capable of supporting the growth of crustacean species for a minimum of five generations. Of these, only one (*Monochrysis lutherii*) supports a copepod culture for an extended period of time, and in all cases the condition of the culture markedly improves when a mixture of algal foods is offered. No single alga has been shown to be acceptable to all copepod species.

Several generalizations may be warranted on the basis of the data in Table 7.2. First, although a large number of highly diverse algae have been tested as foods, by far the majority of those which support good growth are naked flagellates of small size; few are diatoms. This may suggest an important role for chrysophytes as feeds for herbivorous zooplankton. Second, as evidence suggests strongly that single algal species will seldom support long-term crustacean (zooplankton) cultures, the chemical composition of algal species must differ. Thus, algal mixtures likely would overcome the nutritional deficiencies of the constituent species. Finally, that several algal species, employed singly, will support short-term zooplankton culture suggests that many nutritional deficiencies are compounds required by zooplankton only in trace amounts.

TABLE 7.2   *Acceptable Foods for Species of Zooplankton*

Species	Acceptable foods
Acartia tonda	*Isochrysis galbana, Rhodomonas* sp.
Artemia salina	*Stephanoptera, Brachiomones, Rhodomonas, Dunaliella salina*
Tibriopus japonicus	*Monochrysis lutherii, Rhodomonas lens, Isochrysis galbana, Chroomonas*
Acartia clausi	*Isochrysis galbana, Rhodomonas* sp.
Idya furcata	*Isochrysis galbana, Rhodomonas* sp.
Pseudocalanus minutes	*Isochrysis galbana*
Eurythemora hirundoides	*Isochrysis galbana*
Temora longicornis	*Isochrysis galbana*
Pseudocalanus elongatus	*Isochrysis galbana*
Eurythemora affinis, Eurythemora herdmani	*Isochrysis galbana, Cyclotella nana, Platymonas* sp., *Skeletonema costatum*

These possibilities can be investigated by the cultivation of the crustacean species upon fully defined foods and under sterile conditions. Preliminary studies to develop such foods have been completed for *Artemia salina* [21–23]. Table 7.3 shows nutritional requirements of this species to the extent that they are known. Although some dissolved organic materials are absorbed from the medium, death results when particulates are omitted. This confirms the need for particulate foods, but may indicate that the nutritional quality of particulate material is supplemented by the dissolved constituents of seawater. If true, the organic chemistry of seawater may have some influence upon the characteristics of phytoplankton populations.

Nutritional Requirements of Mollusks

All mollusks with aquaculture potential feed upon small particulates throughout their life span. If live feed is ingested, it is predominantly phytoplankton; thus these mollusks may be classified as herbivores. As suspension feeders, however, they collect any particulates of suitable size, and, as the majority of small particulate matter in the ocean is detritus, these mollusks may also be classified as detrital feeders. Finally, evidence suggests

TABLE 7.3 *Nutrient Media for Axenic Culture of* Artemia Salina *in Artificial Seawater*

Particulates	Egg albumin, rice starch, cholesterol	
Solubles	1. Sugars	Glucose, sucrose
	2. Amino acid supplements	L-threonine, L-histidines, L-phenylalamine, L—serine, L-glutamate
	3. Vitamins	Thiamine, nicotine acid, calcium pantothemate, pyridoxine, riboflavin, folic acid, biotin, putriscine
	4. Nucleic acid	Adenylate, guanglate, cytidylate, uridylate, thymidine

that dissolved organic matter may be absorbed directly from solution, indicating yet another nutritional mode [24].

Feeding Process. Mollusks feed by passing a continual stream of water through their digestive systems, filtering off particulate material, and coating it with mucus. Coated particles are passed to labial palps and are either selected as food or rejected. The selection process is not fully understood but apparently relies upon complex interactions of chemoreceptors, filtering apparatus, phagocytes, and mucus-secreting elements of the digestive system. The phagocytes are individually selective in their collection of particulates. Although observed to ingest all particles, they later expel those of nonnutritive value.

Planktonic Foods. The most important nutritional mode available to mollusks is the herbivorous mode. Oysters, clams, and mussels can be raised exclusively upon phytoplankton foods, with naked nannoplankton flagellates being nutritionally more acceptable than other phytoplankton [25]. Adult oysters and clams can be grown on diatom foods, but again growth rates improve significantly when naked flagellates are offered. Free-swimming larvae are less adaptable, and many have obligate requirements for naked phytoplankters [24].

Table 7.4 shows planktonic foods suitable for culturing mollusk species. These foods, and particularly the ones specific to the larval stages, possess some characteristics in common with the zooplankton diet. In short, suitable foods generally lack cell walls, are of small size, and most commonly are chrysophytes. Mixtures of these organisms improve growth response by making up for the nutritional deficiencies of single foods. Since various individual algal species can support mollusk growth for limited time periods, it would appear that, as for zooplankton, species mixtures provide required trace nutrients rather than major constituents.

Dissolved Organics. The assimilation of dissolved organics from seawater may provide mollusks with nutritional components lacking in their particulate foods. Dextrose is rapidly absorbed from seawater solution and stored as glycogen by oysters; glucose solutions produce increased weight and survival in laboratory-grown oysters. Amino acids (especially glycine), vitamins (riboflavin, calcium pantothenate, thiamine, pyridoxine), and polysaccharides are extracted from solution by a variety of mollusks. Calcium and phosphorus are preferentially concentrated by mollusks, suggesting that some minerals also may be absorbed from the sea.

TABLE 7.4 *Suitable Food for Cultivation of Mollusks*

Mollusk	Stage	Food

Clam
Mya arenaria
Mercenaria mercenaria
	Larvae	*Chlorella* sp.;
		Chlorococcum;
		Isochrysis galbana;
		Monochrysis lutherii;
		Mixture of
		  M. lutherii, I. galbana,
		  Dunaliella euchlora, and
		  Platymonas sp.

	Juveniles	As above and *Skeletonema* sp., Chryptomonads, *Actinocyclus* spp.

Oyster
Crassostrea virginica
Ostrea edulis
	Larvae	*Isochrysis galbana;*
		Pyraminomonas grossi;
		Dicrateria mornata;
		Chromulina pleiades;
		Hemiselmis rufesceno;
		mixture of
		  I. galbana,
		  M. lutherii,
		  Dunaliella euchlora, and
		  Platymonas sp.

	Juveniles	*Skeletonema,* Cryptomonads, *Actinocyclus* spp.

Natural seawater is a complex mixture of organic and inorganic materials, derived in part from coastal runoff as well as from the natural decay and excretions of marine organisms. Coastal-zone waters naturally are more heavily loaded with dissolved organic materials than the open ocean, although both areas are characterized by considerable variation in concentration and composition, depending upon location. If a benefit is derivable from dissolved organics by mollusks, it would seem likely that locations having higher organic concentrations would be favorable for mollusk cultivation. However, not all dissolved constituents of seawater are potentially beneficial. Heavy metals, pesticides, herbicides, oil solvents, and the like, may occur in sufficient quantities to adversely affect any organism that absorbs dissolved organics to meet (partially) its nutritional needs. These contaminants are most common in the coastal zone near industrial and urban areas and may be deterrents to aquaculture activities there.

Detritus Feeding. It is very likely that the concentration of dissolved organics in seawater is too low to support mollusk growth entirely. Furthermore, indications are that an obligate requirement exists for particulate foods, and that solutes are utilized only to supplement nutritive needs. The question remains: can detritus fulfill the need for particulates?

Studies to define the role of detritus in mollusk nutrition have shown that significant increases in weight occur when artificially prepared detritus is offered, but not when natural marine detritus is used. The chemical compositions of the detrital sources used in these experiments were entirely different, explaining the discrepancy in the results to some degree. In the first study, the detrital preparation was made from ground plant and animal materials, rather than collected from the marine environment. In the second instance, the detrital samples were collected from the seawater at several locations. It seems likely that higher concentrations of nutrients were present in the former preparation.

Detritus, especially of colloidal size, is highly charged and will form aggregates in the presence of dissolved organic compounds. It is suggested that the absorption of dissolved organics by detritus may improve its natural nutritional properties and provide a more efficient mechanism for supplementing the nutritional needs of mollusks. A plausible hypothesis might be that phagocytes engulf and digest these particulates, thus decreasing the dependency upon active transport of materials from solution.

Nutritional Requirements of Fishes

The nutritional requirements of fishes cannot be generalized easily. Tremendous diversity in feeding behavior, mechanisms, and habitats is demonstrated among species and may exist within a species from one developmental stage to another. For instance, a carnivorous adult may produce larvae with herbivorous requirements, or herbivorous adults may develop from carnivorous larvae. Fishes can collect particulate foods by filtering the water through which they pass, or capture prey through a biting attack, or both. They may sift sediments through their gills, collecting the organic materials therein. In short, fishes display all the conceivable nutritional modes, with requirements being species specific and dependent upon the developmental stage.

Larval Fishes

Phytoplanktonic Foods for Larvae. There have been many documented attempts to use phytoplankton as food for marine fish larvae, but only a few have proved successful [26]. Fabre-Domerque and Bietrix offered the naked green flagellate *Dunaliella salina* to freshly hatched sole larvae (*S. solea*) and observed that it was eaten in preference to zooplankton for the first few days [27]. Similar results were obtained by Kasahara et al. using the unarmored dinoflagellate *Oxyrrhis* supplemented with other marine organisms [28]. The fish larvae actively sought the algae for the first few days of feeding and then became entirely carnivorous. In both cases all larvae were offered a choice of food organisms. The rearing of anchovy larvae (*Engraulis mordax*) in the laboratory, using a phytoplankter as the only source of food, was accomplished by Lasker et al. [29]. The larvae maintained good growth and survival when offered the naked dinoflagellate *Gymnodinium splendens*. Further work indicated that growth and survival could be improved when a dietary choice was offered between the dinoflagellate and veligers (gastropod larvae) of *Bulla gouldiana*.

In general, throughout all the laboratory culture attempts in which phytoplankters were used as the sole nutritional source, fish larvae were observed to feed heavily upon them, but finally were found to survive no better than starved controls. All successful experiments have utilized unarmored, naked phytoplankters; there are no reports of a successful technique for rearing marine fish larvae on algae with true cell walls. It may be that the larvae are unable to digest the thick cellulose walls of the algae.

Zooplankton Foods for Larvae. The usefulness of zooplankton as a food source for larval fishes has been amply demonstrated. During early post-yolk-sac periods, teleost larvae readily consume copepod nauplii and eventually, as they grow, capture larger zooplankton. More than 40 species of fish, from mullet to mackerel, have been reared successfully through metamorphosis on diets of wild zooplankton.

These results would seem to indicate that the large majority of fish larvae are carnivorous and that zooplankton are the mainstay of their diet. However, laboratory experiments have shown that at least two to four edible organisms must be present per cubic centimeter of the rearing medium for these larvae to survive [30]. Open-ocean densities of this magnitude are not reported to occur even in the vicinity of upwellings, which suggests that the observations are a laboratory phenomenon modified from nature.

Stomach-content analyses of wild fish larvae indicate that zooplankton constitute more than 90 percent of the visible ingested food. Interestingly, larvae that capture zooplankton frequently contain more than one. However, 60 to 95 percent of the larvae do not contain particulate food of any kind.

Many explanations have been offered for this observation, including sampling error, regurgitation of ingested foods caused by collection, presence of invisible food supplies, and so on. An additional explanation that may merit further investigation is based upon the spawning habits of fishes and the distributions of zooplankton in the sea.

Zooplankton provide a classical exmaple of clumped or patchy distributions. High concentrations of any given species occur in small, spatially separated volumes. Between these organismic aggregations few individuals of the species are found, thus creating a patch-like network of planktonic organisms. These patches are observable whether sampled with a pipette or a plankton net, and indiciations are that the distributions within a patch are themselves patchy. Data are presently insufficient to indicate the maximum densities within these "microclumps," but it is possible that they might provide pockets of plenty for larval fish.

The fecundity of pelagic spawning fishes is tremendous; approximately 500,000 eggs are released per mature female at each spawning. Subsequent to release and fertilization, the eggs develop at the surface, are transported by currents, and separated by surface turbulence. By the time of emergence, some 24 to 72 hours later, the larvae are dispersed over a fairly large area. Emergent larvae are nutritionally supported by yolk materials for another 36 to 72 hours, during which time their mouths open and their eyes become pigmented. They have limited mobility and are capable of quick motion only for short distances. It is absolutely necessary that they find food quickly after the yolk has been fully absorbed.

Laboratory experiments with herring larvae indicate that successful capture of a suitable food organism must occur within the first three or four attempts. If it does, further feeding is stimulated; if not, the larva discontinues further attempts and ultimately dies. Thus, the likelihood of survival is improved by high food densities in the vicinity of the larvae; distances traveled to obtain food are shorter and less energy is expended. If a larva is located near a dense, even though small, clump of edible plankton, its probability for survival is considerably higher than for one that is not. The larva possibly could maintain its proximity to the package of food and gradually expand its hunting area as it developed and grew. It may well be, then, that survival is a matter of probability, and larvae fortuitously existing near dense clumps may survive when others do not. If larval fishes indeed utilize "microclumps" of zooplankton in this way, many of the interpretive discrepancies between laboratory and field observations could be resolved.

In the laboratory, where predation and environmental parameters can be controlled, larval fish survivals to the time of metamorphosis have been reported as high as 40 percent when wild zooplankton feeds have been offered at high densities. This could mean production as high as 200,000 juveniles from one average pelagic fish, which brings up the question of why commercial aquaculture has not succeeded when using zooplankton as food. There are several reasons for this: the quantities of zooplankton required, their undependable nutritional quality, and the variability in species composition and abundance. These are discussed next.

Effect of Natural Plankton Growth Cycles on Aquaculture. Consumer growth responses are related directly to food supply. Phytoplankton growth kinetics are exponential, and their population densities initially increase more rapidly than those of their predators. Then, owing to this abundance of food, zooplankton growth increases until grazing activities exceed phytoplankton production, and a decline in the algal population follows. Zooplankton densities decline shortly thereafter from lack of food. When the grazing pressures are significantly reduced, the cycle repeats.

This produces two effects of consequence to aquaculture: (1) variation in the zooplankton supply, and (2) changes in the nutritional value of the zooplankton. Healthy animals thrive when the food supply exceeds their requirements, a condition which produces large numbers of zooplankton that possess high nutritional value for fish larvae. When food becomes scarce, zooplankton must mobilize internal food reserves to survive, thereby reducing the amounts of carbohydrates, amino acids, lipids, and the like, available to their predators. At that time the zooplankton are poorer food sources. The significance to aquaculture is obvious: variable quantities and qualities of foods will produce variable yields of cultured products.

Adult and Juvenile Fishes. For aquaculture to be successful, dependable feeds must be developed. Current research is concentrating upon methods for the continuous cultivation of dependable zooplankton supplies, the development of artificial foods, and the testing of algal diets. Although intensive research on artificial feeds prepared by drying and other methods has just begun, several promising methods and concoctions have been developed. Most notable are freeze-dried feeds, the discovery of chemical stimulants for feeding, and methods for encapsulating nutritional mixes.

Current Use of Artificial Feeds. A number of suitable feeds are available for growing marine fishes in captivity. Although many of the artificial feeds were developed for fresh-water fishes (e.g., trout, carp, and catfish), they also work well with marine species. Fishes that grow well on Purina Trout Chow, for example, include California flying fish, sardine, and jack mackerel.

The natural feeding habits of these various species differ dramatically. Sardine generally feed upon plankton and are omnivorous in habit; jack mackerel are entirely carnivorous, feeding upon other fishes such as anchovy and sardine. That both will adapt to feeding upon pelletized artificial mixes is encouraging for finfish culture.

Table 7.5 shows the composition of an artificial feed used to maintain gray mullet in captivity. As is true of all commercial preparations, the nutritional composition is imprecisely defined.

Probable Nutritional Requirements. The specific nutritional requirements for the salmonids are the best defined. Table 7.6 lists the known components and quantities required for normal growth of these fishes [31]. It can be seen that informational deficiencies still exist, particularly for fatty acid and trace-metal requirements. Although salmonids are not truly marine fishes, oceanic species can grow upon feeds formulated for salmonids, which suggests that certain generalizations might be drawn from these data.

Marine fishes probably have large dietary protein requirements. Experiments have shown that the quantities required by salmonids decrease with age and increase with temperature, yielding a requirement range from 40 to 55 percent protein of dietary dry weight. Arginine is required in greater quantities than the other amino acids, followed closely by lysine. As a consequence of the materials used to construct commercial fish diets, it is likely that amino acid deficiencies would first be observed for lysine.

Adequate quantities of crystalline vitamins can reduce the need for protein and allow for an increase in carbohydrates. Vitamin deficiencies hamper the biosynthesis of amino acids from organic acids, increasing the requirement for amino acid supplementation within the diet. For example, nicotinic acid is

TABLE 7.5 *Composition of Artificial Feed Used for Gray Mullet (*Mugil cephalus*)*

Material	Quantity
A. Dry ration	
Fish meal	400 g
Soybean meal	250 g
Chicken starter mash	100 g
Fish bone meal	50 g
Dairy whey	50 g
Wheat germ	50 g
Wheat middlings	50 g
B. Liquid ration	
Water	900 ml
Choline chloride	0.5 g
Urea	0.5 g
Propylene glycol	5.0 g

Mix A and B separately, then combine.

TABLE 7.6 *Nutritional Requirements of Salmonids*

A. General
   Protein/carbohydrates/fats and oils = 50:37:9

B. Amino acid requirements

Essential amino acids	*% Total dry ration*
Arginine	2.5
Lysine	2.1
Phenylalanine	2.0
Leucine	1.5
Valine	1.5
Isoleucine	1.0
Threonine	0.8
Histidine	0.7
Methionine	0.5
Tryptophane	0.2

C. Vitamin requirements

Vitamin	*mg vitamin/kg body wt/day*
Choline	50–60
Inositol	18–20
Nicotinic acid	5–7
Ascorbic acid	2.0–3.0
Pantothenic acid	1.3–2.0
Riboflavin	0.75–1.00
Pyridoxine	0.38–0.43
Thiamine	0.13–0.20
Folic acid	0.1–0.15
Biotin	0.03–0.04
B_{12}	0.0002–0.0003
E (tocopherol acetate)	Required but amount not specified (0.4% of dry ration used in test diet)

D. Carbohydrates (in order of decreasing availability)

   1. Glucose, fructose, mannose
   2. Sucrose, maltose
   3. Short-chain polysaccharides
   4. Long-chain polysaccharides
   5. Processed starches
   6. Alpha-cellulose

E. Fats: a requirement for polyunsaturated fatty acids exists;
   oxidized fats deleterious.

required for lysine biosynthesis. Dietary amino acid deficiencies will probably accompany vitamin deficiencies.

Components rich in simple sugars will likely provide the best sources of carbohydrates. Mannose, glucose, and fructose are more readily utilized by salmonids than are disaccharides and polysaccharides. Starches are not readily used.

Nutritional Requirements of Macrocrustacea

The Macrocrustacea (e.g., lobsters and shrimps) are generally carnivorous as larvae, but may display a preference for phytoplankton during zoeal substages of larval development. Adults are scavengers, feeding on freshly settled organic debris, and, when crowded and hungry, may become cannibalistic. Unfortunately, very little is known about their nutritional requirements.

Lobsters. Our limited knowledge of lobster (*Homarus americanus*) feed requirements has been developed largely by Hughes [*32, 33*] and Stewart [*34*]. Suitable foods for larvae include finely ground quahog (*Venus mercenaria*), frozen adult brine shrimp (*Artemia salina*), and macerated clam. Adult lobster will grow when provided with herring and beef liver, clams (*Mya arenaria*), bay scallop (*Pecten irradians*) viscera, or freshly killed fish, primarily *Pomolobus pseudoharengus*. Adults also accept artificial foods, such as commercial trout pellets or dog food, but since growth data related to these foods are not available, their suitability cannot be evaluated.

Mixtures of feed produce more normal colors and better growth than do single foods. It may be surmised from the nutritional knowledge gained for other marine organisms that single-food diets are deficient in trace nutrients, rather than in major components.

Shrimp

Feeding Patterns of Larvae and Juveniles. Penaeid shrimps are benthic ocean dwellers and the adults are scavengers. The larvae are pelagic, remaining in the ocean waters where they have hatched until they reach postlarval stages. During postlarval development, which continues for several months, the shrimp inhabit estuaries, later migrating back to offshore waters as juveniles. The larvae can be herbivorous during protozoeal substages and become carnivorous during later planktonic stages. Hudinaga [*35, 36*] and Cook and Murphy [*37, 38*] are largely responsible for developing our present understanding of penaeid shrimp nutrition. Protozoea feed heavily upon diatoms: *Pemaeus japonicus* prefers *Skeletonema costatum,* and *P. aztecus* grows best upon *Thalassiosira* sp. Also suitable are *Dunaliella* sp., *Gymnodinium splendens,* and *Exuviella* sp. The presence of thick cell walls apparently does not affect suitability, probably because the larvae possess strong mouth parts with which to crush their food. In all cases, food mixtures yield better survival rates than do single-species diets.

Nauplii of *Artemia salina* are used extensively as food for shrimp in the mysis substage of larval development. Ground fish flesh also supports growth, but at a reduced rate and with decreased survival. The larvae are entirely carnivorous after the protozoeal substages of development.

Adult Nutrition. Few data are available for assessing the nutritional requirements for adult penaeids. In ponds they feed upon small benthic plants and animals, both dead and alive. The relative proportion of each is unknown, as is its chemical composition. None of the supplemental foods tested have produced increases in growth and survival beyond what is normally recorded within a pond.

Some Biological Implications of Nutrition for Open Sea Mariculture

Solely on the basis of nutritional considerations, mollusks presently offer the greatest potential for open sea mariculture. Suitable foods have been developed for their cultivation from larva to adult, the same foods can be used for all life stages, and the foods can be easily distributed.

Feeds for mollusks are closely similar to the requirements of zooplankton, which, in the open production system, will probably occur as contaminants and competitors for the food supply. Harvesting these may provide an open sea mollusk farm with a useful by-product.

The variable success with raising larval marine fishes and the paucity of knowledge about their nutritional needs argue against their culture in the open sea at this time. In addition, their fragility and inability to withstand environmental change may suggest a greater potential for land-based activities, where environmental conditions can be more easily controlled and manipulated.

Insufficient development of an acceptable food for the intensive production of young penaeids, and lack of information regarding lobster nutrition, suggest that attempts at this stage to establish their cultivation would be premature and accompanied by a high degree of risk. However, the ability of these species to utilize as food a variety of waste materials, both plant and animal, may indicate a greater culture potential in the future when nutritional questions have been answered.

A Hypothesis for Developing an Optimum Nutritional Environment for Mollusk Culture

The data reviewed in the preceding sections indicate that suitable foods for mollusks have the following characteristics in common:

• They are live phytoplankton of small size.
• They lack a thick cellulose cell wall.
• They are generally chrysophytes.

In all cases, mixtures of food organisms produce better growth than does any individual component. Among algal organisms, many chrysophytes have the capacity to phagotrophically ingest bacteria. This suggests a role in bacterial control that may indirectly reduce the incidence of disease, a possibility of interest to the aquaculturist. The question remains: how does one practically produce the consistent quantities of high-quality food required by a mollusk farm?

The available data suggest that mollusk production might be enhanced by developing an environment which favorably influences chrysophyte growth relative to diatom and dinoflagellate production. However, the existing knowledge of phytoplankton ecology does not allow such an environment to be constructed at this time. On the following pages, a hypothesis for the development of this technology is presented. The major chemical characteristics and nutrient requirements of each of the three major phytoplankton categories are described and their differences noted. These data also provide inferences of the dietary suitabilities of these phytoplankters for aquaculture.

Chrysophyceae

Chemical Characteristics. The most extensively analyzed chrysophytes in terms of chemical composition are *Monochrysis lutherii* and *Syracosphaera carterae*. During periods of exponential growth, *Monochrysis* contained 49 percent protein, 31 percent carbohydrate, 11.6 percent fat, and 6.4 percent ash; *Syracosphaera* reportedly contained 56, 18, 4.6, and 36.5 percent, respectively. Over 71 percent of the organic materials in *S. carterae* was protein; the observed difference was due to a high ash content associated with its coccolith coating (calcium deposits on the outer cell wall) [39].

The carbohydrates of *Monochrysis* included glucose (22.1 percent), galactose (4.4 percent), ribose (1.3 percent), and xylose (3.5 percent); *Syracosphaera* contained glucose (9.2 percent), galactose (7.1 percent), ribose (1.5 percent), xylose (0.8 percent), and arabinose (1.9 percent). Both stored excess photosynthate as chrysolaminarin, a relatively short chained glucose polymer. These carbohydrates are readily used by predators.

Four amino acids made up almost 80 percent of the total protein fraction; these were aspartic acid (25.3 percent, 35.5 percent), alanine (26.0 percent, 11.4 percent), lysine (19.6 percent, 17.8 percent), and glycine (8.6 percent, 11.4 percent) for *M. lutherii* and *S. carterae,* respectively [39, 40]. Glutamic acid, threonine, and valine each contributed approximately 5 percent to the total protein fraction. Present in lesser amounts were arginine, phenylalanine, leucine, isoleucine, and methionine. Serine, proline, tyrosine, and methionine sulfoxide were represented only as traces. Both chrysomonads lacked histidine. In view of the dietary requirements of zooplankton and fishes, it is likely that some of the abnormalities that arise when chrysophytes alone are offered as food may result from histidine deficiencies. The glycine requirements of mollusks would appear to be amply met.

It must be realized that the organisms used for the determination of amino acid composition were grown on purely mineral media and are therefore only indications of the organisms' synthetic capabilities. Yet it is well documented that many chrysomonads can utilize amino acids as nitrogen sources [41, 42,

43]. Arginine is among those assimilated by *Syracosphaera,* and histidine is utilized by *Coccolithus huxleyi.* Before utilization these compounds are concentrated in intracellular pools and are available to predators. This might suggest that chrysophytes cultured on amino nitrogen sources, particularly histidine, would have improved nutritional characteristics.

Perhaps one reason the chrysophytes frequently are suitable feeds is their unusual fatty acid composition. Both the alpha form (characteristic of animals) and the beta form of linolenic acid have been found in all chrysophytes thus far investigated. Also present is arachidonic acid, previously reported only in metazoa [40]. All are unsaturated fatty acids, for which the need has been stated previously.

Nutrient Requirements. No data are available on the vitamin content of chrysophytes, but there is considerable information about their vitamin requirements. It is reasonable to assume that these organisms have the capacity to synthesize most of the vitamins they need to function, but data need to be developed to assess their contributions to the nutritional requirements of their predators.

Of the 22 species investigated, only *Stichochrysis immoblis* does not require added vitamins. The remaining 21 require added thiamine, B_{12}, and/or biotin. Of these, thiamine is the most frequently needed, but the majority of the species require at least two, particularly thiamine and B_{12}. Only two species have demonstrated a requirement for all three [40].

We might note at this point that Provasoli et al. [44] demonstrated the importance of algal enrichment with vitamins. Their experiments showed that vitamin availability to the zooplankton was dependent upon the assimilative capacity of their algal food; the zooplankters did not receive direct benefits from vitamins in solution. This finding has implications concerning the quality of algal food organisms, making it probable that the organic composition of seawater influences characteristics of the food chain.

The concentrations, as well as the compositions, of dissolved organics, then, can have a direct influence upon chrysophytes. Many of these nannoplankters can utilize certain sugars, organic acids, and organically bound phosphates, in addition to vitamins and amino acids, as supplements to their photosynthetic fixation of carbon dioxide or as an alternative source of nutrients [40]. Significant growth stimulation is observed when *Hymenomonas* sp. has access to lactic acid and pyruvate, and *Pavola gyrens* to acetate. Although the magnitude of the response varies, all Chrysophyceae examined to date have the capacity for heterotrophy.

Present data indicate that suitable carbon sources are those which occupy entry locations to the Embden–Meyerhof pathway and the Krebs and glyoxylate cycles. (Information on these pathways is available in a number of standard biochemistry texts.) Table 7.7 identifies the compounds and their points of entry into the biochemical systems.

Chrysophytes apparently are able to tolerate high nitrogen concentrations,

TABLE 7.7 *Organic Compounds Stimulatory to the Growth of Chrysophytes*

Compound	Pathway entered
Sucrose	Embden–Meyerhof
Glycerol	Embden–Meyerhof
Pyruvate	Krebs cycle
Lactate	Krebs cycle
Acetate	Krebs cycle, glyoxylate cycle
Propylene glycol	Glyoxylate cycle
Malate	Krebs cycle
Glycerol–PO$_4$	Embden–Meyerhof

which can be satisfied with amino acids as well as with nitrates. Ammonia at concentrations of 10 parts per million (ppm; milligrams per liter) is toxic to *Coccolithus huxleyi;* although some chrysophytes can utilize it, only *Pavlova gyrens* and *Hymenomonas* sp. can tolerate concentrations of 200 ppm.

Phosphate is required only at low to moderate concentrations; *Syracosphaera carterae* utilizes 1 phosphorus atom for every 17 nitrogen atoms assimilated. Actual proportionate requirements of these nutrients depend upon the species and the past history of individual cells.

Chrysophytes can tolerate a wide variation in nutrient concentrations before their growth is adversely affected. *Isochrysis galbana* and *Monochrysis lutherii* exhibit growth to 280 mg percent nitrate–nitrogen. Both reach maximum cell densities in a laboratory media containing 210 μg of iron/liter.

To recapitulate, it appears that chrysophytes should grow best in waters having high nitrogen, phosphorus, and iron concentrations, a high nitrogen-to-phosphorus ratio (10 to 15:1), low ammonia levels, and high organic acid concentrations. It is further hypothesized that chrysophyte growth would be favorably influenced if a large proportion of the nutrients was organically bound. Under these conditions, it is not yet known if chrysophytes would dominate a phytoplankton population.

Bacillariophyceae

Chemical Characteristics. Diatoms contain less protein as a fraction of total cell weight than do chrysophytes when grown under identical conditions; *Coscinodiscus* sp., a centric diatom, had only 17 percent protein as compared to 34 percent for the three pennate forms investigated [39].

The four amino acids that yield the majority of chrysophycean protein do the same for diatoms, with two important exceptions. First, glutamic acid contributes twofold more to the total protein nitrogen than it does in chrysophytes, although the nutritional significance to the predator is lessened by the reduction in total protein concentration. Second, the percentage of lysine is less in diatoms, particularly for *Skeletonema costatum,* where it is present in only trace amounts.

In common with chrysophytes, diatoms apparently lack the power to produce histidine during exponential growth. An exception is *S. costatum,* which contains only qualitatively detectable amounts. Tyrosine was produced by *Coscinodiscus* sp., but not by other species investigated. These results suggest that amino acid deficiency symptoms for lysine, histidine, and/or tyrosine would be the first to develop in organisms receiving a diatom diet.

The percentage of ash is higher in diatoms, probably because of the high silica content in their cell wall. Diatom cell walls are composed almost entirely of this element, making silica an obligate requirement for normal diatom growth.

Although the major storage product of diatoms is oils, during rapid growth, differences in fat content are not found between diatoms and chrysophytes. That differences are likely to occur in nature is demonstrated by the results of Ackman et al. [45]. On the average, fatty acids make up 20 percent more of the total lipid fraction in diatoms than in chrysophytes after 14 days of culture. Diatoms were characterized by high levels of C_{14} and C_{16} fatty acids, and low levels of C_{18} and C_{22} fatty acids. The chrysophytes were richer in C_{20}, C_{22}, and, most particularly, C_{18} fatty acids.

Crude fiber is taken as an indication of the amount of cell-wall material. As a percentage of total carbohydrates, crude fiber is generally a large part of a diatom cell. *Phaeodactylum tricornutum,* which lacks a siliceous cell wall, is an exception. This measurement, however, most generally demonstrates the relationship of food suitability for mollusk and zooplankton culture; as much as 29 percent of the total diatom carbohydrate may be cell-wall material. Naked cells, such as chrysophytes, generally provide crude fiber levels of less than 4 percent.

Glucose is apparently the most important of the principal sugars for algal organisms. The composition of algal sugars is reasonably constant, with the exception of mannose. This sugar is found in measurable concentrations only in diatoms.

Nutrient Requirements. The principal vitamin requirement for diatom growth is B_{12}, replacing the status of thiamine in Chrysophyceae. The status of thiamine, however, becomes that of a supplement to B_{12}. (The reverse is true for chrysophytes.) Biotin has not been demonstrated as a requirement for diatom growth.

The heterotrophic capabilities of diatoms are considerably more variable than for chrysophytes and are highly species dependent. A major difference is that some marine diatoms are capable of substantial growth in the dark, a capability not demonstrated for marine chrysophytes.

Organic nitrogen sources are widely utilized but seldom produce the same yields as nitrate. As a single nitrogen source, amino acids generally permit only limited growth; as a supplement to nitrate, however, they markedly stimulate growth. Diatoms can more readily utilize urea and uric acid than can the Chrysophyceae.

In common with all other algal groups, diatoms have the capacity to utilize organic phosphates. These include glycerophosphate and adenylic, quanylic, and cytidylic acids.

Littoral pennate diatoms display a capacity for heterotrophic growth more frequently than do centric forms or oceanic Pennatae. Organic carbon sources include several of those utilized by chrysophytes, but glucose and lactic acid appear most suitable for diatom growth. The ready utilization of glucose seems specific to this group.

Lewin demonstrated that diatoms require both reduced carbon and nitrogen sources for heterotrophic growth [46]. This is not the case for chrysophytes, suggesting that diatom growth might be impaired, to the chrysophytes' advantage, if only inorganic nitrogen were available.

More chrysophyte species can tolerate high nutrient concentrations than can diatoms. Under high nutrient conditions, the species diversity of diatoms generally diminishes, but high cell number persists. In contrast to the chrysophytes, however, diatom growth is not adversely affected by high ammonia concentration. As is practiced in Israeli fish ponds, ammonia might be used to limit chrysophyte growth and favor diatoms.

Differences in cell chemistry suggest that diatom growth may be favorably influenced by lower nitrogen and phosphorus concentrations than those preferred by chrysophytes. The lower protein concentration in diatom cells indicates that less nitrogen is necessary to produce identical dry weights. As intracellular nitrogen-to-phosphorus ratios are similar between the two organismic groups, it seems logical that more diatoms could be produced from the same amount of phosphorus than could chrysophytes of similar size.

In view of the cell composition and chemical requirements of diatoms, chrysophyte production might hypothetically be favored under the following conditions:

1. High concentrations of nitrate–nitrogen, either organic or inorganic phosphorus, and iron (the widespread capabilities of marine algae to utilize organic phosphates suggest no advantage in distinguishing its chemical form).

2. High organic nutrient concentrations, excluding glucose and amino acids not required for chrysophyte and/or predator growth.

3. Low concentration of silica.

Due to their taxonomic positions, many characteristics of diatoms and chrysophytes are similar; chemical and nutritional dissimilarities should be more apparent between chrysophytes and dinoflagellates.

Photosynthetic Dinoflagellates

Chemical Characteristics. In contrast to the previous algal groups, carbohydrates are the principal components of Dinophyceae. *Amphidinium carteri* contained 30.5 percent carbohydrate, 28 percent protein, and 18 percent fat; *Exuviella* sp. showed values of 37, 31, and 15 percent, respectively.

Glucose is the dominant carbohydrate, and only two other sugars, galactose and ribose, are present in measurable amounts. The lack of hexuronic acids, which are common to all other phytoplankters, is notable, but the nutritional significance of this is obscure.

The large majority of dinoflagellate protein is composed of aspartic acid, glutamic acid, glycine, and alanine, and in this respect dinoflagellates are similar to diatoms but different from chrysophytes. Dinoflagellates contain variable quantities of lysine, threonine, and tyrosine, ranging from barely detectable to significant amounts, an apparent characteristic of phytoplankton in general. Histidine was not detected in rapidly growing cultures, but arginine levels were among the highest for algal organisms.

Compared with other algal organisms, dinoflagellates contain more fats, are particularly rich in palmitic acid, and have only trace amounts of C_{16} fatty acids. The majority of their longer-chained fatty acids are unsaturated, and the most abundant is docosahexaeoic acid. This pattern apparently is unique among phytoplankton.

As is generally true for diatoms, dinoflagellates contain little linoleic, linolenic, or arachidonic acids, suggesting a possible requirement for fatty acid supplementation for organisms feeding primarily upon these algae.

Nutrient Requirements. The large majority of dinoflagellates display a requirement for vitamin B_{12}, and several also require biotin and thiamine. It will be recalled that the dominance of B_{12} is a characteristic in diatoms and that the need for biotin is representative of some chrysophytes, suggesting that a variety of vitamin supplements is more frequently required by the Dinophyceae.

photosynthetic dinoflagellates are limited in their heterotrophic capabilities although, in common with the other algae, organically bound phosphates permit good growth. Amino acids can also be utilized as nitrogen sources, but do not stimulate growth, in contrast to chrysophyte and diatom responses. More importantly, available data indicate an inability among phytosynthetic dinoflagellates to utilize the organic intermediates of glycolysis or the Krebs cycle. It would appear that these organisms are obligate autotrophs and obtain almost all the carbon they need from dissolved carbon dioxide. This is a significant divergence from the capabilities presented for diatoms and chrysophytes.

The phagotrophic capabilities of photosynthetic dinoflagellates have not been investigated, but certain colorless forms are known to ingest bacteria and nannoplankton in this way. *Oxyrrhis marina* and *Noctiluca* spp., for example, are obligate heterotrophs that feed upon *Nannochloris* sp., *Phaeodactylum tricornutum, Isochrysis galbana,* and *Monochrysis lutherii,* to name a few.

It is possible that the indications of limited heterotrophy among the photosynthetic dinoflagellates arise from a greater divergence and distinction of physiological forms within the Dinophyceae. It is certain, however, that colorless dinoflagellates will contaminate and compete for food produced in open culture for a mollusk farm. Aside from the competition for food, it is unlikely that these organisms will produce any adverse effects upon mollusks.

Autotrophic dinoflagellates are very sensitive to ammonia. Toxicity has developed at the lowest concentrations tried in laboratory culture (5 ppm), but it is not known if similar results occur in the open sea where natural concentrations are approximately 100 times lower.

Perhaps as a result of low protein content, dinoflagellates tend to have a low nitrogen requirement. As suggested by an internal nitrogen-to-phosphorus ratio near 13, phosphate requirements may also be proportionately low. This suggests that optimal media for dinoflagellate culture would have a high ratio of nitrogen to phosphorus atoms. However, dinoflagellates are known to tolerate high concentrations of nitrate—nitrogen and phosphorus, with proportionate increases in cell number. This range of tolerance suggests difficulties in decreasing dinoflagellate contributions to the biomass by inorganic enrichments.

It is hypothesized that chrysophytes will dominate organically enriched seawater. In inorganic media containing vitamins, chrysophytes and small dinoflagellates (e.g., *Amphidinium carteri*) grow at about the same rate. The growth of chrysophytes, however, can be enhanced two to three times in the presence of organics, and under these conditions chrysophytes might successfully compete with dinoflagellates for inorganic nutrients. Thus, it is possible that the conditions arrived at for limiting diatom growth may also favor chrysophyte production in the presence of dinoflagellates. Experimental verifications of this are not available.

Chrysophyte Culture for Mollusk Farming

Closed Production System. Any nutrient source used in open sea mariculture must be available in sufficient quantity to maintain continuous growth of chrysophytes since, if their growth rate changes, so will the quantity and quality of the mollusk food supply. The relationships that enter into this consideration are shown in Figure 7.9.

Nutrients arriving from their sources are taken up in required proportions by the chrysophytes, with the resulting growth supplying food for mollusks. Mollusks harvest only a small fraction of the chrysophytes in the water, digest and assimilate substances for their growth, and excrete the remainder. Some nutrients are returned with the wastes, many in a chemical form different from the original source. Thus, in a closed production system the original conditions will become highly modified as the various components seek equilibrium.

Open Production System. In an open production system, as illustrated in

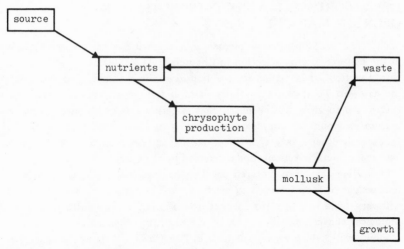

FIGURE 7.9 *Nutrient flow in a mollusk farm.*

Figure 7.10, the cycle can be broken as waste materials are removed from the algal production site by ocean currents. This process would be similar to chemostatic production systems in which growth is controlled by a continuous inflow of nutrients strict proportions, accompanied by a continuous outflow of unused nutrients. Excess nutrients and waste products are not recycled.

Although an open production system is not as efficient or as environmentally acceptable as a closed recycling unit, its advantages are simplicity and predictability. If a cheap and ready source of nutrients is available in large supply, it may also be practical.

For example, a very large and potentially useful source of these materials is the nutrient-rich Deep Water layer. Within this layer, peak nitrate–nitrogen and phosphate concentrations occur and the nitrogen-to-phosphorus ratio is high. It is likely, but not certain, that dissolved organics are also present in large quantities. Although the organic chemistry of the Deep Water layer has not been studied, a number of chrysophytes and diatoms, as well as dinoflagellates, are found in it; and since no light penetrates to these depths, it is almost certain that some of these organisms are supported by their heterotrophic capabilities. These considerations are discussed next.

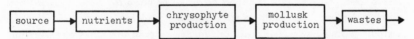

FIGURE 7.10 *Nutrient flow in an open production system.*

DESIGN-ORIENTED ASPECTS OF NUTRITION IN OPEN SEA MARICULTURE

The concepts of natural, supplemental, and provisional feeding are explored in the context of crop selection in Chapter 8. As well as being influential to that function, these concepts are important in the design of nutritional mechanisms. To state them briefly, *natural feeding* involves allowing a cultured crop to grow on the nutritional materials that occur naturally in the culture environment; *supplemental feeding* involves supplementing natural foods with other foods by artificial means; and *provisional feeding* supplies through artificial means all food consumed by the crop.

In feeding any given cultured species, the *conversion rate*, defined as the unit weight of food required to produce one unit weight of crop at harvest, will vary from one food to another. (This is not to be confused with the biological conversion efficiencies of the organisms themselves at the several trophic levels, which were dealt with in Part One.) The lower the conversion rate, that is, the lower the quantity of food needed to achieve a given growth rate, the more effective the feed. Conversion rates in natural and supplemental feeding necessarily are approximations, since it is difficult to determine the rate of consumption of natural foods. Absolute conversion rates can be measured only in cases of provisional feeding. Moreover, other factors, such as population density, health, temperature, and feeding methods, strongly influence the conversion rates of both natural and synthetic foods. Nonetheless, if these factors are held more or less constant, the concept can be a useful one by enabling a rough comparison between one artificially supplied feed and another in terms of their relative effectiveness in increasing crop weight [47].

A meaningful extension of this concept would be an *economic conversion rate*, defined as the amount of money required to establish, maintain, and operate a feeding system per unit weight of crop harvested at any point in the life of the system. Such costs would of course include the initial capital investment, maintenance, operating costs, and cost of feed. Clearly, the economic conversion rate (ECR) would tend to become more favorable with time for longer-lived systems, representing a significant one-time capital investment, since that investment gradually would be amortized. The formula

$$\text{ECR} = \frac{\text{total weight of product at time } t}{\underset{\text{investment}}{\text{capital}} + \underset{\text{costs at } t}{\text{maintenance}} + \underset{\text{costs at } t}{\text{operating}} + \underset{\text{at } t}{\text{feed costs}}}$$

seems appropriate in that it allows credit to be given for systems having high initial costs but low maintenance, operating, and feed costs, as well as for systems having low initial costs but higher costs in the other factors. Thus, the design objective for nutritional systems is to minimize the average dollars per pound (economic conversion rate) expected over the operational life of the system.

But nutrition is a complex question. For example, with any given food and set of environmental conditions, there is some maximum rate at which feed can be consumed by stock to produce rapid growth; beyond this rate, uneaten food may remain and putrify, causing oxygen depletion and hydrogen sulfide production and increasing the likelihood of disease. Another factor is the digestibility of the feed. Food may irritate digestive tracts and/or be difficult to digest even though it is high in nutrients; it may contain indigestible and irritating components, or constituents that the cultured organisms convert into self-toxic excrement; and it may contain pathogenic constituents. Yet another concern is attractiveness; food does no good at all unless consumed, and some difficulties have been encountered with the acceptability of synthetic feeds. Finally, there is the question of feeding method. In addition to being dependable and economical, the feed-dispensing process must be efficient and must not alter the acceptability of the feed in any way. If the process is manual and labor intensive, its operation will be costly. If it is mechanized, it must be reliable and relatively maintenance free. If automated, monitoring must be provided.

This section examines terrestrial sources, intrasystem sources, and artificial-upwelling as possibilities for meeting the nutritional requirements for open sea mariculture. The question of nutrient dispersal is then briefly considered.

Terrestrial Sources of Nutrition

The work of Ryther et al. at Woods Hole [48] and others leaves little doubt that beneficial marine food chains can be based upon either urban effluent or commercial fertilizers. In fact, the only other long-term source of terrestrially derived nutrient that seems to hold promise for large-scale open sea mariculture lies in commercial pelletized feeds. The status of both developments is reviewed briefly here.

Urban Effluents for Primary Production. During the early stages of this project, we made an attempt to determine the volume and general composition of urban wastes produced by major coastal communities in the United States. We spare the reader the title that the staff attached to this subproject. Formatted questionnaires were sent to representative cities, and in several cases a communications history developed between the city and our office. Our naive hope was to develop even a rough approximation of per capita effluent volume and constituents; our aim, to form a basis for projecting the potential utilization of sewage as a primary-production stimulant and to sense some of the problems that might be involved. We hoped to gain some appreciation for total volume, phosphates, nitrates, metals, salts, bases, inert solids, organic components, and coliform counts introduced by coastal urbanites into their nearby marine environment. Probably it is a manifestation of the hubris of the species that, although humanity is deeply concerned with what it extracts from the environment, it records very little information about what it injects into it.

Some cities did not respond to the survey. Some that did respond recorded very little concerning effluent constituents. There was no way, within our limited resources, to correlate effluent with population; this would be a major study in itself. Wide variances in the data caused suspicion about differing sensitivities in the measurement techniques employed. Adequate quantities of data were received from nine cities and arranged in a matrix with cities on one axis and constituents on the other. No positive correlation between constituent ratios was apparent among the cities, nor even among sewage treatment plants within a single city. Neither could any correlation among volumes of constituents be found among cities: in some cities phosphates exceeded nitrates, for example, and in others the reverse was true. There was a slight tendency for treatment plants that reported a high value of any one constituent to report high values for all.

There was no choice under the circumstances but to conclude that there were unknown differences in data sensing, recording, and reporting at work, and that the data were therefore of little value. We thus must take refuge in generalities unlikely to startle any reader. The combined total effluent production of U.S. coastal cities runs into trillions of gallons annually; phosphates and nitrates run into billions of tons; and metals, salts, and bases into billions of tons, with organic compounds and inert solids somewhere in the same range. In our data, coliform bacteria counts ranged from 2/100 ml to 150,000,000/100 ml, with an average around 10,000/100 ml.

Assuming that there are no other limiting factors, Isaacs and Schmitt state that the conversion ratio of phosphorus weight to dry phytoplankton weight can be as high as 100 to 1 [49]. Thus, even when reduced by several orders of magnitude through a food chain and by other limiters and inefficiencies in conversion, the effluents of U.S. coastal cities alone contain sufficient nutrients to stimulate the production of billions of tons of animal protein annually.

How these effluents would be detoxified, collected, concentrated to remove fresh water, transported to mariculture sites, and distributed effectively, on an economic basis, would be a subject for a study much more extensive than our entire project. Such a study should be valuable. The prospective returns from the endeavor—mitigation of coastal sewage impact and the subsequent reclamation of resources to apply to mariculture—could justify intensive research along these lines.

Later in this section artificial upwellings designed to exploit deep nutrient water are considered. Perhaps it is foolish to be concerned yet about possible eventual depletion of deep nutrients and the effects this might produce in the global marine ecosystem. Yet 100 years ago or less it would have been considered foolish to be concerned about shortages of oil, fresh water, copper, or reserves of hundreds of other natural resources that worry humanity greatly today. Maybe now is not too early to consider how deep-lying oceanic nutrients might be replaced. One avenue that seems potentially feasible lies in the injection of urban wastes below the surface-water thermocline, where they can be contained in, absorbed, and processed by the dynamics of the oceanic Intermediate Water (see Chapter 4). The global oceanic dynamics involved—and the need for dehydration or densification, detoxification, or the recovery of mineral resources from the effluents prior to injection—could form the basis of a highly valuable research program, as pointed out in Chapter 15.

Commercial Fertilizers. The prospect of adding commercially prepared phosphates, nitrates, iron, and other elements to a culture medium to stimulate primary production is certainly a possibility. According to Department of Agriculture statistics, commercial fertilizer consumption in the United States approximately doubled, from 20 to nearly 40 million tons, in the 20 years from 1950 to 1970. Apparently, the fertilizer industry had no trouble keeping up with demand during this period. Could a quadrupled demand over the next three or four decades be met as well?

Solid artificial fertilizers currently in production have a specific gravity greater than that of water, and many of them dissolve at controlled rates. So the solid products now available would be appropriate to contained, shallow-water culture only. There are also many artificial fertilizers produced in liquid form. Given controlled application to a culture medium, some of these might be effective as aquatic, primary-production stimulants. But large-scale employment by mariculture probably would require, and could stimulate, development of new commercial fertilizer blends with particular physical and chemical characteristics.

One attraction of this highly controlled approach is brought out in the second section of this chapter: if phytoplankton population mixes can be controlled chemically, artificial fertilizer blends that promote desirable population mixes might be an avenue to high productivity.

However, the price of any commercial fertilizer must include the delivered cost of the raw materials, the processing and production cost, and the cost to deliver it to the site of use. Although it cannot be said yet that commercial fertilizers must be more expensive per unit of primary production than other approaches, it must be concluded that they will be expensive. Moreover, since the fertilizer per se is purchased, the expense relates directly to the volume applied; this is not strictly true of artificial upwellings and even of some means of employing urban wastes. The positive side of this coin is that the application of artificial fertilizers could and would be tightly controlled.

All this indicates that artificial fertilizers might be attractive for two types of marine applications: in contained systems such as atoll lagoons, wherein full benefit could be derived from fertilizer expenditures, and as a supplement to the upwelling nutrient source.

Relative to the second category, it should be pointed out that one of the main constraints of the upwelling theory lies in its *suspect ability* to deliver enough nutrients to the surface to stimulate phytoplankton densities to the point where subsequent grazing is efficient enough to permit aquaculture. But the upwelling—essnetially free fertilizer—could provide the lion's share of the nutrients required, and commercial fertilizers could supplement this supply, serving merely to enhance phytoplankton concentrations to make the production of the final crop considerably more economical.

Pelletized Feeds. Dry, concentrated, and pelletized feed mixtures have been in use for salmonids since 1960 and are now being used increasingly for other varieties of cultured finfish. Recently, work has begun on developing similar feeds for shrimp and lobster [50].

Several considerations are spurring the development of these feeds. They can be produced in predictable volumes from various raw materials, thus freeing the culturist from vagaries in feed supply. Theoretically at least, they can also be blended to provide complete diets with advisable variations for each development stage of the fishes. They can be made disease free, and of uniform size to facilitate use by automatic feed dispensers. All this tends to reduce labor costs, maximize equipment utilization, and thus enhance the economic conversion rate. In some cases, even preventative medications are included [47].

Pelletized foods are blended from mixtures of packing-house wastes, sea-food-processing wastes, fresh low-value fish, vegetable meal, grains, yeast, minerals, pure vitamins, and so forth. Thus, they are manufactured foods and, as such, represent finished products derived from raw materials; the usual manufacturing and transportation costs therefore are included in their price. Research continues, and no doubt quality and supply will increase (and costs decrease) as a function of demand by aquaculturists. However, until delivered costs can be brought to very low levels, the use of pelletized foods in open sea mariculture will very likely be limited to special cases of high-value crops in comparatively tightly controlled cultures.

Intrasystem Nutrition Sources

Intrasystem sources of nutrition are foods that might be produced within an open sea mariculture system as an integral part of its operation. Two possibilities are: (1) the use of on-site processing wastes, and (2) the production of some form of natural food within the culturing mechanism.

On-Site Processing Wastes. Depending on the product species and on the degree of on-site processing (Chapter 11), the volume of on-site processing waste might range from 20 to over 50 percent of the harvested crop.

Generally, it can be assumed that there are three major constituents: (1) digestible animal fats and proteins, (2) indigestible bones, and (3) calcium carbonate in shells and exoskeletons.

For carnivorous finfish and for larger detritus-feeding finfish and crustacea, it often is possible to feed on-site processing offal from a harvested crop directly back to immature members of the same species. For species on other trophic levels, a polyculture system would be necessary.

Monoculture. Fish and crustacea offal might be returned "as is" to the culture environment of carnivores and detritus feeders. Alternatively, it might be finely ground and employed as fry feed. There is some danger in this, since any disease-bearing microorganisms in the offal would either be returned to the large culture or innoculated into the fry culture. The offal could be heat processed, of course, but care would be needed not to "overcook" it and thereby make the proteins less digestible. This might argue in favor of fine grinding and sterilization by chemical means. Then the finely ground product might be used as fry feed or even agglutinated into wet-pellet feed. The removal of bony, indigestible irritants might also be necessary, particularly if the material was to be employed as fry feed.

Polyculture. From one perspective, a polyculture system appears to be an expeditious method to minimize wastes and increase product per dollar. The intrinsic character of the aquatic habitat provides for a fine division into natural niches. Since no single organism can effectively avail itself of the space and nutritional benefits of all niches, polyculture may provide a provident means of capitalizing on otherwise unexploited natural behavior characteristics.

To illustrate, consider a large-volume culture system containing a population of nekton. If abundant pelletized food is supplied so that food is nonlimiting, the production of the nekton species per unit of food depends upon the growth, excretion, respiration, and mortality rates of the organism, where

net production = gross production − (excretion + respiration + mortality)

As established earlier, attractive economic conversion rates (ECR's) are achieved by maximizing the net production per unit food *and* minimizing capital expenses for space, feed, dispersal, and so on. Thus, from

$$ECR = \frac{\text{Net production}}{\text{dollars spent}}$$

it is obvious that increases in production at no, or proportionately less, expense will increase the ECR.

In a typically enclosed nekton monoculture system, uneaten food, fecal pellets, and natural mortalities fall to the bottom and represent immediate system losses. Moreover, the culturist is faced with frequent cleaning expenses (additional system losses) to avoid increases in mortality or decreases in growth caused by bacterial proliferation in the settled material. Here the addition of a co-crop, which derives nutrition from the nekton losses, could have a doubled positive effect by increasing system productivity and decreasing operating expenses. The ECR of the entire system would thus be increased by the culture of compatible and marketable organisms by the exploitation of the nekton system's losses. As an example, a co-culture of benthic detritivores (shrimp or lobster) would derive nutrition, in part, from fecal pellets and moribund nekton. The relative success of an actual polycultural design, of course, depends in part upon the facility of culture and the compatibility of the two populations.

Polyculture systems may take several forms. The first, of course, is the culturing of feed organisms for the primary cultured crop—plankton for plankton feeders, and plankton feeders for carnivores. Second, as mentioned in Chapter 10, some varieties added to a polyculture system to act as controls for fouling organisms or other "weeds" might constitute secondary crops if they can be harvested and processed economically, and if they have sufficient market value. If they should not possess adequate market value, they still might be harvested, at a rate that does not interfere with their primary function, to be used as feed supplements for the primary crop. Finally, the existence of an open sea mariculture operation will probably attract a variety of uncultured species to the site. Some of these likely will be consumed by the cultured crop directly. As such they will constitute a form of natural food. Where such organisms are attracted in numbers, it might be worthwhile to harvest and process them into prepared feeds on site. These opportunities are explored further in Chapter 11.

Artificial Upwelling

Provident harvesting of marine crops by and large is dependent on phytoplankton density, which in turn depends on the rate of primary production. Since marine photosynthesis is limited most often by nutrients, economical fertilization is essential for any successful mariculture based on primary productivity.

It is common knowledge to oceanographers that a vast resource of plant nutrients lies just below the photic zone. This was discussed in Chapters 4 and 5. These dissolved nutrients, the products of the mineralization and ionization of biological material, show a marked increase in concentration at about 200 to 500 m and reach a maximum at approximately 700 m. Typical concentrations of nitrate, phosphate, and silicate, for example, are 40, 2, and 100 μg atoms/liter, respectively; these are 10 to 100 times greater than complementary surface values.

Clearly, this nutrient-rich water, if raised to the photic zone and kept there, could provide the basis for drastically increased primary production. This, in turn, could support greatly increased production of marine animals suitable for human consumption. Research conducted in Hawaii by Gundersen and Bienfang [51] showed that phytoplankton production in elevated Intermediate Water, at surface conditions, may exceed normal surface production by sixty-fold. In this work, carbon fixation values in excess of 200 mg of carbon/m^3/h were common. By analyzing these carbon fixation data, it is possible to predict the potential benefits of an artificial upwelling system in terms relevant to the mariculturist. If the aforementioned productivity is maintained in a 100,000-m^3 "pond" having a 1-day residence time, the resulting annual productivity will be about 70 metric tons of carbon/pond. Since carbon comprises approximately 40 percent of total organic matter, this translates to 175 metric tons (dry weight) of primary organic matter/pond/year. Assuming organic matter to be 25 percent of the second trophic level and assuming the usual 10 percent conversion ratio between the first and second trophic levels, the result is 70 metric tons (wet) of herbivore/pond/year. Admittedly, these figures, based on many contingencies, can only represent approximations.

The feasibility of actually producing the projected quantity of protein is confirmed by the analysis of nutrient availability in the volumes of water under consideration. Calculations of nitrate, for example, demonstrate that the amount made available by such a system is in excess of the amount required. This allows for the culturing of particularly beneficial, high-protein crops.

Nature itself provides evidence for the validity of such a design. Many of the world's fishing areas are, in fact, regions of nutrient-rich, Intermediate Water upwellings. The California, Peru, and Benguela current systems, for example, result from a particular interaction of atmospheric and oceanic circulation relative to the adjacent landmass. The subsequent Coriolis current causes very slow upwelling rates, which are on the order of 2 m/day. But because of the extensive areas involved, millions of cubic meters of water are brought to the photic zone each second.

The concept of biostimulation was discussed by Stommel et al. over 15 years ago [52]. He described a mechanism for the production of small artificial upwellings based on water-density dynamics. Since salinity is lower in deeper water, a pipe extending from the surface to a desired depth with the deep water pumped slowly to the surface would induce the flow of water, once started, to continue in accordance with the physical laws of diffusion and density. The slow motion through the pipe would allow the water inside the pipe to equilibrate its temperature with the water outside the pipe, and because of salinity differences the warmed deep water would be less dense and would rise. Since Stommel's "Salt Fountain" design rests on temperature equilibration through the pipe, obvious restrictions are placed on the maximum pipe radius that will result in the upward motion of water. Nonetheless, flow-rate estimates by Groves [53] indicate that 3,000 liters/s could be transported upward by a pipe with a 1-m radius if temperature equilibration could be achieved in practice.

Another means of raising Intermediate Water is to mechanically pump it up to surface holding tanks. This would be feasible in coastal aquaculture ventures located adjacent to a steep-sloping ocean bottom, because the intake pipe could rest on natural substrate that could also support the holding tanks. Columbia University's Lamont-Doherty Geological Observatory currently has such a system operating in St. Croix in the Virgin Islands. Under the direction of O. A. Roels, deep water is being pumped into surface aquaculture ponds to stimulate primary production. In this particular research, mollusks are used as the secondary trophic level. This experimental work to date has shown that dense cultures of phytoplankton result and the mollusks grow at remarkable rates [54].

In an open-ocean environment, however, holding upwelled water at or near the surface is a significant problem. In fact, there are only three ways to achieve surface retention of the relatively dense Intermediate Water: by physical containment or by lowering its density through heating and/or dilution. Containment would require sizable and survivable tanks and would amount to moving to sea the shoreside practice just described. Such a move would demand floating enclosures that could survive the most severe weather conditions experienced in an area, and so very likely would involve considerable expense. It does, however, seem possible that buoyant flexible containers might be developed that would achieve high survivability at nominal cost.

Dilution of upwelled water with fresh water seems an unattractive alternative. Not only is fresh water a scarce commodity in the offshore environment, but significantly lowered salinities would very likely produce deleterious effects on the vigor and growth of primary- and secondary-trophic-level organisms. Gradual dilution and warming by diffusion of upwelled water through surface water might be a feasible alternative; but it imples significantly lowering the primary-nutrient input rate.

Direct heating is probably the most attractive alternative. But Isaacs and Schmitt [48] point out that the energy required to heat upwelled water to surface-retention temperature is on the order of 1,000 times as great as the pumping power required to raise it to the surface in the first place. Thus, unless clever designs make efficient use of energy fluxes already available, the ongoing costs of heating will be considerable. The available energy fluxes might be either natural or man-made. Among the natural fluxes are solar radiation, winds, waves, and the relatively high temperature of Surface Water as compared with Intermediate Water. Man-made sources of heat would be limited to the sites of offshore nuclear or conventional power plants (see Chapter 13). Artificial upwelling systems might employ wind and wave energy as the primary source of pumping power (which has relatively low power requirements) and a combination of solar radiation and higher surface water temperature as the source of heat. The maze of low-power pumps and low-cost heat exchangers that would appear to be required to make this approach work in any practical fashion would be considerable. Its description is not attempted here. Perhaps multiple "Stommel Salt Fountains" would be preferable.

Until something along the lines of the natural-energy-flux systems suggested is contrived and proved, it seems that uncontained artificial upwellings will depend either upon high-cost heat or upon association with offshore power production. The latter approach, employing Intermediate Water as the power plant coolant and thereby warming it sufficiently for surface retention, is discussed further in Chapter 13.

The bioengineering aspects of converting Intermediate Water nutrients into organic matter are influenced by the physical nature of these nutrients. Because the nutrients are a dissolved constituent in the water, the continual flow of water is explicit in the continual supply of nutrients. The design of this continuous-flow culture must satisfy logistic, kinetic, and economic criteria.

The location of such an operation should be in an area where surface currents are somewhat predictable and of moderately slow velocity. There will be a definite lag between the introduction of deep water into the surface and subsequent primary production. During the time that nutrient assimilation is occurring, the phytoplankton and nutrients will be subject to dispersal by the currents. The duration of this interval will be a function of the phytoplankton density and the nutrient concentration, and the dispersion area will be related to the current velocity and the interval duration. Thus, the concentration and orientation of production and grazing activities require a clear description of current direction and velocity. Moderate current velocities are prerequisite so that the assimilation, primary, and secondary production areas can be as localized as possible, permitting more expeditious and economical harvesting.

The flow-through production system has advantages that are reflected in nearly all facets of mariculture. Continuous culture ensures that phytoplankton are a viable and nutritious food source, waste materials from all trophic levels are continuously diluted, and disease-causing pathogens are less likely to ravage the dense crop population.

The specific growth rate of the phytoplankton will depend upon the supply rate of the limiting substrate. For a given set of flow conditions, the utilization rate of nutrients and the phytoplankton growth rate, density, and nutritional content will remain reasonably constant. Thus, the overall primary production is constant and provides the consistency and stability required by such feeding operations.

The performance of such a design can be evaluated by considering two facets of the system: the production rate and necessary phytoplankton density. The production rate (phytoplankton per hour) is given by the density (phytoplankton per volume) times the flow rate (volume per hour). The maximum production rate per volume is achieved by maintaining the highest possible flow rate without washing out the phytoplankton. The dependence of herbivorous grazing upon phytoplankton concentration was discussed earlier in this chapter. It is the strategy of mariculture to provide maximum production rates at sufficient concentrations to permit maximum production and density at the herbivorous level. Because the intitial nutrient concentration and upwell rate are constants of the system, the dependent criteria are the area involved (current velocity multiplied by time) and the phytoplankton density required for efficient grazing by the herbivore.

As the phytoplankton concentration flowing through the "pond" becomes reasonably constant, it can be assumed that the production of plankton equals the output, with the output represented by phytoplankton that are grazed upon, as well as those which flow out of the system directly. Clearly, efficiency considerations dictate that grazing comprise the vast majority of output. So the time required to effectively graze the transient phytoplankton population places additional constraints on the "pond" size.

At manageable dilution rates, the concentration of phytoplankton will be directly proportional to the limiting nutrient concentration in the deep water, and this is a constant, as already mentioned. Thus, with an adequate knowledge of grazing–growth kinetics, the mariculturist has the ability to ascertain whether and to what extent additional nutrients must be added to produce phytoplankton concentrations that will support the profitable culture of the grazing population.

Market value notwithstanding, it appears that herbivores, or at least low-trophic-level organisms, represent the most feasible culture organisms for the type of system considered here. In the face of a multitude of unknowns, this must certainly be true at least for initial ventures. For this reason the suspension-feeding mollusks, such as clams, oysters, cockles, scallops, and mussels, should receive initial attention.

However, one possibility that might circumvent the necessity of warming the deep water would be the culture of abalone. A temperate species of high market value, the abalone grazes on benthic algae. Thus, the provision of copious surface area within a large pen could provide substrate for the benthic algal food. Because benthic algae are capable of remarkable rates of production and assimilation of nutrients from the environment, this concentrated primary production could be very effectively grazed by a dense abalone population.

Baitfish, such as the anchovy, represent another prime candidate for production under artificial upwelling conditions. Natural populations of these highly desirable, low-trophic-level organisms are critically low, and fishing industries are constantly looking for new sources. The substantial open sea production of such a crop would have several ramifications, including the availability of more finfish and/or the production of fish meal.

CONCLUSIONS

The ultimate choice of the product(s) to be cultured will depend on the natural abundance of larvae, market value, rate of growth, facility of culture, and amenability to control throughout the life cycle. In any event, fertilization of the phytoplankton growth environment by enrichment with Intermediate Water is especially advantageous for many reasons. The deeper water contains a seemingly optimum balance of nutrients and essential growth elements, since it is a reservoir of the detritus from natural living systems. Possible contamination by toxic substances is not a matter of concern here, as it often is when sewage is used; nor is there any psychological or esthetic opposition to consuming food produced in this way, as there is to food organisms nourished (directly or indirectly) on sewage.

The prospective economic feasibility of any aquacultural venture is obscured by an unquantifiable array of technical and biological unknowns. The forecast for aquaculture must describe not only the potential benefit but also the extent of the difficulties that must be overcome; and the investment necessary to solve the technical unknowns most certainly must await a favorable cost—benefit prognosis. The biological questions require the union of basic research and on-site trial-and-error efforts.

REFERENCES

1. Strickland, J. D. H. 1958. Solar radiation penetrating the ocean: a review of requirements, data, and methods of measurements, with particular reference to photosynthetic productivity. *J. Fisheries Res. Board Can.* 15:453—493.
2. Ryther, J. H. 1964. Potential productivity of the sea. In W. E. Hazen, ed., *Population and community ecology.* W. B. Saunders Company, Philadelphia, pp. 275—285.
3. Yentsch, C. S., and R. W. Lee. 1966. A study of photosynthetic light reactions, and a new interpretation of sun and shade phytoplankton. *J. Marine Res.* 24:319—337.

4. Steemann-Nielsen, E., and E. G. Jorgensen. 1968. The adaptation of plankton algae. I. General part. *Plant Physiol.* 21:401–413.
5. Steemann-Nielsen, E. 1966. Productivity, definition and measurement. In M. H. Hill, ed., *The sea,* Vol. 2. Wiley-Interscience, New York, pp. 129–164.
6. Carlucci, L. 1973. Unpublished manuscript.
7. Droop, M. R. 1968. Vitamin B-12 and marine ecology. IV. The kinetics of uptake, growth and inhibition in *Monochrysis lutherii. J. Marine Biol. Assoc. U.K.* 48:689–733.
8. Caperon, J., and J. Meyer. 1972. Nitrogen-limited growth of marine phytoplankton—I and II. *Deep Sea Res.* 19:601–632.
9. Droop, M. R. 1961. Vitamin B-12 and marine ecology: the response of *Monochrysis lutherii. J. Marine Biol. Assoc. U.K.* 41:69–76.
10. Eppley, R. W. 1973. Phytoplankton physiology: a guide to the literature for modelers. NATO Conf. on Modeling of Marine Systems, Lisbon. June 4–8.
11. Hardy, A. C. 1956. *The open sea.* Houghton Mifflin Company, Boston. 322 pp.
12. Friedrich, H. 1969. *Marine biology.* University of Washington Press, Seattle, Wash. 477 pp.
13. Wimpenny, R. 1966. *The plankton of the sea.* Elsevier Publishing Company, Amsterdam. 426 pp.
14. Barnes, R. D. 1968. *Invertebrate zoology.* W. B. Saunders Company, Philadelphia. 742 pp.
15. Lasker, R. 1966. Feeding growth, respiration and carbon utilization of a euphausid crustacean (*E. pacifica*). *J. Fisheries Res. Board Can.* 23: 1291–1317.
16. Rigler, F. H. 1961. The relation between concentration of food and feeding rate of *Daphnia magna. Can. J. Zool.* 39(6):857–868.
17. Burns, C. W., and F. H. Rigler. 1967. Comparison of filtering rates of *Daphnia Rosea* in lake water and suspensions of yeast. *Limnol. Oceanogr.* 12:492–502.
18. Parsons, T. R., K. Stephens, and R. J. LeBrasseur. 1969. Production studies in the Strait of Georgia. P. I. Primary production under the Fraser River plume, Feb. to May, 1967. *J. Exp. Marine Biol. Ecol.* 3(1):27–38.
19. Betzlemeshem, C. W. 1962. Superfluous feeding of marine herbivorous zooplankton. *Rapp. Proc. Verb. Cons. Intern. Explor. Mer.* 153:108–113.
20. Mullin, M. M. 1963. Some factors affecting the feeding of marine copepods of the genus *Calanus. Limnol. Oceanogr.* 8:238–250.
21. Provasoli, L., and A. D'Agostino. 1969. Development of artificial media for *Artemia salina. Biol. Bull.* 136(3):434–453.
22. Provasoli, L., D. E. Conklin, and A. S. D'Agostino. 1970. Factors inducing fertility in aseptic crustacea. *Helgolander wiss. Meeresunter.* 20:443–454.
23. Provasoli, L., K. Shiraishi, and J. R. Lance. 1959. Nutritional idiosynchroses of *Artemia* and *Tigriopus* in monoaxenic culture. *Ann. N.Y. Acad. Sci.* 77(2):250–261.

24. Ukeles, R. 1970. Nutrition requirements in shellfish culture. In K. S. Price, and D. L. Maurer, eds., *Proc. Conf. on Artificial Propagation of Commercially Valuable Shellfish.* College of Marine Studies, University of Delaware, pp. 43–64.

25. Martin, G. W. 1928. Experimental feeding of oysters. *Ecology 9*:49–55.

26. May, R. C. 1970. Feeding larval marine fishes in the laboratory: a review. *Calif. Marine Res. Comm. CalCOFI Rept. 14*:76–83.

27. Fabre-Domergue, P., and E. Bietrix. 1905. Developpement de la Sole (*Solea vulgaris*). Introduction a l'etude de la pisciculture marine. Travail du Laboratoire de Zoologie Maritime de Concarneau. Vuibert et Nony, Paris. 243 pp.

28. Kasahara, S, R. Hirano, and Y. Oshima. 1960. A study on the growth and rearing methods of black porgy, *Mylio macrocephalus* (Basilewsky). *Bull. Jap. Soc. Sci. Fishery 26*:239–244.

29. Lasker, R., H. M. Feder, G. H. Theilacker, and R. C. May. 1970. Feeding, growth, and survival of *Engraulis mordax* larvae reared in the laboratory. *Marine Biol. 5*:345–353.

30. O'Connell, C. P., and L. P. Raymond. 1970. The effect of food density on survival and growth of early post-yolk sac larvae on the northern anchovy (*Engraulis mordax*) in the laboratory. *J. Exp. Marine Biol. Ecol. 5*:187–197.

31. Halver, J. E. 1968. Nutrition in marine aquiculture. In W. V. McNeil, ed., *Marine aquiculture.* Oregon State University Press, Corvallis, Ore.

32. Hughes, J. T., and G. C. Matthiessen. 1962. Observations on the biology of the American lobster, *Homarus americanus. Limnol. Oceanogr. 1*(3): 414–421.

33. Hughes, J. T., J. J. Sullivan, and R. Shleser. 1972. Enhancement of lobster growth. *Science 177*:1110–1111.

34. Stewart, J. E., G. W. Horner, and B. Arie. 1972. Effects of temperature, food and starvation on several physiologic parameters of the lobster *Homarus americanus. J. Fisheries Res. Board Can. 2*:439–442.

35. Hudinaga, M. 1942. Reproduction, development and rearing of *Penaeus japonicus* Bate. *Jap. J. Zool. 10*(2):305–393.

36. Hudinaga, J., and Z. Kittaka. 1966. Studies on food and growth of larval stages of a prawn *Penaeus japonicus,* with reference to the application of practical mass culture. *Nihon Purankuton Kenkyu Renrakukaiho. 13*:88–94.

37. Cook, H. L., and M. L. Murphy. 1966. Rearing penaeid shrimp from eggs to postlarvae. *Proc. 19th Ann. Conf. S.E. Assoc. Game and Fish Comm. 19*:238–288.

38. Cook, H. L., and M. L. Murphy. 1969. The culture of larval penaeid shrimp. *Trans. Amer. Fisheries Soc. 98*(4):751–754.

39. Parsons, T. R., K. Stephens, and J. D. H. Strickland. 1961. On the chemical composition of eleven species of marine phytoplankters. *J. Fisheries Res. Board Can. 18*:1001–1016.

40. Allen, M. B. 1969. Structure, physiology and biochemistry of the Chrysophyceae. *Ann. Rev. Gen. Micro. 23*:93–101.

41. Oppenheimer, C. H., ed. 1963. *Symposium on marine microbiology.* Charles C Thomas, Publisher, Springfield, Ill.

42. Provasoli, L., and J. J. A. McLaughlin. 1963. Limited heterotrophy of some photosynthetic dinoflagellates. In C. H. Oppenheimer, ed., *Symposium on marine microbiology.* Charles C Thomas, Publisher, Springfield, Ill., pp. 105–113.

43. Thomas, W. H. 1966. Surface nitrogenous nutrients and phytoplankton in the northeastern tropical Pacific. *Limnol. Oceanogr. 11*(3):393–400.

44. Pintner, I. J., and L. Provasoli. 1968. Heterotrophy in subdued light of three Chrysochromulina species. *Bull. Misaki Marine Biol. Inst., Kyoto Univ. 12:*25–31.

45. Ackman, R. G., C. S. Tocher, and J. McLachlin. 1968. Marine phytoplankter fatty acids. *J. Fisheries Res. Board Can. 25*(8):1603–1620.

46. Lewin, R. A. 1962. *Physiology and biochemistry of algae.* Academic Press, Inc., New York.

47. Huet, J., and J. A. Timmermans. 1970. *A textbook of fish culture.* Fishery News (Books) Ltd., London.

48. Ryther, J. H., K. R. Tenore, W. M. Dunstan, J. C. Goldman, J. S. Prince, V. Vreeland, W. B. Kerfoot, N. Corwin, J. E. Huguenin, and J. M. Vaughn. 1972. The use of flowing biological systems in aquaculture, sewage treatment, pollution assay, and food-chain studies. *Prog. Rept. NSF-RANN GI-32140.*

49. Isaacs, J. D., and W. R. Schmitt. 1969. Stimulation of marine productivity with waste heat and mechanical power. *J. Cons. Int. Explor. Mer. 33:*20–29.

50. Deshimaru, O., and K. Shigeno. 1972. Introduction to the artificial diet for prawn *Penaeus japonicus. Aquaculture 1:*115–133.

51. Gundersen, K., and P. Bienfang. 1970. Thermal pollution: use of deep, cold, nutrient rich sea water for power plant cooling and subsequent aquaculture in Hawaii. Tech. Conf. on Marine Pollution and Its Effects on Living Resources and Fishing, Rome, Italy. *FAO Rept. E-84.*

52. Stommel, H., A. B. Aarons, and D. Blanchard. 1955. An oceanographic curiosity: the perpetual salt fountain. *Deep Sea Res. 3:*152–153.

53. Groves, G. W. 1959. Flow estimate for the perpetual salt foundation. *Deep Sea Res. 7:*209–214.

54. Roels, O. A. 1970. *New Scientist 47:*378.

Additional Readings

Aaronson, S., and H. Baker. 1959. A comparative biochemical study of two species of Ochromonas. *J. Protozool. 6*(3):282–284.

Ackman, R. G., C. S. Tocher, and J. McLachlin. 1968. Marine phytoplankter fatty acids. *J. Fisheries Res. Board Can. 25*(8):1603–1620.

Alexander, M. 1971. Microbial ecology. John Wiley & Sons, Inc., New York, 482 pp.

Amos, A. F., C. Garside, K. C. Haines, and O. A. Roels. 1972. Effects of surface-discharged deep sea mining effluent. *J. Marine Tech. Soc. 6*(4):40–45.

Arthur, D. K. 1956. The particulate food and the food resources of the larvae of three pelagic fishes, especially the Pacific sardine, *Sardinops caerulea,* (Girard). Scripps Inst. Oceanogr., La Jolla, Calif.

Bishai, H. M. 1961. Rearing fish larvae. *Bull. Zool. Soc. Egypt 16:*4–29.

Blaxter, J. H. S. 1962. Herring rearing. IV. Rearing beyond the yolk-sac stage. *Marine Res. Scot. 1:*18.

Blaxter, J. H. S. 1969. Experimental rearing of the pilchard larvae, *Sardinia pilchardus. J. Marine Biol. Assoc. U.K. 49:*557–575.

Blinks, L. R. 1951. Physiology biochemistry of algae. In G. M. Small, ed., Manual of phycology—an introduction to the algae and their biology. *A New Series of Plant Science Books 27*(14):263–291.

Boney, A. D. 1970. Scale-bearing phytoflagellates: an interim review. *Oceanogr. Marine Biol. Rev. 8:*251–305.

Bruce, V. R., M. Knight, and M. W. Parke. 1939. The rearing of oyster larvae on an algal diet. *J. Marine Biol. Assoc. U.K. 24:*337–374.

Budd, P. L. 1940. Development of the eggs and larvae of six California fishes. *Calif. Dept. Fish and Game Fishery Bull. 56:*50.

Caperon, J. 1968. Population growth response of *Isochrysis galbana* to nitrate variation at limiting concentrations. *Ecology 49:*866–872.

Clarke, G. L. 1934. Further observations on the diurnal migration of copepods in the Gulf of Maine. *Biol. Bull. Woods Hole 67:*432–448.

Conover, R. J. 1966. Factors affecting the assimilation of organic matter by zooplankton and the question of superfluous feeding. *Limnol. Oceanogr. 11*(3):346–354.

Conover, R. J. 1968. Zooplankton—life in a nutritionally dilute environment. *Amer. Zool. 8:*107–118.

Corkett, C. J. 1968. La Reproduction en laboratoire des *Copepodes marins, Arcartia claussi* and *Idya furcatu. Bull. Inst. Oceanogr. Alger. 10:*77–90.

Corkett, C. J., and I. A. McLaren. 1970. Relationships between development rate of eggs and older stages of copepods. *J. Marine Biol. Assoc. U.K. 50:*161–168.

Corner, E. D. S., C. B. Covey, and S. M. Marshall. 1967. On the nutrition and metabolism of zooplankton. V. Feeding efficiency of *Calanus finmarchicus. J. Marine Biol. Assoc. U.K. 47:*259–270.

Corner, E. D. S., and B. S. Newell. 1967. On the nutrition and metabolism of zooplankton, IV. The forms of nitrogen excreted by *Calanus. J. Marine Biol. Assoc. U.K. 47:*113–120.

Darnell, R. M. 1968. Animal nutrition in relation to secondary production. *Zoologist 8:*83–93.

Davis, A. G. 1970. Iron, chelation and the growth of marine phytoplankton. *J. Marine Biol. Assoc. U.K. 50:*65–86.

Delmonte, P. J., I. Rubinoff, and R. W. Rubinoff. 1968. Laboratory rearing through metamorphosis of some Panamanian gobies. *Copeia:*411–412.

Detwyler, R., and E. D. Houde. 1970. Food selection by laboratory-reared larvae of the scaled sardine *Harengula pensacolae* and the bay anchovy *Anchoa mitchilli. Marine Biol. 7*(3):214–222.

Droop, M. R. 1958. Requirement for thiamine among some marine and supralittoral Protista. *J. Marine Biol. Assoc. U.K. 37:*323–329.

178   NUTRITIONAL CONSIDERATIONS

Droop, M. R. 1960. Some chemical considerations in the design of the synthetic media for marine algae. *Botanica Marine II:* 230–245.

Eltone, C. 1927. *Animal ecology.* Methuen & Company, Ltd., London. 207 pp.

Fleming, R. H. 1957. General features of the oceans. *Geol. Soc. Amer. Mem. 67,* V. 1, pp. 87–108.

Forrester, C. R. 1964. Laboratory observation on embryonic development and larvae of Pacific cod (*Gadus macrocephalus* Telesius). *J. Fisheries Res. Board Can. 21:*9–16.

Fuhs, G. W. 1969. Phosphorus content and rate of growth in the diatoms of *Cyclotella nana* and *Thalassiosira fluviatilis. J. Phycol. 5:*312–321.

Gerard, R. D., and O. A. Roels. 1970. Deep ocean water as a resource for combined mariculture, power and fresh water production. *J. Marine Tech. Soc. 4*(5):69–79.

Gillespie, L., R. M. Ingle, and W. K. Havens. 1964. Glucose nutrition and longevity in oysters. *Quarterly Fla. Acad. Sci. 27:*279–288.

Gordon, D. C., Jr. 1970. Some studies on the distribution and composition of particulate carbon in the North Atlantic Ocean. *Deep Sea Res. 17:*235–243.

Gross, F. 1937. Notes on the culture of some marine plankton organisms. *J. Marine Biol. Assoc. U.K. 21:*753–768.

Hedgpeth, U. W., ed. 1957. Treatise on marine ecology and paleoecology, Vol. I. *Geol. Soc. Amer. Mem. 67.* 1296 pp.

Hempel, G. 1970. Fish-farming including farming of other organisms of economic importance. *Helgolander wiss. Meeresunters. 21:*445–465.

Heinle, D. R. 1969. Culture of the calanoid copepods in synthetic sea water. *J. Fisheries Res. Board Can. 26:*150–153.

Herbert, D., R. Elsworth, and R. C. Telling. 1956. The continuous culture of bacteria; a theoretical and experimental study. *J. Gen. Microbiol. 14:*601–622.

Hertling, H. 1962. Die Zuchtung von Meresfischen für wissenschaftliche und praktische Zwecke. In E. Abderhalden, ed., *Handbuch der biologoische Arbeitsmethoden, Abt. 9, Teil6 Heft 2:*195–366.

Hickling, C. F. 1970. Estuarine fish farming. *Adv. Marine Biol. 8:*119–213.

Holling, C. S. 1965. The functional response of predators to prey density and role in mimicry and population regulation. *Mem. Entomol. Soc. Can. 45.* 60 pp.

Ivlev, V. S. 1961. Experimental ecology of the feeding of fishes. Yale University Press, New Haven, Conn. 302 pp.

Johnston, R. 1962. Sea water, the natural medium of phytoplankton. *J. Marine Biol. Assoc. U.K. 43:*427–456.

Johnston, R. 1964. Sea water, the natural medium of phytoplankton. *J. Marine Biol. Assoc. U.K. 44:*87–109.

Jorgensen, C. B. 1966. Biology of suspension feeding. Pergamon Press, Inc., Elmsford, N.Y. 357 pp.

Katona, S. K. 1970. Growth characteristics of the copepods *Eurytemora affinis* and *E. herdmani* in laboratory cultures. *Helgolander wiss. Meeresunters. 20:*373–384.

Katona, S. K., and C. F. Moodie. 1969. Breeding of *Pseudocalanus elongatus* in the laboratory. *J. Marine Biol. Assoc. U.K. 49:*743–747.

Ketchum, B. H. 1953. Mineral nutrition of phytoplankton. *Woods Hole Oceanogr. Inst. 687:*55–71.

Kimball, J. F., Jr., E. Corcoran, and E. J. Ferguson Wood. 1963. Chlorophyll-containing microorganisms in the aphotic zone of the oceans. *Bull. Marine Sci. Gulf and Carib. 13*(4):574–577.

Krauss, R. W. and W. H. Thomas. 1954. The growth and inorganic nutrition of *Scenedesmus obliquus* in mass culture. *Plant Physiol. 29*(3):205–214.

Kurata, H. 1956. On the rearing of the larvae of the flatfish, *Liopsetta obscura,* in small aquaria. *Bull. Hokkaido Reg. Fishery Res. Lab. 13:*20–29.

Kurata, H. 1959. Preliminary report on the rearing of the herring larvae. *Bull. Hokkaido Reg. Fishery Res. Lab. 20:*117–138.

Lasker, R. 1960. Utilization of organic carbon by a marine crustacean: analysis with C-14. *Science 131:*1098–1100.

Lasker, R. 1962. Efficiency and rate of yolk utilization by developing embryos and larvae of the Pacific sardine *Sardinops caerulea* (Girard). *J. Fisheries Res. Board Can. 19*(5):867–873.

Lasker, R., H. M. Feder, G. H. Theilacker, and R. C. May. 1970. Feeding, growth, and survival of *Engraulis mordax* larvae reared in the laboratory. *Marine Biol. 5:*345–353.

Lindeman, R. L. 1964. The trophic dynamic aspect of ecology. In W. Hazen, ed. *Readings in population and community ecology.* W. B. Saunders Company, Philadelphia, pp. 206–266.

Linder, M. J., and H. L. Cook. 1968. Progress in shrimp mariculture in the U.S. *Proc. Symp. on Investigations and Resrouces of the Caribbean Sea and Adajacent Regions,* Willemstad, Curacao, Netherlands Antilles, pp. 1–4.

Loosanoff, V. L. 1970. Development of shellfish culture techniques. In K. S. Price and D. L. Maurer, eds., *Proc. Conf. on Artificial Propagation of Commercially Valuable Shellfish.* College of Marine Studies, University of Delaware, pp. 9–40.

Lotka, A. J. 1956. *Elements of mathematical biology.* Dover Publications, Inc., New York. 465 pp.

Lucas, C. E. 1947. The ecological effects of external metabolites. *Biol. Rev. 22:*270–295.

Marshall, S. M., and A. P. Orr. 1955. Studies on the biology of *Calanus finmarchicus.* VIII. Food uptake, assimilation and excretion in adult and stage V. *Calanus. J. Marine Biol. Assoc. U.K. 34:*495–529.

Marshall, S. M., and A. P. Orr. 1955. The biology of a marine copepod, *Calanus finmarchicus.* Oliver & Boyd Ltd., Edinburgh, 188 pp.

Mayrer, D. L. 1970. Introduction to development of culture techniques for a pilot shellfish hatchery. In K. S. Price and D. L. Maurer, eds., *Proc. Conf. on Artificial Propagation of Commercially Valuable Shellfish.* College of Marine Studies, University of Delaware, pp. 5–9.

Menzel, D. W., and J. H. Ryther. 1960. Nutrients limiting the production of phytoplankton in the Sargasso Sea, with special reference to iron. *Deep Sea Res.* 7:276–281.

Menzel, D. W., and J. Goering. 1966. The distribution of organic detritus in the sea. *Limnol. Oceanogr.* 11:333–337.

Menzel, D. W. 1967. Particulate carbon in the deep sea. *Deep Sea Res.* 11:757–765.

Mock, C. R., and M. A. Murphy. 1970. Techniques for raising Penaeid shrimp from egg to postlarvae. In J. W. Avault, ed., *Proc. First Ann. Workshop, World Mariculture Soc.,* pp. 143–155.

Monad, J. 1942. *La Croissance des cultures bacteriennes.* Hermann & Cie, Paris. 210 pp.

Morris, R. W. 1956. Some aspects of the problems of rearing marine fishes. *Bull. Inst. Oceanogr. Monaco.* 10:82. 61 pp.

Myers, J. 1947. Culture conditions and the development of the photo-synthetic mechanism. V. Influence of the composition of the nutrient medium. *Plant Physiol.* 22:590–597.

North, B. B., and G. C. Stephens. 1967. Uptake and assimilation of amino acids by *Platymonas. Biol. Bull.* 133(2):391–400.

North, W. J., and C. A. Hubbs. 1968. Utilization of kelp-bed resources in southern California. *Calif. Dept. Fish and Game Fishery Bull.* 139:9–33.

Nybakken, J. W., ed. 1971. *Readings in marine ecology.* Harper & Row, Publishers, New York. 543 pp.

Odum, E. P. 1959. *Fundamentals of ecology.* W. B. Saunders Company, Philadelphia. 543 pp.

Parsons, T. R., R. J. LeBrasseur, and J. D. Fulton. 1967. Some observations on the dependence of zooplankton grazing on the cell size and concentration of phytoplankton blooms. *J. Oceanogr. Soc. Jap.* 23:10–17.

Peterson, C. G. J., and P. B. Jensen. 1911. Valuation of the sea. Annual life of the sea bottom, its food and quantity. *Rept. Danish Biol. Sta.* 20:1–81.

Pintner, I. J., and L. Provasoli. 1963. Nutritional characteristics of some Chrysomonads. In C. H. Oppenheimer, ed., *Symp. on Marine Microbiol.* Charles C Thomas Publisher, Springfield, Ill., pp. 114–121.

Provasoli, L. 1969. Eutrophication: causes, consequences, correctives. In Natl. Acad. Sci., *Algal nutrition and eutrophication,* pp. 574–590.

Provasoli, L., and A. D'Agostino. 1962. Vitamin requirements of *Artemia salina* in aseptic culture. *Amer. Zool.* 2(3):439.

Provasoli, L., J. J. A. McLaughlin, and M. R. Droop. 1957. The development of artificial media for marine algae. *Archiv. fur Mikgiogie Bd.* 25 S.392–428.

Qasim, S. Z. 1955. Rearing experiments on marine teleost larvae and evidence of their need for sleep. *Nature (London)* 175:217–218.

Qasim, S. Z., P. M. A. Bhattathiri, and V. P. Devassy. 1972. The effect of intensity and quality of illumination on the photosynthesis of some tropical marine phytoplankton. *Marine Biol. 16:*22–27.

Rashevsky, N. 1959. Some remarks on the mathematical theory of the nutrition of fishes. *Bull. Math. Biophys. 21:*161–183.

Reeve, M. R. 1964. Feeding of zooplankton with special reference to some experiments with *Sagitta. Nature 201:*211–213.

Russell-Hunter, W. D. 1971. Aquatic productivity. Collier Macmillan, Publishers, London. 306 pp.

Ryther, J. H., and R. R. L. Gullard. 1959. Enrichment experiments as a means of studying nutrients limiting to phytoplankton production. *Deep Sea Res. 6:*65–69.

Schwimmer, D., and M. Schwimmer. 1964. Algae and medicine. In D. F. Jackson, ed., *Algae and man.* Plenum Publishing Corporation, New York.

Shelef, G., W. J. Oswald, and C. G. Golueke. 1969. The continuous culture of algal biomass on wastes. In I. Malek et al., eds., Continuous cultivation of microorganisms. *Proc. 4th Symp., Prague,* pp. 601–629.

Sieburth, J. M. 1968. The influence of algal antibiosis on the ecology of marine microorganisms. *Adv. Microbiol. Sea 1:*63–89.

Slobodkin, L. 1961. *Growth and regulation of animal population.* Holt, Rinehart and Winston, New York. 184 pp.

Soleim, P. A. 1942. Arsaker til rike og fattige arganger av slid. *Fiskeriderek toratets sfrifter, Ser. Havunderspkelser.* 7(2):39 pp.; English summary, 2 pp.

Sorokin, J. I. 1971. On the role of bacteria in the productivity of tropical oceanic waters. *Inst. Revue ges. Hydrobiol. 56:*1–48.

Spicer, C. C. 1955. The theory of bacterial constant growth chambers. *Biometrics:*225–230.

Strickland, J. D. H. 1965. Production of organic matter in the primary stages of the marine food chain. In J. P. Riley and G. Skirrow, eds., *Chemical oceanography.* Academic Press, Inc., New York, pp. 433–477.

Taub, F. B. 1970. Algal cultures as a source of feed. In J. W. Avault, Jr., E. Boudreaux, and E. Jaspers, eds., *Proc. First Ann. Workshop, World Mariculture Soc.,* pp. 101–117.

Taylor, J. E. 1970. Introduction to nutritional requirements in shellfish culture. In K. S. Price and D. L. Maurer, eds., *Proc. Conf. on Artificial Propagation of Commercially Valuable Shellfish,* College of Marine Studies, University of Delaware, pp. 41–42.

Theilacker, G. H., and M. F. McMaster. 1971. Mass culture of the rotifer *Brachionus plicatilis* and its evaluation as a food for larval anchovies. *J. Marine Biol. 10*(2):183–188.

Urry, D. L. 1965. Observation on the relationship between food and survival of *Pseudocalanus elongatus* in the laboratory. *J. Marine Biol. Assoc. U.K. 45:*49–58.

Wheeler, R. S. 1967. Experimental rearing of postlarval brown shrimp to marketable size in ponds. *Com. Fisheries Rev. 785:*49–52.

Whittaker, R. H. 1971. *Community and ecosystems.*Collier Macmillan Publishers, London. 158 pp.

Zein-Eldin, Z. P. 1963. Effect of salinity on growth of postlarval Penaeid shrimp. *Biol. Bull. 125*(1):188–196.

Zein-Eldin, Z. P., and D. V. Aldrich. 1965. Growth and survival of postlarval *Penaeus Aztecus* under controlled conditions of temperature and salinity. *Biol. Bull. 129*(1):199–216.

Zein-Eldin, Z. P., and G. W. Griffith. 1966. The effect of temperature upon the growth of laboratory-held postlarval *Penaeus Aztecus. Biol. Bull. 131*(1):186–196.

Zein-Eldin, Z. P., and G. W. Griffith. 1969. An appraisal of the effects of salinity and temperature of postlarval Penaeids. *FAO Fisheries Rept. 57*(3):1015–1024.

Zillious, E. J., and D. F. Wilson. 1966. Culture of the planktonic Calanoid copepod through multiple generations. *Science 151*(3713):996–998.

8

CROP SELECTION ISSUES

C.E. Nash

To make a selection of the aquatic species most suited for culture in the open sea, it is necessary first to assume that economic feasibility of the practice will be achieved. The two major components on which feasibility can be evaluated are bioeconomics and the economics of production. The first of these is concerned directly with the biology of the species, the technological level of culture presently developed, environmental compatibility, and so on. The second depends directly on production logistics: capital costs, production costs, marketing, and labor.

But the selection of species for any particular mariculture venture also will be sensitive to the geographic location of the site, as well as to applicable technology. Since most successful aquaculture to date has been of the "green-thumb" variety, one might expect the historical predilection for local species over imported species to continue for a time at least; consequently, the generalities drawn here and the examples used should be viewed as guideposts rather than blueprints. Also, this chapter is not consistent in the phylogenetic level to which it attends. Aquaculture evolution to date has not been characterized by simplistic logical consistency; to force this chapter into such a mold could result in a distorted picture.

BIOECONOMICS

The *culture* of aquatic species has been described by Hickling [1] as the production of finfish and shellfish, by whatever means, at higher population levels than found naturally. The objectives of culturing aquatic species artificially are therefore the same as for agriculture—to use science and technology to improve selected species, control disease, reduce predation, and improve the environment (creating entirely artificial habitats if necessary) to condition and accelerate growth. In other words, it seeks to achieve the domestication of aquatic species as has been accomplished with terrestrial species throughout recorded history.

The biological suitability of species for intensive culture or other farming practices depends ultimately on a combination of several associated factors. The most important of these appear to be the following:

- Gregarious nature
- High fecundity
- Short reproductive life cycle
- High survival rate
- Natural trophic level
- Fast growth rate, or large biomass production
- High available protein content of tissues
- Inherent behavior

Gregarious Nature

The gregarious nature of many animals fulfills at least two main functions in the fight for survival: it facilitates reproduction and affords protection. Gregariousness is inhibited by a lack of food more than other factors, and it is not surprising that in the many intensive culture experiments where food has not been limiting, the results in terms of behavioral response to the confined environment and to stock density have been most encouraging.

The natural territorial behavior of some crustaceans reduces the suitability of these species for intensive culture systems. Although continuous-feeding experiments show that aggressive tendencies can be minimized, continual moulting still makes them vulnerable to cannibalism. Rearing in separate compartments or provision of a multitude of protected places to hide are possible solutions currently under test, but such facilities are proving expensive to construct and conducive to a complex, labor-intensive system [2–4].

However, many fish species instinctively herd or shoal, and the majority of likely candidate species for farming have been reared intensively (at least experimentally) with little difficulty. Providing that the age and size ranges of individuals in a tank or pond are almost identical, no cannibalism among carnivorous fish species has been recorded. Populations of some of these fish in captivity have also been maintained at densities far above what is known to occur naturally.

Many of the mollusks amenable to culture methods and principles are sessile and normally gregarious. Intensive raft culture systems for oysters [5] and mussels are therefore extremely practical and approximate natural conditions for the species. Some naturally active species, for example, the abalones and scallops, have been shown to accept intensive rearing methods under certain circumstances [6].

Fecundity

High natural fecundity to overcome the disadvantages and exploit the advantages of external fertilization of water-borne gametes is characteristic of the majority of aquatic species. Potential larval and juvenile resources for culture purposes are therefore extremely high. However, high fecundity is usually associated with very small ova. The egg yolk is therefore small, and the emergent larva has little food on which to survive before finding external food organisms. This factor has consequently been one of the major restrictions to aquaculture throughout its history. Table 8.1 illustrates the importance of fish egg size to successful aquaculture, to date.

Even with high fecundity, of course, it is highly desirable to gain control over the reproductive process. Only when such control is in hand can eggs and larvae in large quantities be assured over the long term. At present, adequate control over the reproductive processes exists for only a few species of finfish and shellfish [7].

Life Cycle

The reproductive cycle length and the number of stages in the life cycle are both key factors in the suitability of candidate species for culture. The majority of aquatic species can reproduce to a degree in their first or second year; with the larger species, full maturity and maximum fecundity may not be reached until some years later. One more inhibitor to aquaculture is the number of developmental stages that may have to be completed before reproduction can occur. This is particularly true for crustaceans for which the number of metamorphoses can range between 1 and 30, depending on species; each stage is a potential hazard for culture operations and management. Mollusks may have a few critical stages in their life cycle, particularly if the early larvae are free swimming and must subsequently attach themselves to the substratum and metamorphose to a sessile form. A few finfish, such as the flatfish, also undergo a major metamorphosis early in their life history, which presents an added problem to culturists [8].

The greater the number of metamorphoses, the higher the risk of loss, both in nature and in the hatchery. During the process of metamorphosing, the individual is very vulnerable; and its new form often requires a different diet, so problems associated with the first feeding or weaning are repeated. The timing of metamorphosis and age are also important facets of culture control

TABLE 8.1 *Relationship of Egg Size to Mass Culture of Fishes*

Species	Egg diameter (mm)	Larval length (mm)	Yolk usage (days)	Mass culture
Salmon	4–5.0	25.0	21–60	Achieved
Trout	4.0	20.0	21–50	Achieved
Plaice	2.0	6.5	5–7	Achieved
Dover sole	2.0	6.0	5–7	Achieved
Turbot	1.0	3.0	3–5	Difficult
Mullet	0.9	2.4	3–5	Difficult
Rabbitfish	0.5	2.0	2–4	Difficult

as metamorphosed individuals can be either cannibalistic or injurious to those not yet metamorphosed. Ideally, aquaculture exponents prefer species with no, or very few, separate life stages and a short time to sexual maturity.

Survival Rate

Natural survival rates of marine organisms are always low. The natural loss from predation alone is horrendous. For example, it has been found that in nature the laying of 10,000 eggs of the plaice results in one marketable fish. Although there are some losses due to congenital abnormalities caused or simply revealed by high fecundity, it must be assumed that almost all fertile eggs have the inherent capability of becoming mature adults. Predation at all stages of the organisms' life cycle is probably the dominant control on survival and must, of course, be minimized in both hatcheries and rearing stations.

Plankton is the fulcrum of all life in the sea, and the majority of species contribute in some way to it during a part of their life cycle. It is in its planktonic stages that the greatest flux in the dynamics of a species occurs. Apart from predation, survival is obviously affected by the availability of natural food resources and suitable environmental conditions. Mobile species for the most part can deal with serious shortages of food or changes in conditions by ranging elsewhere. Sessile species obviously cannot, but often have the compensating ability to tolerate difficult situations.

Trophic Level

The natural trophic level of a species directly influences its candidacy for farming. The simplicity and economy with which appetites can be satisfied are a key to a practical culture enterprise. Detritus feeders and herbivores are, in theory, the most suitable species. Food resources are usually plentiful, and food conversion is direct from plant to animal protein. The farther the food chain progresses through the primary and secondary carnivores, the greater the demand on the basic resources. The efficiency of conversion at each trophic level is only 10 to 15 percent of the preceding level.

Growth Rate

Fast individual growth rates and/or large biomass production are another essential for financially successful culture practices. The time required for a population to reach marketable size is a significant contributor to economic evaluation. Alternatively, for certain cultures of smaller organisms that rely on rapid reproduction, the time the population requires to reach a volume economical to harvest is equally important. The sooner a species is fattened and dispatched to market, the greater the saving on feed. Furthermore, the chances for higher survival are increased, the facilities are utilized more efficiently, and turnover is increased.

In temperate waters most commercial finfish and shellfish reach marketable size in their third and fourth years. Rearing in the thermally enriched waters of an electrical generating plant can reduce this time by at least one half [9–12]. At present, culturists involved in developing techniques are concentrating on species that can attain marketable size within their first year or can be fattened within 1 year after leaving the juvenile rearing facilities.

Protein Content

The available animal protein is usually the end product of mariculture, and efficiency in producing it varies among the different forms. Mollusks utilize some energy in shell production, and crustaceans expend energy each time they change their exoskeleton. Per capita, fish are probably the most efficient producers of readily available animal protein.

Inherent Behavior

The inherent behavior of a species and its environmental compatibility further contribute to all the bioeconomical associations that have just been briefly outlined. The unpredictable fisheries of the herring and the anchoveta, in spite of years of research, illustrate the complexity of the population dynamics of these species. Some culture practices might be based on the collection of juveniles from the sea, but our present uncertainty about their behavior makes such efforts unreliable and potentially uneconomical.

ECONOMICS OF PRODUCTION

Production economics are primarily affected by local environmental advantages and disadvantages as manifested in the numerous topographical, physical, chemical, and biological properties of a proposed site. It is assumed that the selectors of a location for the practice of aquaculture will make an evaluation of these factors before construction and operation begin. It remains then to consider the following production factors in terms of costs.

Availability of Juveniles

The availability of juveniles as and when required is of vital importance and, unless the conditions for the natural occurrence and collection of juveniles are exceptionally favorable, all juveniles for an aquaculture farm must be obtained from a hatchery. This will guarantee production annually. At the present time, knowledge about intensive culture of a variety of finfish, crustaceans, or mollusks in either marine or brackish water is severely limited [13]. The marine species cultured most successfully throughout the world on a commercial scale is the oyster.

The number of juveniles required for farming is substantial, as we can see by the following random examples: (1) In France, the stock density for juvenile oysters is 4 million/hectare; (2) the production of postlarval shrimp for a commercial farm is estimated at over 200 million/year; (3) the Japanese government has restricted the collection of yellowtail juveniles to 31 million/year to conserve natural stocks; and (4) for a commercial flatfish farm producing 908 metric tons/year, a total of 25 million fertile eggs are required by the hatchery each year.

Food Costs

The logistics and cost of the food supply are the most important economic criteria for the success of aquaculture. The larval stages of some species cultured in captivity have quite specialized nutritional requirements, and these can be relatively costly to supply. The major expense of fattening either finfish, crustaceans, or mollusks is in fulfilling their gross carnivorous or herbivorous needs, however.

Herbivorous filter feeders (mussels and oysters), reared intensively, may be satisfied by the naturally high level of food in the water, but only in exceptional areas of the world (e.g., in the Galician estuaries of northwest Spain). Alternatively, available food can be increased by increasing the surface area within the compound, as is ingeniously practiced with mullet in Hawaii. Or biological productivity can be increased by the addition of organic and inorganic fertilizers, as is widely practiced in Asia.

These relatively modest costs are nothing compared to those of supplying the needs of carnivorous finfish or crustaceans, and the logistics of the purchase, preparation, and presentation of food influence the choice of a species more than anything else. To illustrate the potential costs of feeding, realistic conversion rates for aquatic carnivorous species (i.e., the weight of food required to produce a unit weight of flesh) are between 7 to 1 and 10 to 1, both wet weights; or between 2 to 1 and 4 to 1 for dry pelleted foods. The cost of catfish feed at present is about $0.05/lb and salmon feed (moist pellets) is $0.17/lb. Both formulations have been refined by many years of research and development.

Labor Costs

Labor costs are significant in farm management; therefore, operation of a farm relies on a minimum of permanent staff, but allowing for additional help at certain seasons. In Japan only low labor and material costs and an inflated market value make some farming practices profitable. In Asia, traditionally a stronghold of aquaculture, labor requirements average between two and four persons per 50 acres of freshwater ponds. The vagaries of salt water and the marine environments will undoubtedly tend to increase this number. In France it is estimated that oyster production per worker is between 2.26 and 9.08 metric tons/year. Present production in France exceeds 68,085 metric tons, valued at over $30 million.

Containment Costs

The cost of the facilities for impounding stock also can be significant if the farming method is sophisticated. Facility requirements of posthatchery juveniles depend on a number of factors—available water or ground space, stock density, operating efficiency, and feeding system, to name a few. In consequence, there are a variety of recognized types of facilities for retaining stock, from a simple net empoundment in a bay, to a raceway using pumped seawater, to, of course, nothing more than an abundance of food in an area to encourage the population to remain. Each system has advantages that depend on the species under consideration and must therefore be evaluated in context.

Processing Costs

The preparation of the harvested crop for sale in the market can add a considerable increment to the total cost, since it entails capital investment for processing equipment and operating expenses. The processing facet of a mariculture industry is discussed in detail in Chapter 14, and some indication of potential cost is given here. It may suffice, here, to point out that the forms in which the product is marketed vary from species to species, ranging from simple chilled fresh seafood to frozen, canned, filleted, or packaged precooked convenience dishes, or fish meal or flour, to name a few. Since costs will vary according to the type and degree of processing required, the choice of crop species will be influenced by the forms in which each species is acceptable to the market, the attendant processing costs, the combinations of species that can be processed by a single facility and set of equipment, and the magnitude of this increment relative to the total expense.

Potential Market Value

Obviously, the choice of a species depends on its demand by the market, which in turn determines its value. With high expenses for feed and labor, it is necessary to choose species of high commercial value. It is unfortunate that, until cultured species can establish their own level of quality and command a premium price, production costs will have to compete with those of the hunted species, which are sold chilled or frozen at dockside prices or processed in floating factory ships.

FARMING METHODS

Culture practices are limited by the proposed methods for feeding each candidate species; in turn, these are influenced by the commercial value of the species concerned, cost of feed and expected yield, and annual capital investment.

There are four basic types of feeding practice on which to base any farming technique. These are

- Natural feeding
- Provisional feeding
- Supplemental feeding
- Increasing basic productivity

Natural feeding involves only the careful selection of a site where the nutritional demands can be abundantly and continuously met by the existing environment. *Provisional feeding,* by contrast, undertakes to supply all the nutritional demands artificially. *Supplemental feeding* is a combination of provisional (artificial) and natural feeding. It may be unintentional through the colonization of a facility by other organisms that may be consumed by the cultured species. Or it may be by design through the intentional introduction of live organisms into a facility to provide nutrients that may be lacking in an artificially prepared food or seasonally unavailable. In either case, this supplemental activity is confined to foods that can be directly ingested or filtered by the cultured stock. It is not to be confused with "increasing basic productivity," which is accomplished by adding fertilizers or using nutritionally enriched water from deep upwellings to stimulate primary production.

Obviously, an important economic relationship exists between type of feed, method of presentation, and stock concentration. However, unless a given area is extremely and consistently rich in marine life, there are usually only two approaches to profitable aquaculture that are practical in commercial terms. These are (1) high stock density with full nutritional support, or (2) low stock density with minimal nutritional support. Provisional feeding must be allied to fast growth and high yields from heavy stock concentration. (It could, however, be used in conjunction with natural feeding under appropriate circumstances for particularly suitable species.) Relative to the second approach, a combination of natural feeding and increased basic, or primary, productivity is highly desirable, but restricted in application.

POTENTIALLY SUITABLE SPECIES

A great number and variety of species are potentially suitable for mariculture. They are, in fact, far too numerous to be individually described in an overview of this length and level of abstraction. However, when we attempt to deal with them in terms of groups, or communities, with common characteristics, we are handicapped by the fact tha many species occupy a number of environmental niches during their life cycles. So they cannot be functionally grouped by phyla, mode of feeding, or water-quality requirements.

For convenience only, then, in this very general discussion of marine, freshwater, and brackish-water species, we have arbitrarily settled on four broad categories:

- Herbivorous finfish and shellfish
- Carnivorous finfish and shellfish
- Planktonic species
- Miscellaneous detritus feeders

Herbivorous Finfish and Shellfish

Herbivorous finfish figure largely as supplementary foods in many Asian countries [14]. A great deal of information is available about their general husbandry and about the construction and care of ponds, but farmers still rely on capturing the juveniles to stock the ponds each year. However, in spite of being a subsistence crop and cheap compared with most other species of fish, even herbivorous fish are a luxury for some nations.

The milkfish (*Chanos chanos*) is the most cultured fish in Southeast Asia. Much expertise has been developed in stock collection and husbandry, but as yet no one has been able to breed this species in captivity. The need to collect and keep the very big parent stock (over 15 kg in weight) has been neglected, and laboratories have made little progress on breeding methods.

For the mullet (Mugilidae), like the milkfish, a great deal is known about their husbandry, but present farming methods rely on the natural availability of juveniles to stock the ponds. Progress in Hawaii and Taiwan in breeding adult mullet in captivity, together with some advances in knowledge about larval rearing, signifies a major breakthrough in rearing important herbivorous species. Mullet can be omnivores, too, accepting and growing well on an animal protein diet supplement in addition to vegetable matter.

Included with these important food fishes are the rabbitfish (Siganidae), which eat filamentous algae and contribute to the protein needs of many Pacific island communities but about which very little is known, and certain carp (*Tilapia* spp.).

Herbivorous shellfish include both mollusks and crustacea. The juvenile stages of many crustaceans require a herbivorous diet for some period of their development (e.g., the Malaysian prawn, *Macrobrachium rosenbergii*). The brine shrimp (*Artemia salina*), on the other hand, requires phytoplankton all its life. This species is important to culture of many other organisms and is thereby of value to culturists. Beyond that, rapid growth and tolerance to a range of environmental conditions make it intrinsically worthy of consideration for mass production of protein food. Sources of brine shrimp eggs are being restricted by reclamation of their salt marsh habitations and contamination with pesticides, but it is conceivable that they can be grown in association with artificial upwellings or the effluents of desalination plants.

Most commercially raised mollusks feed passively on phytoplankters or benthic algae. Seven species of oysters are suitable for farming throughout the world for both flesh (*Ostrea* and *Crassostrea* spp.) and cultured pearls (*Pinctada* spp.). All are extremely well suited to polyculture because of modern techniques for producing cultchless seed and for raft culture. Oysters have always remained in high demand in the markets of most countries. In all probability they are the most widely cultured aquatic organism in the world, and the state-of-the-art of this industry is considerably advanced. Mussels (*Mytilus* spp.), by contrast, hold the world record for per acre equivalent production (150 metric tons) and, although the hatchery technique is well established, the natural production of juveniles is so vast in areas of commercial production that artificial propagation is unnecessary. Such rich areas are the exception and not the rule, but production of mussel protein concentrate is worthy of consideration.

Abalones (*Haliotis* spp.) are temperate and subtropical species for which culture techniques have recently been established. Good market value and the fact that they are herbivores make them serious contenders for immediate culture. They could be marketed as young abalones, when they are more tender, or as adults.

Scallops, certain limpets, clams, and sea urchins have some importance to aquaculture both for their market potential and for producing larvae to feed to higher organisms. Rudimentary rearing techniques have been devised for some species.

The green turtle (*Chelonia mydas*) has been reared for some time in the Caribbean. The eggs, which are laid in sand, are incubated in a hatchery. The turtle's diet is algae and the animal may therefore be suitable for ranching on seagrass, or on artificial algal pasture that encourages additional plant growth in confined areas. Problems to date have been in general husbandry and disease control.

Carnivorous Finfish and Shellfish

Carnivorous finfish and shellfish require such costly feeding procedures that at present culture is limited to the most valuable commercial species.

An industry is well established for many of the salmonids and much is known about their husbandry, nutrition, disease control, and breeding. Recent work with the culture of migratory salmon and trout has developed a major new potential for aquaculture, and the successes are so promising that no marine fish farm in a temperate region should be without facilities for raising salmonids [15].

The pompano (*Trachinotus carolinus*) is of great interest to culturists in America. A full description of the biology of the pompano and its potential for culture is provided by Finucane [16]. This species seems well suited to fish farmers operating along the Gulf Coast; consequently, a great deal of interest has been shown in pompano by several large organizations. Much of the work with pompano has relied on catching large numbers of juveniles in the surf during the summer months. Although it is possible to net many thousands of fish, the method is not efficient and cannot guarantee sufficient numbers to support a commercial industry. It is necessary to culture juveniles in a hatchery.

Other tropical and semitropical species are worth reserach and development because of high market value, a close relationship to fish for which rudimentary breeding techniques already exist, or the local importance to a community. Groupers, snappers, dolphin, and a few other fish belong in these categories, but to date only isolated and small efforts in aquaculture have been made.

In temperate regions work is well advanced on many flatfish species, including plaice (*Pleuronectes platessa*), Dover sole (*S. solea*), and turbot (*Scophthalmus maximum*) [13, 17, 18]. Culture techniques have been developed for all of them, but more work is required on nutrition and disease control. The Japanese work with yellowtail (*Seriola quinqueradiata*) is at about this same level of advancement, except that induced spawning has not yet been achieved. Induced spawning has been achieved in the Mediterranean for the bass, *Dicentrarchus labrax,* and the bream, *Sparus auratus.* For these two, larval rearing has been achieved also.

Crustaceans are at the center of most of the present work in aquaculture because of high commercial value and consumer appeal. The major difficulty found in raising certain shrimps and prawns is that the many stages of larval development require a variety of diets, and the larvae are susceptible to disease. A complete culture technique is now available for the Malaysian prawn (*Macrobrachium rosenbergii*), a large freshwater species that is therefore suitable for culture in many areas of the world. This particular species seems capable of displacing the marine species in the lucrative commercial markets. Further work is necessary to rear the species more intensively.

A great deal of success is attributed to the Japanese in the rearing of the Japanese shrimp (*Penaeus japonicus*). Although a market is well established, there are certain weaknesses, and the farming of shrimp is possible only by virtue of their very high market value in Japan. Every effort to raise the same shrimp in other parts of the world, using the same techniques, has been economically unsuccessful [19]. Better species are *P. occidentalis, P. vannamei, P. stylirostris,* or *P. schmitti* [20].

Finally, the lobster (*Homarus vulgaris*) is still considered the El Dorado of aquaculture, and research continues, but with little progress in spite of the considerable work that has been done. Nutrition and disease seem to be the major barriers to progress—as well as the length of time it takes the lobster to reach marketable size. Culture techniques have been known for many years.

Planktonic Species

Planktonic species offer a somewhat confusing picture. It is considered that, at present, microscopic plankton plants (the phytoplankton) cannot be harvested economically enough to make a significant contribution to the food of man. Although techniques are available, mechanical inefficiency makes the operation totally impractical for naturally occurring populations. It is conceivable, though, that intensive rearing might someday produce populations of sufficient density to warrant economical harvest [21].

However, many planktonic animals (the zooplankton) are macroscopic, and their populations reach enormous proportions in certain regions of the oceans. Harvesting techniques are known and practiced by the Japanese. Among the most common zooplankters of possible commercial value to mankind are the euphausids, mysids, copepods, and galatheids.

A certain cynicism has often been aroused by the suggestion of harvesting planktonic organisms; but from a study of the hydrographic factors associated with concentrations of these animals, it appears that there is a close relationship between their location and the edges of sinking waters adjacent to nutritionally enriched upwellings. The possibility of inducing or creating artificial upwellings is increasing; therefore, the chances of harvesting zooplankton on a large scale also seem to be increasing [22].

Miscellaneous Detritus Feeders

The detritus feeders mentioned here are in addition to the groups already discussed. Bottom organic detritus is usually the main source of nourishment for most benthic forms. We have already considered the benthic flatfish, prawns, and clams. This section is mainly intended to consider the remaining burrowing forms that can constitute an important commercial fishery. The bloodworm (*Glycera dibranchiata*) of the eastern seaboard of the United States is such a species, as are *Arenicola* and *Balanoglossus*. The echinoderms also feature several detrital feeders of possible commercial value, such as the bêche-de-mer, or sea cucumber, which is prized in China.

The concentration of detritus feeders is, of course, dependent upon the extent of detritus production by plants and nonscavenging animals. This production is first acted upon by marine bacteria and other organisms, such as protozoa, nematodes, and rotifers, before consumption by detrital feeders.

Detritus eaters are mainly confined to relatively shallow seas. With increasing depths the detrital food available on the bottom becomes minimal. The sinking material produced by upper pelagic organisms usually disintegrates in the euphotic zone through bacterial action and decay or consumption by pelagic feeders before reaching the bottom. Commercially utilized detrital feeders presently are shallow-water, if not intertidal, species readily obtained by simple harvesting procedures.

Selection Matrices

Candidate species for aquaculture cannot be evaluated with any high degree of accuracy at the present time because of the many commercial, biological, and technical factors that are as yet unknown. These factors can be reviewed by the use of simple matrices, although without reference to species by name there have to be additional generalizations. For example, there are many octopus species. Some grow fast, others more slowly; some are fecund, others produce few juveniles. The reviews that follow, therefore, refer to groups of species rather than individuals, but they are based on data for those species in the group that have been cultured or studied. Thus, some selection procedures have already been imposed, as certain values have made the species desirable to culture—for example, local economic importance or ease of maintenance in captivity. The analyses are confined to marine and brackish-water species, but include the anadromous salmonids and the catadromous eels.

Bioeconomic Suitability. Table 8.2 summarizes the bioeconomic suitability based on a simple five-unit scale. The main factors outlined in the beginning of this chapter are scored along a range between "yes" (scoring 5) and "no" (scoring 1). Intermediate responses score in between at a level relative to the two extremes.

From the matrix totals of the groups, the majority of the mollusks would appear to present few problems for culture. Basically, their trophic level makes them relatively inexpensive to feed, and they can be produced in large quantities in hatcheries by techniques that are now well established. Crustaceans as a group present the most problems because of their peculiar development phases, which involve a multiplicity of different larval forms. In addition, regular moulting exposes them to cannibalism and increases susceptibility to disease. Finfish have few biological difficulties, and the two diadromous groups, the salmonids and eels, are especially suitable for culture. The additional species not falling into one of these three categories present some of the problems well known by those culturists presently working with them.

State-of-the-Art and Market Potential. Table 8.3 reviews the current state-of-the-art of culture for the same groups; again, ratings are determined by what is achieved for the best individual species within the group. For example, the data on the mullet is for *Mugil cephalus,* and not other species. The scale is the same as that of Table 8.2 and is based on present knowledge rather than theoretical potentials.

Of all the groups, an advanced technology is established only for the mollusks. This is to be expected in view of their placing in Table 8.2, but it is also because considerable effort has been expended on them, particularly the oysters, for very many years. With one or two exceptions, notably brine

TABLE 8.2 *Basic Bioeconomic Matrix*

Trophic efficiency																				
Carnivores	5	5						2	2		2	2	2		2			2		3
Herbivores and browsers			3					3	3					3		3			3	
Plankton and filter feeders	4	4		4	4	4						5	5							
Scavengers and detritus feeders					5	5													4	
Reproduction in nature																				
High fecundity	5	5	4	4	3	2	2	3	4	5	4	4	4	4	4	5	2	1	5
Eggs/spat easy to collect	5	5	2	1	1	1	1	1	1	2	1	1	3	2	1	5	4	5	1
Juveniles easy to collect	5	5	2	2	1	1	1	2	4	4	2	3	4	3	1	4	1	1	1
Reproduction in captivity																				
Adults easy to collect	5	5	5	5	5	5	5	5	4	5	5	5	5	5	5	5	5	5	5
Adults survive well	5	5	4	5	5	4	4	5	5	4	4	4	4	2	2	3	3	4	3
Spawn naturally	5	5	5	4	2	1	3	5	5	1	4	3	1	2	1	2	1	3	4	1
Spawn by inducement	5	5	5	5	2	1	1	5	3	5	3	1	1	1	3	3	1	1	1	1
Growth or biomass production																				
Natural fast growth rate	3	3	3	3	3	3	3	3	4	4	4	4	4	4	3	5	3	1	3
Growth rate can be accelerated	5	4	4	3	4	4	4	5	5	5	5	5	2	2	5	3	4	2	3

196

Survival high in captivity	5	5	3	3	3	5	3	3	3	5	5	5	5	3	3	3	5	2	2	5	5	2	2	3
Mature early	3	3	3	3	3	3	3	3	5	5	3	3	3	3	3	3	3	4	3	3	2	3	1	4
Gregarious in nature	5	5	5	5	2	2	3	2	5	5	4	3	4	4	3	3	5	5	5	5	4	1	1	4
Density high in captivity	5	5	5	5	4	5	3	3	4	5	4	4	4	4	4	4	5	5	4	5	4	3	3	4
Readily available protein	3	3	3	3	3	3	3	3	3	3	4	4	4	4	4	4	4	4	4	4	4	3	2	4
Life cycle																								
Few development stages	4	5	4	4	4	2	4	2	4	4	5	4	5	5	5	5	5	5	5	5	5	4	5	4
Few predators	4	4	4	4	4	3	3	4	1	1	4	4	3	3	4	3	4	2	5	4	3	4	4	3
Noncannibalistic in captivity	5	5	5	5	5	1	2	1	5	5	5	5	5	5	5	5	5	5	5	5	5	3	5	5
Matrix total	82	83	70	67	63	53	50	49	59	74	72	66	69	68	61	60	68	62	63	67	70	51	50	58
	Oysters	Mussels	Clams	Scallops	Abalone	Crabs	Shrimps	Lobster	Krill	Brine shrimp	Salmon	Flatfish	Mullet	Rabbitfish	Dolphinfish	Pompano	Yellowtail	Anchovy	Herrings	Eels	Milkfish	Octopus	Turtles	Bloodworm

Key: yes, 5; no, 1.

TABLE 8.3 *Aquaculture State-of-the-Art and Market Potential*

	Oysters	Mussels	Clams	Scallops	Abalone	Crabs	Shrimps	Lobster	Krill	Brine shrimp	Salmon	Flatfish	Mullet	Rabbitfish	Dolphinfish	Pompano	Yellowtail	Anchovy	Herrings	Eels	Milkfish	Octopus	Turtles	Bloodworm
Controlled spawning possible	5	5	5	5	4	4	2	1	3	1	5	5	5	4	4	1	1	1	1	1	3	1	2	1
Simple larval development achieved	5	5	5	5	4	5	1	4	5	5	5	5	5	1	1	1	3	5	1	3	5	1	4	1
Mass-produced in hatchery	5	5	5	3	3	1	4	4	3	1	5	5	5	1	1	1	1	1	1	2	2	1	1	1
Fast growth rate potential	5	4	4	4	4	3	4	4	4	5	5	4	4	4	5	5	4	4	4	5	5	4	3	3
Satisfactory feeds known	4	3	3	3	3	1	3	3	3	5	3	3	3	5	3	2	5	3	3	5	5	3	3	2
Commercial feeds available	1	1	1	1	1	1	1	2	1	1	5	1	1	1	2	1	1	1	1	2	1	1	1	1
High conversion efficiency	2	2	2	2	2	3	3	3	3	4	4	4	3	2	4	4	3	3	3	4	3	3	2	1
Hardy in captivity	5	5	5	3	3	3	3	3	3	4	5	5	5	4	3	3	4	3	3	5	5	3	3	3
High disease resistance	4	4	4	4	4	5	3	4	5	5	4	4	4	3	3	3	3	3	3	4	4	2	2	4
High density potential	5	5	5	5	4	3	3	3	5	5	4	4	4	3	4	3	4	4	5	5	4	3	3	5
Farming systems developed	5	5	5	3	3	1	4	2	1	1	4	4	4	1	1	3	1	1	1	1	4	1	1	4
High price range	5	2	4	4	4	4	5	4	1	5	4	4	3	4	4	4	5	2	2	5	1	4	4	5
High market potential U.S.	5	1	5	5	5	1	5	5	5	1	5	3	3	1	5	5	4	5	3	2	2	3	5	3
High market potential foreign	5	5	5	5	5	3	5	5	3	5	5	4	4	4	5	4	5	4	4	5	5	5	5	3
Matrix total	62	54	55	52	51	36	49	50	32	65	67	53	42	27	40	40	55	40	37	56	48	38	47	37

Key: yes, 5; no, 1.

shrimp and salmonids, both crustaceans and finfishes are studied less, although some attention has been concentrated on them within the last decade.

Applicable Farming Systems. Table 8.4 summarizes the applicable farming systems for each group based on their main nutritional requirements. In addition, the table indicates the better method of culture in terms of high- or low-density practices, and also makes reference to the groups' usefulness in polyculture.

In the context of open ocean mariculture, those species immediately associated with basic productivity will have a high priority, followed in order by those needing natural and supplemental feeding, respectively. Provisional feeding, that is, supplying all the nutritional requirements for the stock, must be of low priority. This affects the predatory finfishes, crustaceans, and other organisms high on the trophic scale.

CONCLUSIONS

Although many factors must contribute to the successful practice of mariculture, significant evaluation of a species is summarized by the phrase *production per unit volume per unit cost.* Those species closest to the basic elements for life, the natural resources for productivity, are therefore, theoretically, the most suitable. Those species that are well up the food chain, the predators, are theoretically the least suitable.

If species were ranked by their mode of nutrition with respect to open ocean culture, they would be placed in the following order:

1. Primary producers
   Single-cell phytoplankton
   Multicell phytoplankton
   Benthic algae
2. Secondary producers
   Detritus feeders
   Herbivores
   Primary carnivores with herbivorous larvae
   Primary carnivores with carnivorous larvae
   Secondary carnivores

Only primary producers (marine plants) are able to convert light and chemical energy into the energy-containing plant materials that form the base of nearly all food chains. The major portion of primary production occurs in the open oceans and along the coastal fringes, and it involves single-cell organisms, not the higher plants. However, although gross primary-production estimates serve as an index to the total production of living marine resources, production at different trophic levels in the food chain is not directly proportional to primary production. This is due to many factors, including variability in the density of the phytoplankton standing crop and variability

TABLE 8.4 Suitability for Applicable Feeding Systems (A–D) and Rearing Methods Anticipated (E–G)

	Oysters	Mussels	Clams	Scallops	Abalone	Crabs	Shrimps	Lobster	Krill	Brine shrimp	Salmon	Flatfish	Mullet	Rabbitfish	Dolphinfish	Pompano	Yellowtail	Anchovy	Herrings	Eels	Milkfish	Octopus	Turtles	Bloodworm
A. Natural	3	3	2	2	2	1	1	1	1	1	1	1	2	2	1	1	1	1	1	1	2	1	1	2
B. Provision	1	1	2	1	1	2	2	2	1	1	3	2	1	1	3	3	3	1	2	1	3	3	2	1
C. Supplemental	2	2	2	2	2	1	1	1	1	3	1	1	2	2	1	1	1	1	1	1	2	1	2	2
D. Increased basic production	3	3	2	2	1	1	1	2	2	2	1	1	1	1	1	1	1	1	1	1	1	1	1	1
E. High density	3	3	2	2	2	2	2	1	3	3	3	3	1	1	2	3	3	3	3	3	3	2	2	3
F. Low density	1	1	1	1	1	1	1	1	1	1	1	1	2	1	1	1	1	1	1	1	1	1	1	1
G. Polyculture	3	3	2	2	3	1	2	1	1	2	1	2	3	3	1	1	2	2	1	1	3	1	3	3

Key: yes, 5; no, 1.

in the number of successive groups of animals in the chain. In artificial systems variability could be controlled, to some extent, and the proportionalities somewhat better regulated.

Primary production in one of the most fertile marine areas, the Peru Current region, is about 5,000 mg of carbon/m²/day [23]. The energy required to harvest this production directly is so great that harvest is totally uneconomical except by concentrating it through organisms at higher trophic levels. Several steps in the food chain are often required to arrive at desirable secondary producers. It is estimated that for each change in trophic level the efficiency of conversion is between 10 and 15 percent, depending on the species involved. However, the majority of the food resources of the oceans, unlike those of the land, are cycled in microscopic steps through all but the final portions of the food chain, and it is only the final-stage organisms that are directly edible. The codfish, for example, which eats predatory crustaceans, gastropods, and small fish, requires a primary production base of 1,000 times its weight in plant matter and a vast expanse of bottom acreage [24]. By contrast, the flounder, which consumes detritus and phytoplankton-eating invertebrates, has a total direct and indirect requirement of vegetable matter of about 40 times it weight and a considerably lesser acreage in more accessible coastal regions.

Large-scale controlled production can be achieved only in restricted locations where the natural behavior of organisms can be suppressed. Then fertilization of the water with essential nutrients can be accomplished by artificially induced upwelling or by direct introduction, and both nutrients and species can be retained [25].

It is important to note that on land the physical size of the majority of plant life is large compared with that of the fauna, and direct grazing of plants by animals is a common and efficient conversion process. In the marine environment the position is reversed, since the macroscopic flora are relatively indigestible. Here, through microbial activity, organic detritus provides the link between these levels of primary and secondary production, in contrast to the grazing food chain that exists on land. In most freshwater environments and in estuaries, the fauna and macroscopic flora are in closer contact. As on land, the flora is more directly digestible and comparable grazing occurs, thus providing food at all times for the indigenous populations.

Certain steps in the marine food chain can be assisted, thereby increasing the ultimate yield. For example, filter feeders like mussels, clams, or oysters can convert phytoplankton into animal protein in one step. Within limits, the efficiency of their feeding is directly proportional to phytoplankton concentration. Concentration of plankton by natural advection is important in many places in the sea—at boundaries between current systems and around islands and shoals—and this can be a highly efficient mechanism. In some cases it can increase the local concentration of plankton by a factor of 50 or more over the concentrations of the contributing water.

The coral atolls of the Pacific, many of which enclose shallow lagoons hundreds of square miles in area, are natural sites for fish farming, although remote and highly saline. In some of these atolls, phosphate rock—deposits of potential on-site fertilizer—can be introduced directly into the lagoon. It is possible that small power plants can fix the added nitrogen needed to maintain a sustained level of fertility in enclosed waters and increase protein production [26].

From the general types of organisms reviewed in this chapter, then, the following would take precedence:

- Mussels
- Oysters
- Clams
- Scallops
- Krill

Of these five groups, clams and scallops have the lowest priority, as they are bottom dwellers that rely on additional detritus and particulate organic matter to supplement their diet.

Mussels and oysters, although normally fixed firmly to a substratum, have responded extremely well to the artificial practice of suspended culture. As the practice is relatively labor intensive, mussels would be given priority because of an intrinsic ability to attach themselves to ropes, whereas oysters have to be implanted manually.

Free-swimming krill, here typifying a variety of macroscopic plankters such as the euphausids and copepods, have all the immediate advantages of a low trophic level in the food chain and a tremendous potential for bulk or biomass production. The greatest disadvantages are their attraction for predators, since they are the staple food of a host of marine species, and an extremely low market potential.

The final selection of species will be based on additional regional and specific location factors. For example, specific characteristics will include the microoceanographic parameters of salinity, tidal range and frequency, temperature, current movements and strengths, sea states and other environmental facets, as well as any natural containment characteristics, including depth. Regional factors are the more pragmatic ones of labor resources, market proximity, government receptivity, financial aid, and other socioeconomic aspects.

If we assume that phytoplankton itself will remain totally impractical to extract directly, we must conclude that the edible species that will profit most from high concentrations of primary producers are the filter-feeding bivalves, of which the mussel has the greatest potential. Under good conditions the growth rate of the mussel is fast, production can be high (over 45 metric tons of flesh weight/acre equivalent/year), the organisms have a high fecundity, and the technology for culture is available [27]. In addition, the protein content is high (between 9 and 10 percent), and the flesh can be prepared in a variety of ways. Mussels can be eaten raw, cooked, or processed into mussel protein concentrate (MPC), an enriched flour that can be incorporated with cereal produce to make a highly nutritious food. Depuration of mussels is a simple procedure, and the only health hazard that remains after cleaning is the concentration of stable toxic elements during the preparation of MPC.

RECOMMENDATIONS FOR PRIORITY RESEARCH

The highest-priority areas for technological development at present are genetics and reproduction, aquacultural engineering, nutrition, and disease control. In this short commentary, we direct recommendations specifically to these four areas.

1. *Genetics and reproduction.* The genetic improvement of wild stock is imperative and is the basis of all future production of a truly domesticated aquatic livestock resource. Within the last quarter-century, agriculture has been revolutionized by the results obtained from intensive work in animal and plant genetics and selective breeding. This experience has important and obvious implications for the field of aquaculture.

2. *Aquacultural engineering.* To date, the most serious omission from aquaculture development has been engineering support. The field of agricultural engineering is exceedingly broad and active; aquaculture by contrast has been neglected. Engineering involvement is sparse in basic-component design, hatchery technology, and all the anticipated farming facilities, whether ponds, tank systems, raceways, or cages. Until aquacultural engineering technology is improved, the progress of animal husbandry in this field is severely inhibited.

3. *Nutrition.* Interest in nutrition has increased within the last 5 years. This is partly because poor nutrition has been one of the main hindrances to progress in culturing aquatic organisms, and partly because it is the area of commercial production offering the greatest future cost benefits, provided that commercial aquaculture does succeed. Although highly efficient nutrition for specific organisms is important, the greatest emphasis should be placed initially on the production and testing of broadly adequate diets that will do the job for a large number of species.

4. *Disease control.* The control of disease has been an anticipated area of need for some time, but in most intensive culture systems operated to date outbreaks of serious disease have been few. The pathogenic organisms have been recognized and treatment has been effective and cheap. This does not

mean that disease control can be demoted from the list of priority areas for concentrated research, but it currently appears that outbreaks of disease will be much more manageable than at first anticipated. However, the stock being reared at aquaculture sites at present are so close to wild stock that they may have an inherent immunity, and probably will continue to have it for several generations. Greater problems may come from the same pathogens attacking the highly selected and more vulnerable individuals in generations to come.

Aquaculture is almost as old as civilization itself, but until this century it has been operated only at an intensive subsistence farming level. If desired, it can be brought from this efficient but primitive management practice to a productive farming technology by applying modern agricultural principles. The time span within which this can be achieved is directly proportional to the level of effort and funding directed toward this objective. If the goal for making aquaculture an advanced technology is set for 2000 A.D., this field will achieve in 50 years what it took agriculture 100 years to accomplish, and it will do it in the much more complex environment of the earth's coastal waters and open seas.

REFERENCES

1. Hickling, C. F. 1962. *Fish culture.* Faber & Faber Ltd., London. 295 pp.
2. Carlisle, J. C., C. H. Turner, and E. E. Ebert. 1964. Artificial habitat in the marine environment. *Calif. Dept. Fish and Game Fishery Bull. 124.*
3. Stroud, R. H. 1966. Artificial reefs as tools for sport fishery management in coastal marine waters. *Proc. 10th Internatl. Game Fish Conf.* (Miami, 1965). Inter-Oceanographic Foundation.
4. Ogawa, Y. 1967. Experiments on the attractiveness of artificial reefs for marine fishes. *Bull. Jap. Soc. Sci. Fisheries 33*(9):801.
5. Harada, T. 1970. The present status of marine fish cultivation research in Japan. *Helgolander wiss. Meeresunters. 20*:594–601.
6. Furukawa, A. 1970. Present status of major Japanese marine aquaculture. *14th Indo-Pacific Fisheries Council C70/Sym47.*
7. Shehadeh, Z. H. 1973. Controlled breeding of culturable species of fish—a review of progress and current problems. In T. V. R. Pillay, ed., *Coastal aquaculture in the Indo-Pacific Region.* Fishing News (Books) Ltd., Surrey, England, pp. 180–194.
8. Shelbourne, J. E. 1964. Artificial propagation of marine fish. In J. T. Russell, ed., *Adv. Marine Biol.,* Vol. 2. Academic Press, Inc., New York, pp. 1–83.
9. Ansell, A. D. 1963. The biology of *Venus mercenaria* in British waters, and in relation to generating station effluents. *Ann. Rept. Challenger Soc. 3*(15):38.
10. Gaucher, T. A. 1968. Thermal enrichment and marine aquiculture. Paper presented at Conf. on Marine Aquiculture, Oregon State University, Newport, Ore.

11. Gribanov, L. V., A. N. Korneev, and L. A. Korneeva. 1966. Use of thermal waters for commercial production of carps in floats in the U.S.S.R. World Symp. on Warm-Water Pond Fish Culture. *FAO 44.1.*
12. Nash, C. E. 1970. Marine fish farming, I and II. *Marine Pollution Bull. (U.K.) 1*(1):5–6 and *1*(2):28–30.
13. Ryther, J. H., and Bardach, J. E. 1968. The status and potential of aquaculture, Vols. I and II. *NTIS PB 177–767* and *PB 177–768.* Amer. Inst. Biol. Sci., Washington, D.C.
14. Pillay, T. V. R. 1967. Proceedings of the world symposium on warm-water pond fish culture (1966). *FAO Fisheries Rept. 44*(1–5).
15. Mahnken, C. V. W., A. J. Novotny, and T. Joyner. 1970. Salmon mariculture potential assessed. *Amer. Fish Farmer 2*(1):12–15.
16. Finucane, J. H. 1970. Progress in pompano mariculture in the United States. *Proc. First Ann. Workshop, World Mariculture Soc.*
17. White Fish Authority. 1970. *Ann. Rept.* H. M. Stationary Office, London.
18. Shelbourne, J. E. 1963. Marine fish culture in Britain. *J. Cons. Perm. Int. Explor. Mer. 27*(1):70.
19. Fujinaga, J. 1963. Culture of Kuruma shrimp, *Penaeus japonicus. Curr. Affairs Bull., Indo-Pacific Fish. Counc. 36:*10–11.
20. Mistakidis, M. N., ed. 1968. Proceedings of the world scientific conference on the biology and culture of shrimps and prawns (1967). *FAO Fisheries Rept. 57*(1–4).
21. Ansell, A. D., J. Coughlan, K. F. Lander, and F. A. Loosemore. 1964. Studies on the mass culture of *Phaeodactylum.* IV. Production and nutrient utilization in outdoor mass culture. *Limnol. Oceanogr. 9*:334–342.
22. NOAA. 1971. *Publication of U.S. Dept. of Commerce 1*(1):12–13, 25–29.
23. Gilmartin, M. 1969. Sea farming. In F. E. Firth, eds., *Encyclopedia of marine resources.* Van Nostrand Reinhold Company, New York.
24. McLeod, G. C. 1969. Sea farming. In F. E. Firth, ed., *Encyclopedia of marine resources.* Van Nostrand Reinhold Company, New York.
25. Foyn, E. 1965. Disposal of waste in the marine environment and the pollution of the sea. *Oceanogr. Marine Biol. Ann. Rev. 3*:95.
26. Naylor, E. 1965. The effects of heated effluents upon marine and estuarine organisms. In F. S. Russell, ed., *Advanced marine biology,* Vol. 3, Academic Press, Inc., New York.
27. Joyner, T., and J. Spinelli. 1969. Mussels: a potential source of high quality protein. *Com. Fisheries Rev.* (Aug.–Sept.): 31.

Additional Readings

Allen, D. M. 1963. Shrimp farming. *U.S. Fish Wildlife Serv. Fishery Leafleat 551.* 8 pp.
Allen, K. O., and J. W. Avault, Jr. 1970. Effects of salinity and water quality on survival and growth of juvenile pompano, *Trachinotus carolinus. Coastal Studies Bull. 5,* Louisiana State University, pp. 147–155.
Aquatic Products and Research Inc. 1972. Demonstration project to develop

techniques for rearing freshwater shrimp in a controlled environment. U.S. Dept. Commerce Econ. Develop. Admin. 33 pp.

Baughman, J. L. 1948. *An annotated bibliography of oysters with pertinent material on mussels and other shellfish and an appendix on pollution.* Texas A&M Research Foundation, College Station, Texas. 794 pp.

Bellinger, J. W., and J. W. Avault, Jr. 1970. Seasonal occurrence growth, and the length-weight relationship of juvenile pompano, *Trachinotus carolinus. Trans. Amer. Fisheries Soc. 99*(2):353–358.

Bellinger, J. W., and J. W. Avault, Jr. (in review) Food habits of juvenile pompano, *Trachinotus carolinus,* in Louisiana. *Trans. Amer. Fisheries Soc.*

Bente, P. F., Jr. 1971. Mariculture on the move. Address to Proc. First Ann. Workshop, World Mariculture Society (Louisiana), pp. 18–26.

Ben-Tunia, A., and W. Dickson, eds. 1969. Proceedings of FAO Conference on fish behavior in relation to fishing techniques and tactics (1967). *FAO Fisheries Rept. 62*(1–3).

Berry, F. H., and E. S. Iversen. 1967. Pompano: biology, fisheries and farming potential. *Proc. 19th Ann. Gulf Carib. Fishery Inst.,* pp. 116–128.

Birdsong, C. L., and J. W. Avault, Jr. (in press) Toxicity of certain chemicals to juvenile pompano, *Trachinotus carolinus. Prog. Fish-Cult.*

Bretonne, L. W. de la, Jr., and J. W. Avault, Jr. 1970. Shrimp mariculture methods tested. *Amer. Fish Farmer 1(12).* 12 pp.

Broom, J. G. 1966. Shrimp culture in ponds. *La. Conserv. 18*(11–12):7.

Broom, J. G. 1969. Pond culture of shrimp on Grand Terre Island, Louisiana, 1962-68. *Proc. 21st Ann. Gulf Carib. Fishery Inst.,* pp. 137–151.

Broom, J. G. 1971. Shrimp culture. *Proc. First Ann. Workshop, World Mariculture Society* (Baton Rouge, La.), pp. 63–68.

Choudhury, P. C. 1970. Complete larval development of the palaemonid shrimp, *Macrobrachium acanthurus* (Wiegmann, 1836), reared in the laboratory. *Crustaceana 18*:113–132.

Choudhury, P. C. 1971. Complete larval development of the palaemonid shrimp, *Macrobrachium carcinus* (L.), reared in the laboratory (Decapoda, Palaemonidae). *Crustaceana 20*:51–69.

Cook, H. L., and M. A. Murphy. 1966. Rearing penaeid shrimp from eggs to postlarvae. *Proc. S.E. Assoc. Game Fish Comm. 19th Ann. Conf.,* pp. 283–288.

Cook, H. L., and M. A. Murphy. 1969. The culture of larval penaeid shrimp. *Trans. Amer. Fisheries Soc. 98*(4):751–754.

Cook, H. L. 1969. A method of rearing penaeid shrimp larvae for experimental studies. *FAO Fisheries Rept. 57*(3):709–715.

Cordover, R. 1972. Problems and progress in penaeid prawn production. (Shrimp cultivation in Hawaii.) *Amer. Fish Farmer 3*(11):4–7.

Costello, T. J. 1971. Freshwater prawn culture technique developed. *Amer. Fish Farmer 2*(2):8–10, 27.

Delmondo, M. N., and H. R. Rabanal. 1956. Cultivation of 'sugpo' (jumbo tiger shrimp) *Penaeus monodon* Fabricius, in the Philippines. *Proc. 6th Indo-Pacific Fisheries Council,* pp. 424–431.

FAO. 1970. *Yearbook of Fishery Statistics, Catches and Landings,* Vol. 30.

Finucane, J. H. 1969. Summary of pompano mariculture in Tampa Bay, Fla. Paper presented at Ann. Meeting Mariculture Subcommittee, Estuarine Technical Coordinating Committee, Gulf States Marine Fisheries Comm. 8 pp.

Finucane, J. H. 1970. Pompano mariculture in Florida. *Amer. Fish Farmer* 1(4):5–10.

Firth, F. E., ed. 1969. *The encyclopedia of marine resources.* Van Nostrand Reinhold Company, New York.

Forster, J. R. M., and J. F. Wickins. 1972. Prawn culture in the United Kingdom, its status and potential. Fisheries Experiment Station, Conway, Caernorvonshire. *Min. Agr., Fisheries and Food Lab., Leaflet (new series) 27.* 32 pp.

Fujimura, T. 1966. Notes on the development of a practical mass culture of the giant prawn, *Macrobrachium rosenbergii. 12th Indo-Pacific Fisheries Council C66/WP(47).* 3 pp.

Fujimura, T. 1966–1970. Quarter reports of the project "Development of a prawn industry" (culture on the giant prawn, *Macrobrachium rosenbergii*). Fed. Aid Program Work at Keehi Fish Station, Honolulu. *Hawaii Sub-project H-1-D-2.*

Fujimura, T., and H. Okamoto. 1970. Notes on progress made in developing a mass culturing technique for *Macrobrachium rosenbergii* in Hawaii. *14th Indo-Pacific Fisheries Council (Bangkok)* C70/SYM 53. 17 pp.

Fujinaga, M. (Hudinaga). 1969. Kuruma shrimp (*Penaeus japonicus*) cultivation in Japan. *FAO Fisheries Rept. 57* (3):811–832.

Galtsoff, P. S. 1964. The American oyster, *Crassostrea virginica. U.S. Fish Wildlife Serv. Fishery Bull.,* Vol. 64. 480 pp.

Glude, J. B. 1949. Japanese methods of oyster culture. *U.S. Fish Wildlife Serv., Commer. Fishery Rev.* 11(8):1–7.

Hudinaga, M. (Fujinaga). 1942. Reproduction, development and rearing of *Penaeus japonicus* (Bate). *Jap. J. Zool.* 10(2):305–393.

Hudinaga, M. (Fujinaga). 1967. The large scale production of the young Kuruma prawn, *Penaeus japonicus* Bate. *Inform. Bull. Planktonol. Jap.* (Commemoration number of Dr. Y. Matsue's 60th birthday): 35–46.

Hudinaga, M., and J. Kittaka. 1966. Studies on food and growth of larval stages of a prawn, *Penaeus japonicus*, with reference to the application of practical mass culture. *Inform. Bull. Planktonol. Jap. 13:* 83–94.

Idyll, C. P. 1965. Shrimp nursery—science explores new ways to farm the sea. *Natl. Geogr. 127:* 636–659.

Ingle, R. M., and B. Eldred. 1960. Notes on the artificial cultivation of freshwater shrimp. *The West Indies Fishery Bull.* 4(July/Aug.). 5 pp.

Ingle, R. M., and F. G. Walton Smith. 1953. Oyster culture in Florida. *Fla. Board Conserv. Ed. Ser. 5* (Revised). 24 pp.

Iversen, E. S. 1968. *Farming the edge of the sea.* Fishing News (Books) Ltd., London. 300 pp.

Iversen, E. S., and F. H. Berry. 1969. Fish mariculture, progress and potential. *Proc. 21st Ann. Gulf Carib. Fishery Inst.,* pp. 163–176.

Joyce, E. A., Jr. 1972. A partial bibliography of oysters, with annotations. *Florida Dept. Nat. Resources, Marine Res. Lab.* (St. Petersburg, Fla.) *Spec. Sci. Rept. 34.* 845 pp.

Kesteven, G. L., and T. J. Job. 1958. Shrimp culture in Asia and the Far East: a preliminary review. *Proc. 10th Ann. Gulf Carib. Fishery Inst.,* pp. 49–68.

Kinne, O., and H. P. Bulnheim. 1970. Cultivation of marine organisms and its importance for marine biology. International Helgoland Symp. 1969. *Helgolander wiss. Meeresunters. 20:*1–4.

Krenkel, P. A., and F. L. Parker. 1969. *Thermal pollution.* Vol. I, *Biological aspects,* and Vol. II, *Engineering aspects.* Vanderbilt University Press, Nashville, Tenn.

Lewis, J. B. 1962. Preliminary experiments on the rearing of the freshwater shrimp, *Macrobrachium carcinus* (L.) *Proc. 14th Ann. Gulf Carib. Fishery Inst.,* pp. 199–201.

Lewis, J. B. 1966. The breeding cycle, growth and food of the freshwater shrimp, *Macrobrachium carcinus* (Linnaeus). *Crustaceana 10:*48–52.

Lindner, M. J., and H. L. Cook. 1968. Progress in shrimp mariculture in the United States. Symp. on Investigations and Resources of the Caribbean Sea and Adjacent Regions, Nov. 18–26, Willemstad, Curacao, Netherlands Antilles.

Ling, S. W. 1962. Studies on the rearing of larvae and juveniles and culturing of adults of *Macrobrachium rosenbergii. Indo-Pacific Fisheries Council Current Affairs Bull. 35,* 11 pp.

Ling, S. W. 1969. Methods of rearing and culturing *Macrobrachium rosenbergii* de Man. *FAO Fisheries Rept. 57*(3):607–619.

Loosanoff, V. L. 1965. The American or eastern oyster. *U.S. Fish. Wildlife Serv. Fishery Circ. 205.* 36 pp.

Loosanoff, V. L., and H. C. Davis. 1963. Rearing of bivalve mollusks. In F. S. Russell, ed., *Advances in marine biology,* Vol. 1. Academic Press, Inc., New York, pp. 1–136.

Loosanoff, V. L., and H. C. Davis. 1963. Shellfish hatcheries and their future. *U.S. Fish Wildlife Serv. Com. Fisheries Rev. 25*(1):1–11.

Louisiana Wild Life and Fisheries Comm. 1969. Shrimp! *La. Conserv. 21*(7–8):20–23.

MacKenzie, C. L., Jr. 1970. Oyster culture in Long Island Sound 1966–1969. *Com. Fisheries Rev. 32*(1):27–40.

Maghan, B. W. 1967. The Mississippi oyster industry. *U.S. Fish Wildlife Serv. Fishery Leafleat 607.* 12 pp.

May, E. B. 1969. Feasibility of off bottom oyster culture in Alabama. *Ala. Marine Resources Bull. 3,* 14 pp.

Mihursky, J. A. 1967. On possible constructive uses of thermal additions to estuaries. *BioScience 17:*698–702.

Miller, G. C. 1971. Commercial fishery and biology of the fresh-water shrimp, *Macrobrachium,* in the lower St. Paul River, Liberia, 1952–1953. *Nat. Marine Fishery Serv. Spec. Sci. Rept. 626.* 13 pp.

Mock, C. R., and M. A. Murphy. 1971. Techniques for raising penaeid shrimp from the egg to postlarvae. *Proc. First Ann. Workshop, World Mariculture Soc.* (Baton Rouge, La.), pp. 143–156.

Moe, M. A., Jr. R. H. Lewis, and R. M. Ingle. 1968. Pompano mariculture: preliminary data and basic considerations. *Florida State Board Conserv. Marine Lab. Tech. Ser. 55.* 65 pp.

More, W. R., and L. L. Elam. 1970. Salt water pond research, 1970. Job. 1. Pond studies on penaeid shrimp. Coastal Fisheries, Texas Parks and Wildlife Dept., Austin, Tex., pp. 1–15.

Muskie, E. S., Chairman. 1968. Thermal pollution—1968, Part 1. Hearings before Senate subcom. on air and water pollution. 90th congress (1968). Govt. Printing Office, Washington, D.C.

Odum, E. P. 1971. *Fundamentals of ecology.* W. B. Saunders Company, Philadelphia.

Odum, H. T. 1966. Energy transfer and the marine systems of Texas. Paper presented at Conf. on Pollution and Marine Ecology, Galveston, Tex.

Pinchot, G. B. 1970. Marine farming. *Scientific American 223*(6):15–21.

President's First Rept. to Congress on Marine Resources and Engineering Development. 1967. *Marine science affairs—a year of transition.* Govt. Printing Office, Washington, D.C.

Provenzano, A. J. (In press) Some results of a pilot project on freshwater shrimp culture in Jamaica. *Proc. 4th Ann. Workshop, World Mariculture Soc.* (Monterey, Mexico).

Raney, E. C., and B. W. Menzel. 1969. Heated effluents and effects on aquatic life with emphasis on fishes: A bibliography. Cornell Univ. Water Resources and Marine Sci. Center, Ithaca, N.Y. 470 pp.

Ryther, J. H. 1969. Photosynthesis and fish production in the sea. *Science 166*:72–76.

Shang, C. Y. 1972. Economic feasibility of fresh water prawn farming in Hawaii. Econ. Res. Center, Univ. of Hawaii. 49 pp.

Sick, L. V., and J. W. Andrews. 1971. Shrimp mariculture. *Amer. Fish Farmer 2*(11):14–16.

Smith, K. A. 1961. Air-bubble and electrical field barriers as aids to fishing. *Proc. 13th Ann. Gulf Carib. Fishery Inst.*

Subrahmanyam, C. B., and C. H. Oppenheimer. 1971. The influence of feed levels on the growth of grooved penaeid shrimp in mariculture. *Proc. First Ann. Workshop, World Mariculture Soc.* (La.), pp. 91–100.

Tabb, D. C., W. T. Yang, Y. Hirono, and J. Heinen. 1972. A manual for culture of pink shrimp, *Penaeus duorarum*, from eggs to post larvae suitable for stocking. *Univ. Miami Sea Grant Program, NOAA 2-35147.*

Tabb, D. C., W. T. Yang, C. P. Idyll, and E. S. Iversen. 1969. Research in marine aquaculture at the Institute of Marine Sciences, University of Miami. *Trans. Amer. Fisheries Soc. 98*(4):738–742.

Tarzwell, C. M. 1965. Biological problems in water pollution. *PHS Publ. 999-WP-25.* 424 pp.

Turner, C. H., E. E. Ebert, and R. R. Given. 1969. Man-made reef ecology. *Calif. Dept. Fish and Game Fishery Bull. 146.*

University of Miami Sea Grant. 1970. Aquaculture: the new shrimp crop. *Sea Grant Inform. Leafleat 1,* 5 pp.

Vibert, R. 1967. *Fishing with electricity.* Fishing News (Books) Ltd., London.

Wheeler, R. S. 1967. Experimental rearing of postlarval brown shrimp to marketable size in ponds. *Com. Fisheries Rev. 29*(3):49–52.

Wheeler, R. S. 1968. Culture of penaeid shrimp in brackish-water ponds, 1966–67. Paper presented at Southern Div. Amer. Fisheries Soc. Meeting, Baltimore, Md. Oct. 21–23.

White House. 1965. Restoring the quality of our environment. Report of the Environmental Pollution Panel of the President's Science Advisory Committee XII. 317 pp.

9

CONTROLLING THE
BIOLOGICAL ENVIRONMENT

J. A. Hanson and J. M. Collier

In the context of this study the term *biological environment* is used to refer to any organisms coexisting with and affecting a cultured crop in any given area. Our discussion deals primarily with the direct threats to productivity posed by this biological environment—excluding ectocrines produced by plankton (dealt with separately in Chapter 5) and fouling (treated in Chapter 10). The detrimental effects of cohabiting organisms seem to fall into three basic categories: (1) predation, (2) competition, and (3) disease and parasitism. Since our concern is with potential problems and methods for controlling them, we do not examine symbiotic and synergistic relationships in the marine environment [1].

With some notable exceptions that are cited in this chapter, knowledge of organismic interaction in the natural marine environment, to say nothing of large-scale mariculture, is woefully incomplete. Nonetheless, problems have already manifested themselves in marine aquaria, fisheries, live-holding facilities, and in early mariculture ventures, and these provide some insight into the potentially more serious problems that may be expected with large-scale mariculture. Because there is a considerable body of experience with marine aquaria, single-species live-holding facilities and, to a lesser extent, with marine monoculture, and because there is little experience with truly large-scale culturing in the marine environment, knowledge of disease and parasitism is relatively more abundant than knowledge of predation and competition as they affect mariculture. This situation is evident in this chapter. The latter subjects receive brief and general discussion, whereas the former is afforded considerably more authoritative treatment.

PREDATION

Predation is defined as the loss of cultured organisms to other creatures, which prey upon them primarily for food. In the marine food web no organism is without one or more real or potential predators. As we see later, not all predation is necessarily bad. Limited predation can serve to weed out diseased members of a crop, and so help in controlling epizootic infections.

Also, when one contemplates schemes for attracting, versus containing, desire organisms, the attraction of predators through the culturing of their prey might emerge as an important technique. For example, many organisms prey on plankton, which could conceivably be a cultured crop. In the artificial upwelling systems discussed in Chapter 7, it is possible that the stimulation of primary production alone might attract and stimulate the growth of plankton feeders such as herring and anchovy, and they, in turn, may attract carnivores. In this case, it would be the prey that was cultured directly and the predator that was harvested. However, for present purposes, these positive aspects are ignored and attention is devoted wholly to the control of unwanted predators so as to limit their negative economic effect. We divide them into two classes: (1) predators on sessile organisms, and (2) predators on mobile species.

Sessile-Organism Predators

Most sessile organisms of interest to the mariculturist have protective shells. Their principal predators are the various species of sea stars, some fish, octopi, gastropods, crabs, and some birds. Sea stars prey on bivalves by opening the shells; the oyster drill penetrates bivalve shells; others may crush the shells with strong jaws. If sessile-organism cultures were established well offshore, the probability of infestation by all these predators might be reduced, since the crops would not be in the vicinity of naturally occurring predator populations. However, a crop would be an inviting food supply for predators; even though infestation might be slow at the beginning of a project, it could quickly grow serious if a local predator breeding population developed.

Sea stars may, of course, be removed manually by divers if an infestation begins, since most sessile-organism cultures would exist well within diving depths. Dredge mops and quicklime are already being used for sea star control on shellfish grounds. There seem to be no sure controls at the moment other than removal or killing by formalin injection. The latter method has proved somewhat effective in controlling the crown of thorns sea star (*Acanthaster planci*) on tropical reefs. It probably would be pointless in an open sea mariculture operation however, since simple removal would be easier and just as effective. The Japanese oyster farms report low sea star predation associated with their offshore rafts but high predation rates on inshore rafts [2]. Thus, physical removal of a culture from the sea star's natural habitat may be adequate protection.

Removal from natural predator ranges seems likely to be an effective measure for protection against demersal fishes, octopi, and crabs as well. Where infestations did take root, they might be blocked effectively by biological controls, such as disease innoculation and natural predators. If a predator infestation were once eliminated by reducing its density below the density threshold necessary for reproduction, it could be years before a reoccurrence.

Although considerably more mobile than the other predators cited, birds seem to pose a minor problem for all sessile organisms attractive to mariculture. During the growing cycle the cultured organisms are submerged continuously and therefore largely invulnerable to avian predation.

Other possibilities for controlling free-moving predators include the techniques mentioned in the next section.

Mobile-Organism Predators

Free-swimming predators on mobile organisms come in many varieties; among their ranks are marine mammals, sharks, carnivorous fishes, octopi and squid, and some species of crustacea. Their diet range is similarly broad, encompassing the varied nektonic species as well as demersal fishes and benthic crustacea. Chiefly, the problem of control appears to be a matter of excluding the predators from the culture environment, but there are two exceptions to acknowledge before looking further at the problem in terms of exclusion. First, some prospective species for culture are natural "hiders," depending in their wild state upon physical concealment for survival. For these, which include shrimp, lobsters, crabs, and many reef fishes, the provision of artificial reefs and other means for concealment may offer some measure of predation control. It should prove effective, the mariculturist may find this technique more practicable in some cases than an effort to "fence out" predators. The other exception is in the matter of intraspecies cannibalism, which may be serious in some crustacea. Obviously, different techniques will have to be developed to cope with cannibalism. Defensible niches offer one possibility, and the isolation of size groupings within a species is another. Control through the use of aggression-suppressing *pheromones* (chemical signals) is still another approach, explored in recent work at Woods Hole Oceanographic Institution [3]. Recent work with octopi at the Hawaii Institute of Marine Biology indicates that intraculture predation may be controlled through adequate feeding and the provision of defensible niches [4].

Aside from these two considerations, the major question appears to be one of discouraging predator habitation in a culture environment. There appear to be a number of prospective mechanisms for achieving this goal, but all must be classified either as experimental or speculative at the present time.

Mesh Barriers. All free-swimming predators can be controlled, at least theoretically, by exclusion from a culture area through deployment of nets or cages with mesh too small to admit predators or to allow crop escape. This method would not, of course, exclude the very small planktonic larvae of many active predators. However effective the technique may prove to be, its actual use obviously will depend upon considerations of economy and practicability. The Japanese fish cages described in Chapter 10 employ a small-mesh interior net for crop containment inside a stronger, larger-mesh exterior net for predator exclusion. At this time there is little further knowledge or experience concerning the choice of netting materials that would be most effective against any given predators—or, conversely, most favorable to any selected cultured species. The question of practicality clearly is one of mesh strength, durability, and marine fouling resistance versus cost, which probably can be answered only on a case-by-case basis. In Chapter 10 we explore this subject further.

Other more exotic means of controlling the movements of marine organisms are under serious investigation. Among these are air, electrical, and acoustical barriers, and chemical controls.

Air Barriers. Bubble fences are constructed simply by laying perforated pipe around the area to be fenced. Air injected into the pipe under pressure escapes through the perforations to form a fence of densely packed air bubbles continuously rising to the surface. The technique has been employed as a shark deterrent in Australia with some limited success, and it has been used experimentally for finfish herding and containment. If it proved effective, other characteristics of the bubble fence make it an attractive alternative method of control. Chapter 10 contains further discussion of the technique.

Electrical Barriers. It is an accepted principle that in an electrical field finfishes swim (through involuntary skeletal muscle responses) toward the positive pole. Thus, it is theoretically possible to establish a system of electrical fields to turn incoming fish and shark predators away from a culture area. South African investigators using this approach have employed rather weak electrical fields against sharks with considerable success [5]. If the technique can work for predator repulsion, it might also find application in crop containment.

Acoustical Barriers. Acoustical approaches to predation protection may turn out to be varied. Work at the University of Miami [6] indicates that some predatory fish are attracted to sounds resembling their own feeding noises. Whether this type of attraction is general and whether it could be strong enough to lure fish away from real prey remains to be shown. Conversely, beginning work with porpoises by the Naval Undersea Center indicates that some predators may be repelled by the sounds of their natural enemies. Clearly, a good deal more research into the responses of marine organisms to low-frequency radiant energy is indicated.

Chemical Controls. Research on chemical communication in the sea by the chemotaxis group at Woods Hole indicates a high degree of importance for chemical information in the lives of many marine animals. Pheromones have been shown to affect homing, social, and feeding behavior in fishes and crustacea and probably are important for many forms. Although this research is still in its early stages, one application that now appears conceivable is the use of such compounds to lure predators away from stocks, and pests from their hosts. It is not possible to predict the extent of future use of such chemical controls in view of dilution and dispersion problems, but they do offer a promising field for consideration and further research.

Animate Barriers. One final, and somewhat appealing, possibility might lie in the employment of porpoises as watchdogs. That porpoises are highly trainable has been amply demonstrated; what is less well known is that they tend to be highly habitual in feeding behavior. Once trained to feed on dead whole or cut-up fish, they have a strong tendency to ignore live fish [7]. And so we may visualize a small school of porpoises protecting a cultured population against its natural predators, receiving dead fish as remuneration for services rendered, and leaving the cultured stock untouched.

COMPETITION

Competition is defined here as any type of rivalry between cultured organisms and uncultured organisms for any environmental resource that may be limiting. It is immaterial whether the resource in question is food, space, oxygen, or a variety of other environmental constituents. The problem occurs when competition between the cultured and noncultured organisms reduces cultured-organism productivity. At that point, it clearly becomes desirable to control competition.

If little is known about predation in mariculture, it can be affirmed that essentially nothing is known about competition problems with cultured organisms in the open sea environment. Nevertheless, if open sea mariculture operations create attractive ecological niches, it seems reasonable that organisms other than those intentionally cultured will attempt to occupy these niches; in so doing they will almost certainly compete with cultured crops for the available resources.

When competitive forms are similar to cultured forms, they can perhaps be allowed to coexist with the cultured forms if cultured productivity degradation is not excessive. During harvesting these "volunteer" organisms could be culled and diverted to a direct market, converted to fish meal, or processed into feed for the cultured organisms. Such an approach might be more efficient than attempts to effect rigid controls.

In the case of competitive organisms phylogenetically dissimiliar to the cultured forms, biological controls such as diseases and parasites might prove effective in controlling their numbers. Again, this is a subject about which next to nothing is known at present. For phylogenetically separated organisms, chemical control also may be an avenue worth pursuing. There almost certainly are some chemicals that, in some concentrations, would be harmful to the competitors but harmless to the cultured crop.

Finally, of course, there are the physical controls considered for predators, which might be employed to exclude competitors. As with predation, it appears that competition control may prove easier with sessile crops than with free-moving crops. All the foregoing adds up to very little more than speculation and the recognition of an obvious need for comprehensive empirical investigations of the whole spectrum of questions implied by predation and competition.

DISEASE AND PARASITISM

The premise that disease and parasite control must be assigned very high priority in the development of mariculture is supported by examples of the devastating effects of disease outbreaks in fish and shellfish populations under even the limited scales that aquaculture has achieved to date. Many biological factors pose hazards to the successful rearing of healthy marine-animal stocks. Among these, pathogens such as protozoans, bacteria, worms, crustacea, fungi, and viruses are largely responsible for fish mortality. Other causes of debilitation and mortality include dietary deficiencies, wounds, poisons, and environmental factors such as temperature and salinity.

The approach to fish disease necessarily must bear much similarity to the practice of human medicine. The symptoms are documented, the pathogen isolated and studied, and curative and preventive measures are sought through experimentation. For the purpose of mariculture, perhaps the science of epidemiology may prove of prime relevance. That is, where high-volume fish populations are concerned, the humanistic approach of attempting to save and rehabilitate every member of the group, irrespective of inherent individual weakness or handicap, would seem to be impractical if not downright impossible. Instead it probably would be most effective to "play the numbers," seeking to maintain an acceptable percentage of the total population in good health in order to achieve a harvest that renders a satisfactory economic return. In this epidemiological approach, principal emphasis may be placed on disease vectors, environmental factors that either predispose an organism to vulnerability or favor disease resistance, and on removal or isolation of infected individuals. By developing control in these areas it may be possible to hold down the incidence of disease and parasitism in cultures to levels at or below the rate of occurrence in the natural environment or, more to the point, to levels acceptable to a viable mariculture.

The diagnosis and treatment of fish diseases have reached the highest levels of sophistication in the culture of temperate freshwater fishes. Most existing data on marine-fish pathology have been accumulated through the study of marine aquaria, although similar diseases are showing up as problems among the limited number of marine species so far undergoing cultivation. Granted that the infant body of knowledge on marine-animal pathology is largely limited to marine fishes held in captivity or reared under hatchery conditions, and that it may therefore prove to have limited pertinence to fish reared in the somewhat dissimilar environment of open sea mariculture, we are constrained to begin with what knowledge we have.

We start by summarizing some of the known data on bacteria, parasites, fungi, and other pathogens of finfish, including both pertinent experience from studies in marine aquaria and the limited knowledge about diseases in cultured fish populations. We follow that with a review of information concerning cultured shellfish pathology. In so doing we are indebted to the classic work of Carl J. Sindermann, *Principal Diseases of Marine Fish and Shellfish* [8]. We have not made a complete compilation of all the research and findings in the literature; that would be outside the scope of this work. Rather, representative examples of disease problems have been chosen. In general, most of the detail on specific diseases and parasites and their symptoms and cures is omitted here; the focus is placed on their potential importance to a cultured population. From that vantage it is possible to speculate on implications for open sea mariculture. If technical information is desired on individual species and pathogens, this may be found in the supplemental references at the end of the chapter or cited in the discussion, as well as in other technical literature in this field.

Cultured Finfish

Bacterial Diseases. Aquaria studies have been the principal source of data on bacterial infections of marine fish for several reasons:

1. External signs of disease often are visible only when fish are under water. Captive fish are most subject to close scrutiny and continuous surveillance, and most of the bacterial diseases, such as dermatitis and tuberculosis, have in fact been observed in captive populations [9–12].

2. Infectious diseases are easily introduced with fish captured in their natural habitat for addition to aquarium stock. Transmission from fish to fish is expedited in the restricted body of water; in short, the animal and environmental vectors operate more effectively. Moreover, resident stock may not have built up resistance through previous exposure to a newly introduced disease.

3. Environmental factors such as abnormally high temperatures, unusual diet, crowded space, and the like, may simultaneously prove conducive to the rapid growth and proliferation of bacteria, and/or serve to reduce the natural resistance of the hosts. In any event, an unnatural environment per se is likely to place stress on an organism.

4. In most aquaria the absence of predators eliminates one effective mechanism for the removal of infected individuals, thereby prolonging their survival time and operation as disease vectors.

To the extent that a mariculture preserve duplicates the characteristics of aquarium environments, it may create a disease-encouraging environment, just as aquaria do.

Aquarium Stock. Aquaria studies have shown coral fish to be extremely susceptible to a number of microbial infections and apparently incapable of internal defense against invasion. Bacterial tail or fin rot is the most common form of external lesions on marine aquarium fish. This has been observed also in Atlantic herring (*Clupea harengus harengus*) held in captivity over varying periods and in Atlantic cod (*Gadus morrhua*) held in live boxes. In the latter case death occurred usually within 48 hours of symptom appearance. A bacterial dermatitus with similar signs and a progression time of 48 hours was reported from many species of fish held in the Aquarium of Monaco. Preventive and curative measures used on aquaria stock include water sterilization by germicidal lamps, various antibiotic injections, disinfectant dips, and the addition of low-level antibiotics (e.g., Terramycin) to food [8]. The first two measures obviously are unsuited to open sea mariculture; the third might possibly be adapted in some form, but only in a restricted environment; and the last measure, if not too costly, might conceivably find some application where artificial feeding is employed.

Aquaculture Stock. Although there is, except for salmonids, no significant commercial production of marine fish by aquaculture methods in the United States as yet, interest has been expressed recently in developing techniques for culturing pompano, mullets, flounders, and certain other species. Information about pompano culture, specifically, has been summarized, listing recommendations and problem areas, including disease [13]. A preliminary report published in 1968 deals with the role of disease in experimental marine fish farming of flatfish and salmonids [14]. According to this study, flatfishes and salmonids maintained in seawater were affected by a number of diseases. Among these were an outbreak of lymphocystis in a breeding population of Dover sole (*Solea vulgaris*) and a vibrio disease that affected most of the species being held in ponds and tanks. For this last disease, antibiotic therapy was unsuccessful. Another bacterial disease, caused by a myxobacterium, affected rainbow trout held in cages in seawater. Here control was effected by dipping in a copper sulphate solution.

The Japanese, who have made probably the greatest advances in marine aquaculture, have found disease to be one of the major obstacles to further development. Evidence of their work on understanding and controlling finfish diseases may be found in the scientific journal *Fish Pathology* (founded in 1965 to publish the results of research on fish and shellfish disease). A vibrio-caused ulcer disease has been found in cultivated fish of nine species in various parts of Japan, including the most important cultured varieties— yellowtail (*Seriola quinqueradiata*), ayu (*Plecoglossus altivelis*), and horse mackerel (*Trachurus japonicus*) [15]. Mortality rates on the rearing grounds rose to 98 percent for some species, with the disease running its course within a few days. The ulcer disease was seen also in wrasses and mackerel caught in the natural environment, and the bacterium was isolated from seawater samples from the Sea of Japan and along the coast. Cures were achieved by injections of antibiotics and by using sulfisoxazole as a food additive. A phosphate-based chemical developed for fish treatment also was found to be effective on yellowtail infections.

In Mie prefecture, one of the marine fish culture centers in Japan, reports of disease in yellowtail, ayu, and other species described a vibrio disease of ayu, a systemic vibrio disease of yellowtail and other carangids, bacterial dermatitis in a number of fish species, and ulcer disease in puffers [16]. It became apparent that the diseases did not necessarily constitute clear entities; several could have involved the same vibrio, and infections could have been mixed. In 1963 a vibrio disease broke out in ayu cultured in floating net enclosures in Mie prefecture. Fish transferred to fresh water recovered. Symptoms appeared first on the skin and then as inflammation of the viscera. It was similar to an outbreak among cultured ayu in Kurita Bay on the Sea of Japan in 1963, and another among ayu held in net cages in a salt lake on the south coast of Japan, as well as to the diseases of yellowtails and carangids studied earlier in Mie prefecture. It appears that halophilic vibrios are significant pathogens of cultured marine fishes in Japan, but there is uncertainty about the number of disease entities involved. Bacterial dermatitis also is a common and widespread disease among aquaria stocks and cultured fish in Mie prefecture, apparently aggravated by surface abrasions, drastic changes in water temperature, or inadequate diet [8].

Parasites

Aquarium Stock. Among the several parasitic diseases coral fish are subject to in aquaria, a dinoflagellate, *Oodinium ocellatum,* is one of the most widespread. This dinoflagellate, discovered at the beginning of the 1930s and apparently introduced by specimens from Bermuda, caused the "velvet disease" of marine fishes. It quickly assumed epizootic proportions, with heavy mortality rates in London and Singapore aquaria. Primarily a gill parasite, it was known to attack at least 28 species of cold-water and subtropical fish, and so was apparently nonspecific in host selection. It also was reported to be common on many species in the New York Aquarium. In Denmark the *Oodinium* was successfully combated with copper sulfate and lowered salinities [8, 17].

The ciliate parasite *Cryptocaryon irritans* has been another cause of epizootics in marine aquarium fishes in Japan, London, and Singapore, although infections rarely have been heavy in natural populations. For instance, in Fiji only one species, the rock cod (*Epinephelus merra*), was infected out of 36 species on the same coral reef; by contrast, few species are resistant in aquaria, where the mature trophozoite, through multiple divisions, produces large numbers of motile infective stages. Infection was found in 27 species of Indo-Pacific and Atlantic origin in the New York Aquarium. Symptoms include heavy mucous secretion, erosion, and blindness. Heavy infection is fatal, but treatment with formalin, cupric acetate, and tris-(hydroxymethyl) aminomethane in seawater proved effective. At the Amsterdam Aquarium a successful treatment with trypaflavine was devised.

In the 1930s, other parasites found to be responsible for aquaria mortalities included Myxosporidia, Microsporidia, and larval nematodes. Mullet (*Mullus barbatus*) from the Mediterrean frequently became emaciated and died from kidney destruction due to a myxosporidian of the genus *Myxidium*. Other fish apparently were killed by liver degeneration caused by the microsporidian *Nosema ovoïdeum*. Monogenetic trematodes too can reach epizootic proportions in aquaria, infesting the eyes, gills, and nasal cavities. Although these are frequently fatal, survivors usually develop some degree of immunity.

Aquaculture Stock. As might be expected, there is much less information available about parasites in marine culture environments than aquaria. However, the following examples may help to indicate the nature of the problems.

Among the parasitic diseases of cultured marine fishes in Mie prefecture, the greatest damage to yellowtails was caused by monogenetic trematodes [8, 16]. *Benedenia seriolae* and *Axine heterocerca* were the two significant species. The *Benedenia* infestation varied seasonally, the numbers being lowest during winter and spring. The parasites developed in great numbers in regions where there were concentrations of the small floating net enclosures, and heavy infestations caused slow growth, emaciation, and abrasions that made the fish vulnerable to bacterial dermatitis. Fresh water proved to be an effective treatment. In Shizuoka prefecture also, *Benedenia* posed a serious problem to yellowtail fish culture. It was found that the worm reproduced frequently, grew to maturity in only 18 days at 22 to 26°C, and tolerated temperatures up to 28°C before detaching. A chemical, sodium pyrophosphate, provided an effective control [8].

The other species, the *Axine*, has been known to severely affect most of the stock. This bloodsucking trematode attacks the gills; in heavy concentrations, it causes severe anemia and mortality. Even light infestations may cause emaciation. Some control was achieved by short-term brine dips. Other parasites of relatively minor significance to cultured yellowtail include several other species of monogenetic trematodes; the digenetic trematodes *Echinostephanus hispidus* and *Tormopsolus orientalis* in the intestines; several species of ectoparasitic copepods of the genera *Caligus* and *Lernaeopoda* on gills, buccal cavity, and fins; and a parasitic copepod in the gill cavity [8].

In examples involving other species in different geographic areas, Dover sole stocks were found to be severely infested by the monogenetic trematode

Entobdella solaea after 1 year of captivity. Formalin baths controlled the problem in this case. Leeches (*Piscicola* sp.) also were found to flourish in Dover sole and parasitic copepods (*Lepeophtheirus nordmanni*) in turbot. In a 1968 report the only control measure proposed was washing with jets of fresh water [14]. Gray mullet (*Mugil capito*) kept in experimental ponds in Egypt in the 1920s suffered lesions and emaciation from heavy infestations of the copepod *Caligus pageti,* although this parasite had not been seen in wild populations [8]. And, in the course of significant development in the use of brackish-water ponds for fish culture in Israel in the last couple of decades, serious problems with disease were encountered early in the program [18]. Ectoparasites (i.e., monogenetic trematodes and parasitic crustacea) were considered the most harmful, especially to the young fry. In the latter case, control by insecticides proved effective, and the commercial production of carp, mullet, and tilapia in Israel rose to 10,000 tons/year by the late 1960s [8].

Other Diseases. The bacterial and parasitic diseases discussed so far include only the recognized major diseases. A few examples of other pathogens may help to indicate the scope of the disease problem.

Lymphocystis is probably the best-known virus disease of marine and freshwater fishes, and it appears occasionally in marine aquaria. In severe cases much of the body is involved. In aquaria, the disease was observed to appear in midsummer and disappear in late fall and winter; it was seldom fatal and the infected fish usually recovered completely. Documentation is extensive; several reports are included in the reference list for this chapter.

Epizootics in freshwater salmonid hatcheries in the western United States have been caused by the fungus *Ichthyophonus*; in these hatcheries raw marine clupeoid fishes, salmon viscera, infected trout viscera, and infected salt-water forage fishes were used for part of the hatchery diet. A systemic fungus disease probably caused by the same fungus has also appeared in yellowtail cultured in Japan.

A series of studies by Raabe and Guiart (listed in the Additional Readings) notes the appearance in the aquarium environment of pathogens virtually unknown in the natural environment. Two algae, *Leucosphaera oxneri* and *Thallamoebella parasitica,* have been blamed for epizootics and mortalities in a number of marine fish species.

In the case of the turbot, fatalities other than from parasites were occasioned by degeneration of the liver in juveniles due to inadequate diet, which was subsequently corrected. And in the brackish culture ponds of Israel, serious problems have arisen from low concentrations of oxygen that follow algal blooms, and from blooms of toxin-producing algae. Both have resulted in mortalities.

Controls. When modified, the measures recommended [8] for disease control in marine aquaria (and for freshwater aquaria) may be appropriate for application to open sea mariculture. The aquaria recommendations include the following:

1. Prevention of the spread of disease by controlling the water that has passed through tanks containing diseased fish.

2. Periodic sterilization of aquaria with chlorine.

3. Removal of uneaten food before decay.

4. Control of temperature, pH, and water flow to prevent abnormalities that lower resistance.

5. Diet free of pathogens.

6. Quarantine of all new introductions for 2 to 3 weeks.

7. Immediate isolation of any abnormal fish.

In open sea mariculture, items 1, 3, and 4 might effectively be achieved with careful site selection, which would allow the natural characteristics of the open sea site to flush water and uneaten food from the culture area and maintain water temperature and chemistry within normal limits. Pathogen-free diets and the quarantine of new introductions should be possible in open sea mariculture systems also: when juveniles were hatchery reared, they presumably would largely be free of pathogenic infections in any case. The immediate removal of abnormal fish might be achieved in open sea mariculture by predators.

Knowledge is still rudimentary in the field of disease control for captive marine fishes. Methods developed for freshwater purposes, and especially for hatcheries, are not directly transferable to seawater, where the pH and salt content may change the characteristics of antibiotics and other chemicals used in disease therapy. However, there is a growing interest in developing control measures for mariculture. Bacteria, protozoa, and monogenetic trematodes comprise the most serious menaces to captive and cultured species, and outbreaks of each have caused severe mortalities. Other pathogens, and viruses in particular, may be important, but their roles have not yet been determined [8, 19].

Captive and Cultured Shellfish

Disease and parasite problems in shellfish have been recorded in a variety of situations. In general, these situations may be classified under three categories: (1) premarketing impoundment, (2) shellfish culture in a natural environment, and (3) shellfish culture in an artificial environment.

The first category includes adult, market-sized crustaceans (lobster, crabs, etc.) held for a limited time in pounds, tanks, and live cars, prior to marketing, and adult, market-sized mollusks (clams and oysters) held in tanks or inshore beds awaiting marketing.

In the case of culture in a natural environment, examples of epizootic disease and parasitism have occurred in mollusks cultivated in natural water by the capture of seed in their natural habitat and subsequent transfer to growing beds, rafts, or racks, and/or by transfer of partially grown stock to other beds.

In artificial environments, disease has been a problem with both mollusk and crustacea larvae and spat produced in hatcheries in treated or recirculated water, and with mollusks or crustacea reared to market size under artificial conditions.

Recorded instances of disease and parasitism outbreaks under each of the three categories are reviewed in the following sections.

Premarketing Impoundment of Adult Shellfish. In the case of adult shellfish captured in their natural environment and then held for periods of time under artificial conditions prior to marketing, Sindermann cites an instance of parasite infestation in crabs and two infectious diseases of lobsters.

Crabs. In the summers of 1965 and 1966 a serious parasite problem occurred with molting blue crabs from the Chesapeake Bay. Peritrichous ciliates of the genera *Lagenophrys* and *Epistylis* were the pathogens responsible for massive infestations of the gills. Crabs freshly caught from the natural waters were infested heavily enough to interfere with respiration, and fisherman reported some mortalities in their catches. Holding the crabs in shedding tanks or floats during molting aided buildup of the commensal ciliates on the gills, contributing to mortalities, and the losses were most severe in the holding tanks immediately preceding and following molting.

Lobsters. Major losses of American lobsters, *Homarus americanus,* held in pounds or live cars are caused by two bacterial diseases. In each instance infection spreads rapidly in the characteristically crowded conditions. One is gaffkaemia, or red tail disease, caused by the coccus *Gaffkya homari.* It reached epizootic proportions on the Maine coast in 1946–1947, when it was first noted, and again in 1959–1960; it is also known in Canada [8, 20]. Other than the pink coloring, indicated by the name, signs include abnormal changes in the blood. Infected lobsters become progressively weaker and mortality may rise to 50 percent after short periods of storage; losses did in fact reach 58 percent during the Maine epizootics. Water temperature in excess of 15°V increases the mortality rate sharply (stricken lobsters often move to shoal water and die). It was discovered that the pathogen could live and multiply outside the lobster, in the slime on the lobster cars, crates, tanks, and live wells. It was also found in seawater several miles from infected pounds and in the mud of tidal pounds. Healthy lobsters held in seawater containing the bacteria, or fed on infected specimens, died in 2 to 3 weeks. Similar pathogenic organisms were found in European lobsters, *Homarus vulgaris,* in 1962, with mortalities observed in storage tanks in southern England [8, 21]. Losses of impounded lobsters have been reduced by treating tidal pounds with calcium hypochlorite to reduce the pathogen population in bottom mud.

Shell disease, caused by chitin-destroying, gram-netagive bacilli, is the second major cause of lobster losses. The chitin-degrading bacteria were discovered in live American lobsters impounded at Yarmouth, Nova Scotia, in 1937 [8, 22]. Symptoms include pitting and sculpturing of the exoskeleton and white-rimmed lesions on the walking legs, and were observed later in newly caught lobsters from several widely separated Canadian fishing grounds. Although the disease was relatively rare in natural populations, lobsters stored in pounds over the winter suffered severe shell erosion and general weakening. One study, around 1948, found that 71 percent of infected lobsters died from the disease. However, mortality was not correlated with intensity of erosion. The disease was transmitted directly from infected individuals, reaching its advanced stages in 3 months or more. Chitin destruction progresses at a rate directly temperature-dependent; new shell grown after molting was not affected except by reinfection [8, 23].

Shellfish Culture in Natural Environments. An important factor in the diseases of cultured shellfish populations is the introduction and spread of pathogens through the transfer of seed stock to growing areas. This fact is well documented in the case of oysters particularly. They have been transferred by man from place to place probably more freely than any other marine animal, great distances from their natural environments, sometimes with the misguided intention of avoiding diseases occuring in the parent stock. Moreover, this has usually been done under very loose control, or none, under existing regulations.

In Europe and England, the effects and spread of a parasite invasion of cultured mussel stocks have been documented since the beginning of the 20th century. In Japan, evidence of disease in oysters grown both on the bottom and by suspended culture has been accumulating since 1915. On the U.S. West Coast, three diseases are known to be present in oyster stocks: a parasitic copepod, a multiple-abcess bacterium, and an amoeboid organism. All three can cause pathological changes in oysters and all apparently were introduced from Japan. Like important predators, such as the Japanese oyster drill and carnivorous flatworms, they may affect native as well as imported stocks. On the U.S. East Coast four diseases, all apparently unrelated to the Pacific West Coast diseases, seem respectively to predominate in particular geographic areas. Pacific oysters have not been imported to the east in significant quantities for commercial culture, and the importation of foreign seed has not reached proportions comparable to the West Coast. However, oysters have been transferred from one East Coast area to another in significant numbers, and the possibility exists that the practice may have contributed to the spread of the eastern diseases [8].

A short review of some of the documented facts about several severe disease and parasitism problems among cultured shellfish may serve to indicate their potential importance to a developing mariculture industry. Some of the examples given point up the role of stock or seed transfers in the spread of the disease.

Europe and England. A long-existing problem with *Mytilicola intestinalis* invasion of north European sea mussel stocks has provided extensive evidence of the effects of a parasite on cultivated mussel populations. After its initial discovery in Mediterranean mussels, first documented in 1902, and its subsequent appearance near Cuxhaven, Germany, in 1938, the parasite was found to be abundant in some areas of the English coast in 1946. It appeared in mussel stocks of the Netherlands in 1949, where it spread widely, and about a decade later it was studied in mussel populations grown on floats in Spain. Suspected vectors included mussel-encrusted ships, the movement of planktonic larvae, and transplanted seed mussels. Reproduction of the parasitic copepod is accelerated in warm water; during the summer the many young parasites invaded and killed mussels of all sizes, including seed. Young mussels fell from culture racks, and adults died during transport to markets. The mussel population density was thought to affect survival and multiplication of the parasite, infections were lightest where the hosts were scattered thinly, and near the water surface. In Spain, infestation was heaviest near shore where currents were weak. In rope cultures, using 6-m ropes, vertical distribution of the parasite was uniform in the presence of strong currents, but where currents were weak the infestation increased with depth. Suggested control methods included off-bottom culture and the location of beds in fast-moving water or at the brackish water ends of estuaries [8].

Japan. In early research on oyster losses in Japan, specific pathogens usually were not identified, although disease was often suspected. In many cases a direct connection still has not been identified. Several examples of major outbreaks include the following:

1. Large-scale mortalities of oysters in Kanasawa Bay, beginning in 1915 and continuing for several years. Over 80 percent of the populations in that bay died annually during the period.

2. Mass mortalities on the Miura Peninsula beginning in 1927 and continuing for 10 years. Oyster farms along this coast lost 50 to 80 percent of their crop annually.

3. Mortalities of 2-year-old oysters in Hiroshima Bay and adjacent areas, beginning in 1945. Here a 10-year study suggested that a bacterial pathogen caused the losses, but the evidence was not conclusive.

4. Mass mortalities in raft-cultured oysters in Matsushima Bay, Miyagi prefecture, beginning in 1961 and occurring annually in late summer. Mortalities exceeded 60 percent in some areas during 1961–1964.

In this latter instance, research papers published in 1965 described the environmental, physiological, and pathological factors. Mortalities were considered to be related to metabolic changes during fattening and spawning. Although a gram-positive bacterium was present in multiple abcesses in up to

20 percent of the oysters in some samples, this was not directly connected to the mortalities. However, further observation has been recommended on the basis that the abcesses may be only the chronic stage in resistant adults, and that the acute disease might cause mortality. In samples examined at the Oxford Laboratory, it was noted that an ameboid organism, strongly affecting the host, was often present in large numbers in oysters from the mortality area of Matsushima Bay [8]. These two pathogens are the bacterium and amoeboid organism mentioned earlier in connection with the U.S. West Coast oyster stocks.

United States West Coast. Among the examples we are able to give here, nowhere is there clearer evidence of the hazards involved in introducing nonnative shellfish seed stock to a growing area without adequate precautions than in the oyster culture on the U.S. West Coast. This may be seen in the following three instances involving disease problems; however, direct associations between specific pathogens and oyster mortalities on the West Coast have not yet been discovered.

1. The parasitic copepod *Mytilicola orientalis,* found in the digestive tract of the Pacific oyster (*Crassostrea gigas*) in Japan in 1935, was identified as a new species of parasite in the United States in 1938. It is now common on the West Coast. The vector by which it moved across the Pacific ocean was the Pacific oyster host. This species has been imported as seed stock for waters of the Pacific Coast states since the 1930s to offset the decline in populations of the Olympia oyster (*Ostrea lurida*) native to the American waters. In 1946 the parasite was found to be infecting the native Olympia oyster also, where in even small numbers it degraded the condition of the host.

2. Seed oysters imported to the Pacific West Coast from Japan recently were found to be infected with a focal necrosis disease, similar to the multiple-abcess disease outbreaks in Japan, which were described earlier, and caused by the same bacterium. The disease condition also has been detected in adult oysters grown on the Washington coast.

3. In 1966 a disease caused by an amoeboid organism, briefly mentioned earlier in connection with Japanese oyster diseases, was discovered by the Oxford Laboratory of the Bureau of Commercial Fisheries in over 30 percent of parent stocks from a Matsushima Bay area. A possibly similar amoeboid organism was found in oysters grown in Humboldt Bay, California, from Japanese seed imported from Matsushima Bay. Oyster mortalities in Humboldt Bay have appeared to be increasing, annually, since 1960; in 1966 a dredged sample showed 50 percent of the oysters to be recent mortalities [8].

United States East Coast. Severe mass mortalities of cultured oysters have been associated with diseases on the North American East Coast. Of the four known maladies—Delaware Bay disease, *Dermocystidium* disease, Malpeque

Bay disease (in Canada), and seaside disease (caused by a haplosporidian, *Minchinia costalis*)—two are described in some detail by Sindermann and summarized briefly here.

1. Delaware Bay disease, caused by the haplosporidian protozoan *M. nelsoni*, has severely affected planted oyster beds in the Delaware and Chesapeake bays, impacting on oyster industries of Virginia, Maryland, Delaware, and New Jersey. From 1959 to 1961 nearly half the planted beds in the Virginia waters of Chesapeake Bay were forced out of production, and planted beds in Delaware Bay were even more seriously affected. Infection occurred from May through October, with the highest mortality in early summer and early autumn. Death sometimes followed infection within 2 months; and mortality rates ran 50 to 60 percent in each of the first 2 years after initial introduction of the pathogen into a population. In an outbreak of Malpeque Bay disease of similar severity but unknown etiology in Canada some years earlier, the oyster population developed an apparent resistance to the pathogen. That incident would seem to support the approach now being taken by the four afflicted states to solve the problem of restoring beds to production. With the assistance of federal funds, they have been conducting programs to develop disease-resistant stocks. The little information so far available from the program indicates that resistance may be both hereditary and acquired [8, 24]. Additional findings suggest that the survival of oysters in epizootic areas is due more to a resistance developed through early exposure to disease than to an innate resistance derived genetically from resistant parents [12, 25], a fortunate situation for the potential mariculturist since, if true, it would simplify control.

2. *Dermocystidium* disease in planted oysters, caused by the fungus *D. marinum,* also has reached epizootic proportions and is taking an increasing toll. Since it is greatly influenced by the proximity of susceptible hosts, it has been suggested that the modern planting method itself may have precipitated this acute crisis in oyster production. The most intense development of the disease is in the areas of Louisiana and Virginia where crowded planting is the custom. The effects are reduced during periods of lower water temperatures and in lower salinities; for example, the disease is active for most of the year in the Gulf of Mexico, but quiescent for almost half the year in the Chesapeake Bay. In Louisiana the entire crop of oysters can be lost in one summer, but they may be held for two or three summers in Virginia before yield is reduced sufficiently to wipe out profit [8]. The age and size of oysters influence susceptibility and mortality rates. Spat are resistant during the first 3 or 4 months, and mortality is low during the first year. Oyster stocks from different areas also vary in their susceptibility to the fungus pathogen; South Carolina stocks proved more resistant than tested seed from several other states. Some of the proposed control measures include early exposure of spat to build resistance to the disease, planting of the largest seed

possible, and harvesting as soon as the oysters reach marketable size. It has been further recommended that planting be in early autumn to obtain autumn and spring growth before infections occur, and that harvesting be done in late spring [8, 26].

Shellfish Culture in Artificial Environments. According to some schools of thought, culture methods for marine animals seem likely to develop a growing dependence on artificial environments as they become increasingly complex. But open sea mariculture might possibly reverse that trend for the juvenile and maturation stages in marine-stock life cycles. At the same time, mariculture may be expected to increase pressure for improving the sources of replacement stock through controlled breeding in captivity, as opposed to reliance on the economically impractical capture of juveniles. This being the case, the control of disease and parasitism in the hatchery environment will assume added urgency. That there already is a pressing need for disease control in shellfish hatcheries is attested to by the following instances.

● Mass mortalities of bivalve mollusk larvae in hatchery tanks were described in 1965 [8, 27]. The pathogen was gram-negative, motile bacilli believed to be *Vibrio* sp. or *Aeromonas* sp. In this case all the pathogenic serotypes were found to be sensitive to certain antibiotics; Combistrep and chloramphenicol were effective therapeutic agents.

● Other epizootics in hatchery-reared oysters and clam larvae were found, in 1954, to be caused by a fungus that was subsequently identified as *Sirolpidium zoophthorum* [8, 28, 29]. Both juvenile and larval bivalves were killed; in some cultures epizootics wiped out most of the population in 2 to 4 days. Suggested controls included filtration and ultraviolet treatment of seawater, and the use of bacteria-free algal cultures or sterilized artificial diets.

● In the practice of short-term hatching and early larval rearing of lobsters, egg-bearing females held in hatching troughs may die from *Gaffkya* infections before the eggs are hatched. Female lobsters, their eggs, and larvae may be infested with the annelid *Histriobdellea*, or the eggs may be destroyed by the suctorian *Ephelota*. The chitinivorous bacteria described earlier pose another hazard [8].

● Attempts at English prawn (*Palaemon serratus*) culture in seawater tanks support the prediction that long-term cultivation of crustaceans will encounter serious difficulties with parasites and diseases. The prawns were severely affected by a progressive systemic fungus disease first seen in egg-bearing females captured at sea. The infection spread rapidly to most groups of prawns being cultured. In a study of this disease problem [30], the pathogen was tentatively assigned to the Phycomycete genus *Pythium*, and a similarity was noted between this disease and a mycosis caused by *Aphanomyces astaci* in European freshwater crayfish. Brown spot disease, characterized by erosion of the exoskeleton and inflammation of underlying tissues, was another important pathological condition that was widespread among the cultured prawns. The responsible organism was not identified during the study, which alluded to other examples of crustacean disease caused by chitinivorous bacteria. It was also pointed out that mycoses appeared to pose a particular hazard in the crustacean hatchery environment; as far as is known, no effective therapy is available.

Summary

From the study of fish diseases in aquarium environments and to a limited extent in aquaculture enterprises, and from the study of shellfish diseases among impounded and cultured populations, an initial body of data has been accumulated that would seem to support the following general statements.

1. Pathogens and parasites are more rapidly and widely spread among captive marine organisms than among natural populations. This may be due to restricted space (as in aquaria), or to the greater density of populations (as in shellfish cultures), or both.

2. Diseases caused by vibrios have assumed great importance to cultivated fish populations.

3. The microbial parasites, which are characterized by direct water-borne transmission, short generation times, and dependence on a single host during their life cycle, are of major importance in cultivated marine populations.

4. In cultivated stocks of marine fish, pathogenic roles have been assumed by certain organisms that often are rare and innocuous parasites in natural populations.

5. Repeated outbreaks of epizootic disease and consequent mortalities have occurred in the culture of molluscan shellfish, which is the most advanced field of marine aquaculture at the present time. Some counteractive measures have been developed, but limited knowledge about the diseases involved has so far been a roadblock to really effective control.

6. Disease problems in cultured stocks can be aggravated by transfers from one breeding or growing environment to another. Nonresistant stocks may be placed in areas where they are newly exposed to enzootic or epizootic diseases, or new diseases may be introduced into areas by host animals imported from other localities.

In connection with this last statement, it should be pointed out that adequate controls over the introduction of shellfish seed stocks into non-native growing areas are difficult to develop at the present time. The problems involved include the following:

• Even under extensive inspections and testing programs, it is difficult to screen imported stocks for all the possible pathogens that they might be host to, especially the microbes.

- Knowledge about the possible pathogens of both domestic and nonnative shellfish species is incomplete and inadequate.
- Knowledge about pathogen life cycles is lacking.
- The effects of imported pathogens on native organisms that have no history of previous exposure are unpredictable.
- The reactions of imported pathogens to a new environment and/or to unfamiliar host organisms are unpredictable.

Potential controls to avoid the introduction or spread of diseases and parasites might include strict quarantines, the exclusion of imported seed stocks, or closed culture systems using only offspring of previously cultured parents for introductions into uncontaminated areas. Research is strongly needed here, as well as in the entire field of disease, to determine the nature and level of the risks and workable controls.

Several other factors related to aquaculture in its present state of development have a general influence on the incidence rate and severity of pathological conditions. First, any change in the natural environment of an organism tends to place stress upon the animal; and the greater the change, the greater the stress is likely to be. Stress, in turn, lowers the organism's inherent resistance to disease or parasites. Second, the nutrition of cultivated populations may be deficient, thereby lowering their resistance to or actually causing pathological conditions. Moreover, shellfish often are cultivated in areas where natural populations do not exist; some environmental limiting factor or factors may be responsible for their normal absence. Finally, fish and shrimp culture still depends largely upon the capture and impoundment of juveniles. This leads to injuries during capture and to the inclusion of individuals already infected by pathogens [8].

CONCLUSIONS AND RECOMMENDATIONS

The control of predation, competition, disease, and parasitism is a major roadblock in the development path of mariculture. In all probability this challenge will be met only by the application of imaginative thinking that transcends simple extrapolation of the methods in use today.

Predation and Competition Controls

Where culturing is to take place in unenclosed environments, the control of predators and competitors is still problematical. The most attractive mechanism so far apparent is the removal of a cultured crop from the normal ranges of predators and competitors. In this, open sea mariculture seems in some cases to offer an advantage over estuarine mariculture. Although there are a few straightforward methods for controlling relatively immobile predators on sessile organisms, all other methods for dealing with predation in unenclosed cultures so far are experimental at most; yet some of them may prove highly effective. Mesh or other physical barriers offer promise for controlling predation and competition where cultured crops can be enclosed. But these may not be effective in excluding larval and young stages of some predators or competitors. Nonetheless, for all open sea circumstances, effective eradication measures and the geographic location of offshore sites away from natural predator and competitor ranges offer hope of eventually bringing this aspect of the problem under control.

Disease and Parasitism Control

Extrapolating from present knowledge, we are led to postulate several potential means of controlling disease in open ocean mariculture crops, and these means may apply to any form of aquaculture. Obviously, none of these methods has yet been attempted in the open ocean. They all represent avenues for further research and development.

Predators. Selective predators in controlled numbers might be employed to weed out unhealthy cultured organisms and thereby inhibit the spread of disease. This, of course, would work only with free-swimming forms.

Feed Medicines. Preventative and curative medicines might be added to feed or to special supplemental feed. This method might apply to both free-living and sessile forms.

Medication Subsystems. Free-swimming species might be collected or attracted without injury, passed through containers of physically and chemically conditioned water, and then returned to their culture environment.

Medicated Culture Environments. If curative or preventative chemicals were sufficiently inexpensive, they might be simply injected into the culture water either continuously or intermittently. This approach might work particularly well with sessile-organism culture, wherein it seems feasible to include medicated water injectors as components of the attachment apparatus.

Biological Controls. Biological controls are coming to the fore in agriculture and eventually may prove valuable in aquaculture also. Particularly attractive for fish ectoparasites in warm-water cultures might be the "cleaner fishes," *Labroides* spp. The services these small fishes perform for other fish at their cleaning stations is one of the most famous stories in the natural history of coral-reef ecosystems. Other biological controls, of course, would include beneficial commensal organisms, and the isolation and selective breeding of microbiological pathogens that are specific to organisms having pathogenic effects on cultured species.

Immunization. Immunobiology must come to mind also. Superficially at least, it appears that the most promising research path would involve the isolation of heterogenetic antigens and the inoculation of fertilized egg and larval cultures with them. However, inoculation of young animals should be considered also.

Selective Breeding. Although there is evidence of transient immunities in some marine larvae, it appears likely that, except for the few viviparous groups, few antibodies per se are passed on to progeny by nonplacental marine organisms. Nonetheless, resistance to pathogens in the form of morphological or physiological characteristics must surely be a component of the genetic information contained in all organisms. Consequently, selective breeding for disease resistance may be a long but ultimately rewarding path for aquaculture, as it has proved to be for agriculture.

Irradiation. There is evidence that electromagnetic radiation in the ultraviolet range may be effective in controlling pathogens that infect a variety of animals and plants. In addition, viruses have been shown to be sensitive to radiation in the X-band, although we know of no instances in which X-rays have yet been employed to treat viral infections. We do not propose that it is definitely possible to treat marine organism pathogens with these forms of radiation without damaging the host; we suggest only that it is one medical research path that may be worthy of consideration.

Whatever research avenues may be chosen, the focus as it pertains to open sea mariculture should be on the development of economical and non-labor-intensive diagnosis, treatment, and most of all, prevention.

REFERENCES

1. Chang, T. C., ed. 1971. *Aspects of the biology of symbiosis.* University Park Press, Baltimore, Md.
2. Fujiya, M. Oyster farming in Japan. Intl. Helgoland. Symp. *Helgoländer wiss. Meeresunters.*
3. Todd, J. H., J. Atema, and D. B. Boylan. 1972. Chemical communications in the sea. *Marine Tech. Soc. J.* (July–Aug.).
4. May, R. C. 1973. Hawaii Institute of Marine Biology, Univ. Hawaii. Personal communication.
5. Anonymous. 1968. Cheaper anti-shark barrier. *New Scientist, 37*(538).
6. Richard, J. D. 1968. Fish attraction with low frequency sound. *J. Fisheries Res. Board Can.,* pp. 1441–1452.
7. Norris, K. S. 1965. Trained porpoise released in the open sea. *Science 147*(3661):1048–1050.
8. Sindermann, C. J. 1970. *Principal diseases of marine fish and shellfish.* Academic Press, Inc., New York.
9. Oppenheimer, C. H., and G. L. Kesteven. 1953. Disease as a factor in natural mortality of marine fish. *FAO Fisheries Bull. 6:*215–222.
10. Aronson, J. D. 1926. Spontaneous tuberculosis in salt water fish. *J. Infectious Diseases 39:*315–320.
11. ZoBell, C. E. 1946. *Marine microbiology.* Chronica Botanica, Waltham, Mass. 240 pp.
12. Nigrelli, R. F., and H. Vogel. 1963. Spontaneous tuberculosis in fishes

and in other cold-blooded vertebrates, with special reference to *Mycobacterium fortuitum* Cruz from fish and human lesions. *Zoologica 48*:131–144.

13. Berry, F. H., and E. S. Iversen. 1966. Pompano: biology, fisheries and farming potential. *Proc. Gulf Carib. Fisheries Inst., 19th Annual Session,* pp. 116–128.

14. Anderson, J. I. W., and D. A. Conroy. 1968. The significance of disease in preliminary attempts to raise flatfish and salmonids in sea water. *Proc. 3rd Symp. Mond. Comm. Off. Intern. Epizoot. Etude Maladies Poissons* (Stockholm) 5 pp. (mimeo).

15. Kusuda, R. 1966. Studies on the ulcer disease of marine fishes. *Proc. 1st U.S.–Japan Joint Conf. on Marine Microbiol. (Tokyo)* (mimeo. extract).

16. Kubota, S. S., and M. Takakuwa. 1963. Studies on the diseases of marine cultured fishes. I. General description and preliminary discussion of fish diseases in Mie Prefecture. *J. Fac. Fisheries, Prefect. Univ. Mie 6, 1*:107–124. Transl. Fisheries Res. Board Can. Biol. Sta., Nanaimo, B.C. 1966.

17. Højgaard, M. 1962. Experiences made in Danmarks Akvarium concerning the treatment of *Oodinium ocellatum*. *Bull. Inst. Oceanog. Monaco, Numero Spec. 1A, Premier Congr. Intern. Aquariol. A.,* pp. 77–79.

18. Tal, S., and M. Shelubsky. 1952. Review of the fish farming industry in Israel. *Trans. Amer. Fisheries Soc. 81*:218–223.

19. Oppenheimer, C. H. 1962. On marine fish diseases. In G. Borgstrom, ed., *Fish as food,* Vol. 2. Academic Press, Inc., New York, p. 541.

20. Goggins, P. L., and J. W. Hurst, Jr. 1960. Progress report on lobster gaffkyaremia (red tail). Dept. Sea and Shore Fisheries, Augusta, Maine. Unpublished mimeo.

21. Wood, P. C. 1965. A preliminary note on Gaffkaemia investigations in England. *Cons. Perm. Intern. Explor. Mer, Rappt. Proces-Verbaux Reun. 156*:30–34.

22. Hess, E. 1937. A shell disease in lobsters (*Homarus americanus*) caused by chitinovorous bacteria. *J. Biol. Board Can. 3*:358–362.

23. Taylor, C. C. 1948. Shell disease as a mortality factor in the lobster (*Homarus americanus*). *Maine Dept. Sea and Shore Fisheries Circ. 4,* pp. 1–8 (mimeo).

24. Powell, E. H., and J. D. Andrews. 1967. Production of MSX-resistant oysters. *Virginia J. Sci. (N.S.) 18*(4):163 (abstr.).

25. Andrews, J. D. 1968. Oyster mortality studies in Virginia. VII. Review of epizootiology and origin of *Minchinia nelsoni*. *Proc. Natl. Shellfisheries Assoc. 58*:23–26.

26. Andrews, J. D., and W. G. Hewatt, 1957. Oyster mortality studies in Virginia. II. The fungus disease caused by *Dermocystidium marinum* in oysters of Chesapeake Bay. *Ecol. Monographs 27*:1–26.

27. Tubiash, H. S., P. E. Chanley, and E. Leifson. 1965. Bacillary necrosis, a disease of larval and juvenile bivalve mollusks. I. Etiology and epizootiology. *J. Bacteriol. 90*:1036–1044.

28. Davis, H. C., V. L. Loosanoff, W. H. Weston, and C. Martin. 1954. A fungus disease in clam and oyster larvae. *Science 120*:36–38.

29. Vishniac, H. S. 1955. The morphology and nutrition of a new species of *Sirolpidium. Mycologia* 47:633–645.
30. Anderson, J. I. W., and D. A. Conroy. 1968. The significance of disease in preliminary attempts to raise Crustacea in sea water. *Proc. 3rd Symp. Mond. Comm. Off. Intern. Epizool. Etude Maladies Poissons, Separate No. 3.* 8 pp.

Additional Readings

Abeliovitch, A. 1967. Oxygen regime in Beit-Shean fish ponds related to summer mass fish mortalities, preliminary observations. *Bamidgeh* 19:3–15.

Amlacher, E. 1970. *Textbook of fish diseases.* T.F.H. Publications, Inc., Neptune City, N.J.

Andrews, J. D. 1966. Oyster mortality studies in Virginia, V. Epizootiology of MSX, a protistan pathogen of oysters. *Ecology* 47:19–31.

Andrews, J. D., and J. L. Wood. 1967. Oyster mortality studies in Virginia. VI. History and distribution of *Minchinia nelsoni,* a pathogen of oysters, in Virginia. *Chesapeake Sci.* 8:1–13.

Bardach, J. E., J. R. Ryther, and W. O. McLarney. 1972. *Aquaculture: the farming and husbandry of freshwater and marine organisms.* John Wiley & Sons, Inc., New York.

Brown, E. M. 1934. On *Oodinium ocellatum* Brown, a parasitic dinoflagellate causing epidemic disease in marine fish. *Proc. Zool. Soc. London,* pp. 583–607.

Brown, E. M. 1963. Studies on *Crytocaryon irritans* Brown. *Proc. 1st Internatl. Congr. Protozool.* (Prague). Academic Press, Inc., New York, pp. 284–287.

Brown, E. M., and R. Hovasse. 1946. *Amyloodinium ocellatum* (Brown), a peridinian parasite on marine fishes. A complementary study. *Proc. Zool. Soc. London* 116:33–46.

Bullock, G. L. 1971. Identification of fish pathogenic bacteria. In S. F. Snieszko and H. R. Axelrod, eds., *Diseases of fishes.* T.F.H. Publications, Inc., Neptune City, N.J.

Bullock, G. L., D. A. Conroy, and S. F. Snieszko. 1971. Bacterial diseases of fishes. In S. F. Snieszko and H. R. Axelrod, eds., *Diseases of fishes.* T.F.H. Publications, Neptune City, N.J.

Cole, H. A., and R. E. Savage. 1951. The effect of the parasitic copepod, *Mytilicola intestinalis* (Steuer) upon the condition of mussels. *Parasitology* 41:156–161.

Couch, J. A. 1966. Two peritrichous ciliates from the gills of the blue crab. *Chesapeake Sci.* 7:171–173.

Couch, J. A. 1967. A new species of *Lagenophrys* (Ciliatea: Peritrichidae Lagenophridae) from a marine crab, *Callinectes sapidus. Trans. Amer. Microscop. Soc.* 86:205–211.

Davis, H. S. *Culture and diseases of game fishes.* University of California Press, Berkeley, Calif.

DeGraaf, F. 1962. A new parasite causing epidemic infection in captive coral fishes. *Bull. Inst. Oceanog. Monaco, Numero Special 1A, Premier Congr. Intern. Aquariol. A.,* pp. 93–96.

Dijn, C. V., Jr. 1956. Diseases of fishes. *Water-life.* London.

Farrin, A. E., L. W. Scattergood, and C. J. Sindermann. 1957. Maintenance of immature sea herring in captivity. *Progressive Fish Culturist 19*:188–189.

Hepper, B. T. 1955. Environmental factors governing the infection of mussels, *Mytilus edulis,* by *Mytilicola intestinalis. Min. Agr. Fisheries and Food, Fisheries Invest., London 2*(20):1–21.

Hoshina, T. 1968. On the monogenetic trematode, Benedenia seriolae, parasitic on yellowtail, *Seriola quinqueradiata. Proc. 3rd Symp. Mond. Comm. Off. Intern. Epizoot. Etude Maladies Poissons (Stockholm) Separate No. 25.* 12 pp.

Imai, T., K. Numachi, J. Oizumi, and S. Sato. 1965. Studies on the mass mortality of the oyster in Matsushima Bay. II. Search for the cause of mass mortality and the possibility to prevent it by transplantation experiment. (In Japanese with English summary.) *Bull. Tohoku Reg. Fisheries Res. Lab. 25*:27–38.

Kabata, Z. 1970. Crustacea as enemies of fishes. In S. F. Snieszko and H. R. Alexrod, eds., *Diseases of fishes.* T.F.H. Publications, Inc., Neptune City, N.J.

Kan-no, H., M. Sasaki, Y. Sakurai, T. Watanabe, and K. Suzuki. 1965. Studies on the mass mortality of the oyster in Matsushima Bay. I. General aspects of the mass mortality of the oyster in Matsushima Bay and its environmental conditions. (In Japanese with English summary.) *Bull. Tohoku Reg. Fisheries Res. Lab. 25*:1–26.

Kasahara, S. 1967. On the sodium pyrophosphate peroxyhydrate treatment for ectoparasitic trematodes on the yellowtail. (In Japanese.) *Fish Pathol. 1*:48–53.

Korringa, P. 1959. Checking Mytilicola's advance in the Dutch Waddensea. Cons. Perm. Intern. Explor. Mer., 47th Meet., *Shellfish Comm. Rept. 87* (mimeo).

Mackin, J. G. 1962. Oyster disease caused by *Dermocystidium marinum* and other microorganisms in Louisiana. *Publ. Inst. Marine Sci. 7*:132–229.

Muroga, K., and S. Egusa. 1967. *Vibrio anguillarum* from an endemic disease of ayu in Lake Hamana. *Bull. Jap. Soc. Sci. Fisheries 33*:636–640.

Ogasawara, Y., U. Kobayashi, R. Okamoto, A. Furukawa, M. Hisaoka, and K. Nogami. 1962. The use of the hardened seed oyster in the culture of the food oyster and its significance to the oyster culture industry. (In Japanese with English summary.) *Bull. Naikai Reg. Fisheries Res. Lab. 19*:1–153.

Paperna, I. 1964. Host reaction to infestation of carp with *Dactylogyrus vastator* Nybellin, 1924 (Monogenea). *Bamidgeh 16*:129–141.

Reichenbach-Klinke, H. H., and E. Elkan. 1965. *The principal diseases of lower vertebrates.* Academic Press, Inc., New York.

Reichenbach-Klinke, H. H. 1969. *Bestimmungsschlussel zur Diagnose von Fischkrankheiten.* Gustav Fischer Verlag, Jena, Germany.

Roskam, R. T. 1957. Gaffkaemia, a contagious disease, in *Homarus vulgaris*. Cons. Perm. Intern. Explor. Mer. *Shellfish Comm. Rept.,* 4 pp. (mimeo).

Russell, J. 1966. Japan's fish hospital. *New Sci. 31*:150–151.

Schaperclaus, W. 1954. *Fischkrankheiten,* 3rd ed. Akademie-Verlag G.m.b.H., Berlin.

Schaperclaus, W. 1956. Neue Möglichkeiten zur Bekampfung von Infektionskrankheiten bei Aquarienfischen. *Aquar. Terrar. Zeit. 9*:213–215.

Shilo, M. 1953. Prevention of mortality of fish fry caused by *Gyrodactylus* and *Dactylogyrus. Bamidgeh 5*:26.

Sikama, Y. 1962. Study on white spot disease in marine fish. *Agri. Hort. Tokyo 10*(1):29–90.

Sindermann, C. J. 1968. Oyster mortalities, with particular reference to Chesapeake Bay and the Atlantic Coast of North America. *U.S. Fish Wildlife Serv. Spec. Sci. Rept. Fisheries 569,* pp. 1–10.

Sindermann, C. J., and A. Rosenfield. 1954. Diseases of fishes of the western North Atlantic. I. Diseases of the sea herring (*Clupea harengus*). *Maine Dept. Sea Shore Fisheries, Res. Bull. 18*:1–22.

Sindermann, C. J., and A. Rosenfield. 1967. Principal disease of commercially important bivalve Mollusca and Crustacea. *U.S. Fish Wildlife Serv. Fishery Bull. 66*:335–385.

Sparks, A. K., E. J. Robbins, D. Des Voigne, B. C. C. Hsu, and L. Schwartz. 1967. Oyster pathology. Research in fisheries, 1966. Coll. Fishery, Fishery Res. Inst., Univ. Washington, Seattle, Contrib. 240, 37.

Weissenberg, R. 1965. Fifty years of research on the lymphocystis virus disease of fishes (1914–1964). *Ann. N.Y. Acad. Sci. 126*:362–374.

Wood, P. C. 1965. Gaffkaemia, the blood disease of lobsters. *Proc. Soc. Gen. Microbiol.,* p. 14 (abstract).

10

CONCENTRATING AND HARVESTING MARINE CROPS

J. A. Hanson

Implicit in most concepts of culturing is the idea that the biomass of the cultured crop per unit volume will be high. Most frequently this means that the cultured biomass will remain high throughout the growing cycle; but with some open sea mariculture concepts, such is not necessarily the case. For example, anadromous fish as cultured today presumably grow in low concentrations in the open sea and later concentrate themselves as a result of their homing behavior at spawning time. One might also contemplate artificial, reef-based mariculture systems or others wherein stock locations are somewhat loosely controlled by some form of attraction. In such systems the rearing density of the stock is likely to be lower than the density required for efficient harvesting; thus, an additional concentrating mechanism must be employed as a component of the harvesting process.

Harvesting is one area of open sea mariculture technology for which there is some practical precedence, for the fishing industry has been developing equipment and techniques for open sea harvesting for centuries. In fact, it may be that the inertia of traditional fishing customs has acted to inhibit technological advancement in more than a few instances; people who have thorough knowledge of, and sizable capital investments in, a time-honored method are slow to adopt innovations.

For the most part, harvesting in an open sea mariculture operation should pose fewer problems than are found in harvesting naturally occurring open sea stocks. If the crop is sessile, for example, one might design the attachment substrate to facilitate automated harvesting. If the stock is benthic, it probably will require complete or partial containment, and its substrate may be artificial and so designed to facilitate harvesting. If the stock consists of secretive organisms, the concealment places themselves might be so designed as to become harvesting mechanisms. If the stock consists more or less of concentrated free swimmers, at least one usually would know where they are and neither have to search for them nor carry the harvesting system to them. Superimposed on all these examples, of course, is the added promise that, in general, cultured stocks will be vastly more homogeneous in size and species composition than are natural stocks.

In this chapter we first examine the general and organism-oriented functional requirements for concentrating and harvesting marine organisms, after which we present a brief survey of harvesting techniques currently in use by or proposed for the fishing and aquaculture industries. Then we look to the future and speculate on how existing technology might be adapted to new uses and how innovative designs might be applied to concentrating and harvesting mechanisms for use in open sea mariculture. Finally, we offer suggestions for further research and development.

FUNCTIONAL REQUIREMENTS FOR CONCENTRATING AND HARVESTING MARINE ORGANISMS

Any concentrating and harvesting mechanism must meet certain general requirements, regardless of its specific application, if it is to be practicable and economical. For a mariculture operation, as distinguished from conventional fishery harvesting, the following general functional requirements would apply:

1. The systems should provide for harvest of crops at the optimum point in the growth cycle, that is, the point at which the ratio of rearing cost to marketing return is most favorable. Generally, this will mean young adults of most species.

2. The systems must concentrate the crop and transfer it from the rearing area to the on-site processing facility without significantly degrading the market quality of the product. This requirement tends to exclude the more brutal harvesting mechanisms, such as explosives and poisons, and the concentrating mechanisms that favor predation.

3. The systems should concentrate and harvest all or a very high percentage of the crop intended for harvest. Unharvested individuals usually will be wasted and therefore affect the cost effectiveness of the whole operation adversely.

4. Concentrating and harvesting systems for open sea mariculture should be designed so as to be as non-labor-intensive as possible. They also should be designed to minimize capital and operating costs.

5. Harvesting systems should be designed to harvest at rates that are compatible with on-site, harvest-processing rates (Chapter 11) so as to maximize capital equipment utilization and minimize temporary storage requirements.

6. In most cases one would expect cultured crops to be largely homogeneous in species and size. However, it is possible that some polyculture systems will not offer this advantage. Therefore, all harvesting system designs should harvest only those organisms intended for harvest, excluding others.

With these general functional requirements in mind, it is appropriate to consider some important differences among the marine organisms that may be cultured in open sea systems. The distinguishing characteristics of concern are variations in life style or ecological niche rather than phylogenetic differences, although there is certainly some correlation between the two. We suggest that the significant differences among cultured organisms as they may affect harvesting system designs can be conveniently divided into six categories. Our examination of these differences recognizes that harvesting frequently consists of two phases: (1) gathering organisms to the harvest, and (2) the actual harvesting.

Sessile Organisms

Sessile organisms attach themselves to some firm substrate. This implies, first, that the harvesting system must remove them from the substrate without a significant incidence of damage. Second, sessility offers the opportunity to design firm artificial substrates in such a manner that they may facilitate and even become part of the harvesting system.

Benthic Crustacea

These organisms may either be open-bottom dwellers, such as many forms of shrimp, or they may prefer rough-bottom topography (e.g., lobsters and crabs). Benthic crustacean populations that inhabit, or can be concentrated in, open situations lend themselves well to various modes of dredge-type or trawl harvesting. Forms dwelling on rough bottoms must be either attracted to traps or lured (or possibly forced) into the open for dredge-type harvesting; or perhaps one might design systems of artificial hiding places in such a way that the hideouts themselves become the harvesting mechanisms.

Nonschooling Fish

With some notable exceptions, those fish potentially appropriate to culture that do not regularly school are demersal (bottom dwelling) and may also be territorial to greater or lesser degress. The result, of course, is that their natural population densities tend to be significantly lower than schooling fish. Thus, any harvesting system to be employed with them will have to attract or guide large numbers of individuals from and to various places, as opposed to attracting a school from one point to another, in order to harvest them.

Schooling Fish

A school of fish tends strongly to behave as an entity, the school remaining intact or re-forming quickly in spite of a wide variety of outside forces. Thus, where it may be necessary to gather nonschooling fish together to achieve organism densities adequate for harvesting, the harvest of schooling fishes might be initiated through either attracting or herding the entire school.

Anadromous Species

Naturally anadromous organisms, exemplified by the salmonid fishes, return unerringly to their fresh-water hatching site to spawn after achieving sexual maturity in the oceans. Recent work at Woods Hole Oceanographic Institute indicates with little doubt that anadromous species select their hatching site from among all others by highly sensitive chemical discrimination [1]. It has not yet been determined whether this chemically discriminatory homing capability exists with nonanadromous migratory species, such as members of the tuna family. If it does exist, or could be developed through selective breeding, one form of open sea mariculture might rely on seasonal chemical attraction of migratory stocks into enclosures in the open sea— either the same enclosures from which they were released as young or other ones located in the same area and having an identical chemical signature.

Large Marine Animals

Chapters 8 and 14 mention the possibility of culturing large organisms such as sea turtles and whales. Although we do not wish to discount these possibilities completely, such animals do not lend themselves well to either automated harvesting or automated processing, because of the large size of each individual and the comparatively low number of individuals usually involved in each harvest. Therefore, harvesting and processing systems for very large animals would almost certainly remain labor intensive and, as such, comparatively unattractive for large-scale open sea operations other than those located on atolls.

CONTEMPORARY OPEN SEA CONCENTRATION AND HARVESTING METHODS

Most of the current techniques for concentrating and harvesting discussed in this brief survey are employed by, or proposed for, open ocean fisheries. Notable exceptions are the methods for sessile-organism harvesting that presently are employed in estuarine oyster and mussel culture operations, the techniques used in anadromous fish culture, and recent sea cage experiments.

Mollusk Culture

At the present time, oysters and mussels are the only mollusks cultured extensively. In Japan, the occurrence of typhoons may temporarily create heavy seas in the inland sea culture areas; in other parts of the world mollusk farming is confined to protected waters. Present culture methods appear primitive and labor intensive in comparison to techniques that technologically would seem to be possible. Typically, the shellfish are grown on ropes, 3 to 10 m long, that are suspended from rafts or bottom-mounted racks. At harvest time the ropes are taken up individually and transferred to a harvesting facility on shore, where the shellfish are removed manually. Interestingly, the available literature on this subject contains little information on the harvesting mechanisms employed. So far, apparently, problems of production supersede those of harvesting the product [2–4].

Benthic Crustacea

Benthic crustacea may be divided roughly into two categories based on the type of bottom they typically inhabit—sand (or mud) versus rocky terrain.

Smooth-Bottom Dwellers. Penaeid shrimp typically inhabit open sand or mud bottoms. Recent work indicates that the lobster, *Homarus americanus,* also spends much time on such terrain, particularly in deeper or dirtier waters [5]. Shrimp, and possibly lobsters too, lie buried during the day and ascend during the hours of darkness to the surface of the sand or mud and the water somewhat above the bottom in order to feed. Thus, classically, shrimp trawling has taken place at night. Now, however, Electroproducts Incorporated of Pensacola, Florida, has placed research conducted by the National Marine Fisheries Service (NMFS) into commercial application with the production of an electric trawl net [6]. The device employs electrodes to induce low-voltage direct current at four or five pulses per second just ahead of the trawl net. The electrical current causes the shrimp to jump a few centimeters off the bottom, just enough to be caught by the trawl. A roughly similar device has been tested in Rhode Island waters with lobsters, crabs, and benthic fish [5]. The testing of the latter device was not extensive, however, and the results, although encouraging, were not entirely conclusive.

Rough-Bottom Dwellers. For bottoms that feature rough terrain, the only harvesting devices in practical use today are variations of traps, although the NMFS hopes soon to begin development of an off-bottom electric trawl for harvesting shrimp in rough-bottom areas [7]. Organisms are attracted to the traps either by food bait or by the offer of concealment. Once inside, they find egress difficult or impossible, and so remain there until hauled to the surface and harvested. Some modification of the trawling technique might offer utility to open sea mariculture, since it seems potentially productive and non-labor-intensive where dense populations of animals are to be harvested; but it appears that the trapping technique would require major modification to be of value. Yet the fact that simply the offer of concealment can attract marine crustacea in harvestable numbers is extremely interesting. This is explored further in succeeding pages.

Finfish

Concentration or Containment. The initial problem to be solved in harvesting commercial quantities of finfish is to find or attract them in sufficient numbers to a point of harvest. Much of the work in this field is still experimental, but some applications of at least two techniques, attraction to light or shelter, have proven utility. An alternative to attraction is found in containment, or the control of the location and movement of fish during their growth and maturation phases prior to harvest time. Methods for achieving containment are almost entirely in the developmental stages. Here both fields are examined.

Attraction by Light. At night it may be feasible to attract both schooling and nonschooling fish with strategically placed lights. Nocturnal attraction to lights combined with a variety of fishing gear actually has been in use for years. To date most species captured in number with this method have been of the schooling varieties. Methods most recently described by Wickham [8, 9] involve the use of a continuous, 1,000-watt, mercury-vapor underwater lamp in conjunction with either lift nets or purse seines; the fish are attracted to the light and then captured by either type of net. Catches of coastal pelagic fish have reached thousands of pounds per night with this technique, and it seems to be most productive during periods of the new moon. As mentioned, however, the predominant species captured so far have been schooling fish; to the best of our knowledge, no one has yet experimented specifically with the effectiveness of light attraction for nonschooling fish.

Attraction by Sound. Work in the Bahamas [10] indicates that some species of predatory fish can be attracted with randomly pulsed, low-frequency, 25- to 50-hertz (Hz) sound. In the experiments reported, groupers, snappers, margote (grunts), and sharks were attracted to the sound-generation area. These tests were limited in scope, however, and concluded little more than that certain types of predatory fish were attracted by some types of sound, probably because the sounds approximated those made by feeding fish and prey. It is not yet known how many species could be attracted by this device or whether the technique could operate effectively on commercially significant scales. One problem relates to the acoustic receptors of the fish themselves; it concerns the hypothesis that, although fish may be able to detect sounds in both the near and far fields, they can localize sound sources only in the near field [11–13]. Thus, sound sources strong enough to attract fish from a wide area seem likely to be so strong as to repel closely approaching fish. Nonetheless, it seems likely that fish can be attracted by such devices from a radius of about 100 m, which might easily be adequate for some mariculture harvest systems.

Attraction to Shelter. The ability of almost any sort of structure in the marine environment to attract marine life is widely recognized. Artificial reefs constructed of concrete modules, junked automobile bodies, and derelict streetcars have been found to attract dense marine biomasses in both temperate and tropical waters [14–16]. Worn-out automobile and truck tires have been tried as artificial reef bases with considerable success [17]. Offshore petroleum platforms, too, attract marine life in large numbers [*15, 18*]. This phenomenon is discussed further in Chapter 12. Recently, the Exploratory Fishing and Gear Research Base of the NMFS at Pascagoula, Mississippi, has experimented with the effectiveness of various tent-like floating structures deployed at the surface and in mid-water. In these experiments it was found that the "fish tents" attracted hundreds of jacks and thousands of baitfish, although the effectiveness varied greatly with time and mode of deployment [*19*]. (We might note here that our literature surveys indicate that the NMFS facility at Pascagoula is presently at the forefront of research into fish attraction and harvesting; it should, for some time to come, remain an excellent source of current information on these subjects.)

Containment in Net Enclosures. The Japanese have for some years used fairly large, raft-supported net cages for finfish in their Inland Sea, and the Australians, English, and Americans are also experimenting with such structures. All designs presently in use consist of a flotation device forming the surface perimeter, with a net cage suspended from it. The flotation gear frequently acts as a work platform too. Where predation is a problem, the net may be two-layered with a coarse, strong, predator-inhibiting net outside of a finer crop-enclosing net [*4*]. Figure 10.1 provides a generalized schematic of this technique. It has been demonstrated to essentially everyone's satisfaction that this structural approach at its present stage of development will not survive heavy weather in the open sea. For open sea culture we shall either have to find stronger materials and more durable designs, or devise means for protecting cages from extreme seas.

Here the Japanese appear to be in the lead. A large fishing firm, Taiyo Gyogyo, and a shipbuilding firm, Niigota Engineering Company, teamed in

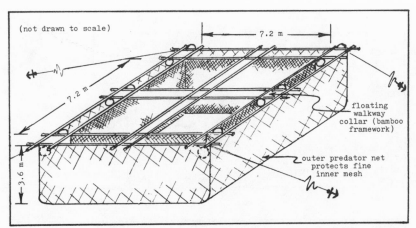

FIGURE 10.1 *Double-layer net cage for unprotected waters. (Adapted from Milne [4].)*

1971 to develop submersible fish enclosures. The design under development consists of an octagonal structure of steel pipe 10 m wide by 5 m deep on which wire mesh is hung. In the center is a flotation buoy with mechanics for blowing and flooding. As Figure 10.2 shows, a taut-wire mooring system is employed in such a way that the cage can be elevated to the surface or submerged well below. The mechanism is radio-controlled. During fine weather the cage remains at the surface; when extreme conditions threaten, it can be lowered to calm subsurface water. At present, experiments are being conducted in 60 m of water; it seems that the design, if practical, could certainly work in 100- to 200-m depths also, thus opening many thousands of square miles of offshore waters to the potential culture of contained finfish [20].

Containment by Bubble Fence. Pneumatic barriers, or bubble fences, have been tried in shallow waters as containers and herders of finfish by a number of investigators, and a few commercial units have been built. Generally speaking, this approach appears to be effective with some species at least so long as the bubble screen rises vertically. Horizontal deflection of the rising bubbles by currents or too rapid movement of the bubble generator (in herding) causes marked loss of effectiveness.

The device is simple in concept; it consists of submerged hose or pipe that contains small holes and is shaped to form the horizontal pattern of an enclosure. It is supplied with air from a compressor. Among the apparent advantages of this technique are automatic aeration of the contained water and a tendency to reduce surface wave action, and so to maintain calmer water within the enclosed space. Some disadvantages include the probably limited effectiveness of the device in excluding predators, potentially dele-

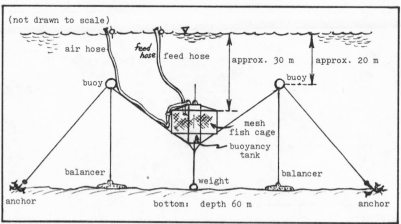

FIGURE 10.2 *Submersible fish enclosure with taut-wire moor. (Redrawn from Sakae [20].)*

terious effects on water temperature, and the possibility of transporting contaminating substances from the bottom into the rearing area.

Other Controls. Electrical barriers have been tried only in fresh water and with limited success. They are expensive to construct and emplace, and their very large power requirements and dubious effectiveness make them unlikely candidates for marine use [4]. In discussing the homing behavior of anadromous fish, we mentioned the speculation that chemical cues might be "programmed" into some other species of finfish. There is little to add to that comment, as yet.

Harvest Techniques. Attention is now given to techniques for harvesting finfish, once sufficient concentration has been achieved. In this necessarily condensed summary we exclude from consideration those methods appropriate only for limited-scale fishing as well as methods with high labor requirements, on the premise that both obviously are of little interest to open sea mariculture. The following techniques do appear to have the potential for meeting the functional requirements for mariculture harvest, which we reviewed earlier.

Purse Seines. Purse seines are a variety of fishnet with a long history of use; they are commonly used for both large and small schooling fish. The net may be more than 1 km in length and several meters deep, with a buoyed top and weighted bottom. One end of the net, usually attached to a skiff, is dropped from the purse-seine boat first. The boat then attempts to encircle a school, paying out the net as it completes a circle back to the skiff. Next the weighted bottom edge is drawn together to close the net into a purse and so prevent the escape of the encircled fish through the bottom. Finally, it is gradually hauled aboard by heavy deck gear and the netted fish are transferred to the hold. Purse seining of an already concentrated fish population could be a relatively non-labor-intensive operation. Thus, this classic method might find application in some forms of open sea mariculture, particularly if the open sea rearing enclosures could operate as purse seines at harvesting time.

Lift Nets. Lift nets are ring-shaped seines having a top cylindrical section over a conical section, as shown in Figure 10.3. They are small in size, usually less than 7 m in diameter. The lift net is submerged until a sufficient number of fish is attracted over it, and then raised so that the perimeter is above the surface, while the main body of the net remains submerged. The fish thus are entrapped and may be hauled aboard with the net by the fishing vessel's deck gear; the bottom of the conical section opens to drop the catch into the hold [8]. Lift nets used more or less continuously over an extended period of time might harvest at rates compatible with on-site processing (Chapter 11) in open sea mariculture operations.

Fish Pumps. Most fish pumps today are employed with baitfish and fry, but at least one model (originally designed for moving fruits and vegetables in

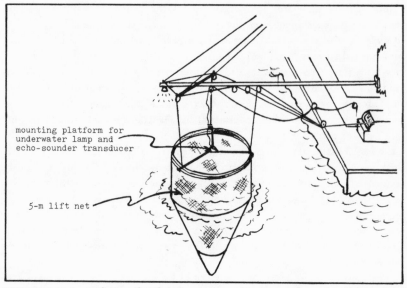

FIGURE 10.3 *Lift net for fish harvesting. (Redrawn from Wickham, [8].)*

canneries) has been used for fish weighing up to ¼ kg. The latter has a spiral impeller, refined to eliminate sharp contours and obstructions and so to allow smooth passage of delicate products. Such pumps have been used with a high degree of success and with low injury and mortality rates in harvesting both active and electrically stunned fish. Since power requirements tend to increase sharply with the volume of water raised a given distance, it seems likely that fish-pump efficiencies may not be competitive with other means for very large fish. Consequently, one would expect to see their use limited to fish of small to moderate size, for the immediate future at least [21–23].

The USSR fisheries are employing an air-lift suction device in the Caspian Sea that seems attractive. In this approach a flexible hose several tens of meters long, with a diameter of 200 mm, is set in a U shape. The collecting funnel, over which light attraction devices are set, faces upward at the end of one leg of the U. Air is injected continuously into the lower portion of the other leg; as the rising air expands, it entrains water and creates a lower-weight mixture in the air-injected leg than in the collecting leg. Consequently, a flow sufficient to suck active small fish into the collector and to transport them through the U to the deck of the collecting boat is produced [24].

Because fish usually have a very slight negative buoyancy, pumping-power demands can be minimized for harvesting with fish pumps if the escape activity is reduced or eliminated. Klima describes three progressive reactions of finfish to electrical fields of increasing strength: (1) fright, (2) electrotaxis (swimming toward the positive pole), and (3) tetanus (stunning). There are problems in creating adequate electrical fields in seawater; but they seem to be surmountable with the employment of capacitors producing pulsed direct current. It has been shown too that fish reactions to electrical currents vary with species and size; a pulse that produces electrotaxis in one group may produce fright or no response in another. These phenomena, from the harvesting viewpoint, may be either beneficial or detrimental or both. In any event, Klima has proposed that pulsed direct-current fields can be employed to concentrate and condition fish sufficiently for harvesting by means of fish pumps [25]. This concept is shown in Figure 10.4.

SPECULATIVE OPEN SEA CONCENTRATION AND HARVESTING SYSTEMS

Having taken an admittedly cursory overview of current concentrating and harvesting methods that may prove applicable to mariculturing, we now are ready to speculate on possible techniques of the future for containing and gathering in marine food crops. We emphasize that they are, so far, speculative indeed; we have not investigated them in any depth, and at most they point to some potential research-and-development avenues.

Oyster and Mussel Culture

The comparatively labor intensive practices in use with shellfish culture today have been mentioned already. These could be improved upon in the design of tomorrow's open sea mariculture systems. If one assumes the achievement of large-scale, nutrient-rich artificial upwellings and the concomitant primary production expected to occur, attention is immediately directed to questions of efficiently employing the planktonic product. Because of high productivity, advantageous position in the trophic chain, and

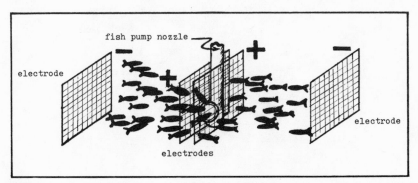

FIGURE 10.4   *Pulsed direct-current electrical field used to concentrate and condition fish for harvesting. (Redrawn from Klima [25].)*

theoretical ease of containment in open sea environments, oysters and mussels come immediately to mind.

To overcome apparent deficiencies in present practices, and to achieve large-scale operation, we consider the scheme in Figure 10.5. Removable growing racks are suspended from horizontal flotation cylinders. The racks are sufficiently thin that mollusk growth occurs for the most part only on the two sides of the large plane. Thus, the attachment surface, although holding organisms on both its sides, is itself two-dimensional. This would enable mechanized removal (harvesting) by drawing the rack through closely spaced scraper blades. An alternative possibility would involve coating the surface with a material that could be peeled off at harvest, taking the attached mollusks with it.

After harvesting, the racks might be placed in containers of settling spat for reseeding and then restored to their open sea flotation cylinders to begin another growing cycle. We shall return shortly to this operation and develop the idea further.

The flotation cylinders might be of either concrete or steel, probably filled with styrofoam. If constructed of concrete, they probably would be a series of short, linked cylinders rather than one continuous cylinder. This might also prove to be the best design for steel cylinders. In fact, connectable modules of, say, 15 m in length might prove to be the most attractive approach from several viewpoints; they would probably prove more economical to build and handle, and could be linked together to form growing rafts of varying lengths as conditions warranted. The growing racks themselves might be made of a flexible rubberized or plastic material. If so, they would require

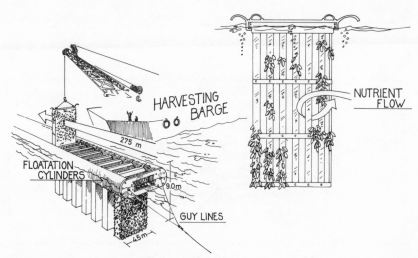

FIGURE 10.5 *Conceptual drawing of an open sea oyster cloister.*

weighting to hold them within a reasonably narrow vertical range and to keep them from colliding in currents and extreme seas.

An immediate question concerns how such racks might be sown with spat or seed oysters or mussels. Current information (for *Crassostrea gigas*) indicates that oyster spat typically settle best on horizontal surfaces and not deeper than 3 [26, 27]. One means of dealing with this situation might involve making the two surfaces of the racks into a series of hinged buoyant flaps or louvers. The flaps would float up to horizontal positions when bare and so be in that position for sowing. As the shellfish matured, their weight would cause the flaps to swing downward into the growing–harvesting position. Of course, this would work only if few or no spat settled on the undersides of the flaps. We do not know if the upper–lower spat-settling dichotomy needed could be achieved in practice. But since there is a good evidence that spatfall is higher on surfaces which contain oyster metabolic products than on those which do not [28, 29], it seems possible that chemical attractants and repellents might be developed to achieve this capability. If not, it might be feasible to design the louvers to rotate through 360° with the edges just missing one another. Thus, the louvers could be set alternately with first one, and then the other, side upward during the settling process. They could then be turned to the vertical position for rearing. This latter concept, however, requires the inclusion of mechanical apparatus in a structure to be immersed in seawater for extended periods. The design would require an inexpensive and reliable louver-turning mechanism that could operate in spite of fouling, which might be difficult to achieve.

Differential attraction of settling spat might hold an answer to the thinning problem, too. At present young oysters are usually thinned by hand, which probably would be a prohibitively labor intensive approach for a large-scale open sea operation. Is it possible that one might space the attractant-coated surfaces in such a pattern that thinning requirements could be reduced or eliminated?

We were able to find little information on the settling of mussel larvae. In the Galician region of Spain in 1971, Nash [30] noted that new ropes were hung in the vicinity of fattening mussels during the spawning period and thereby became so heavily seeded that two or more thinnings were required as the mussels grew. Milne [4] tends to confirm this in his reviews of other culture methods. The ropes were approximately 3 to 10 m long and apparently spatfall occurred over their entire length. With some developmental research, it seems then that we might expect to achieve direct settling of mussel spat in the open sea.

The length of time both oyster and mussel spat may require for settling and becoming reasonably resistant to competition and predation seems to be sensitive to a variety of factors, including temperature, light, and nutrition [28, 29] and so may require from one to several days, depending on conditions. This length of time might argue for temporary enclosures of some plastic material that could be maintained around the rafts and racks during the seeding process as they hung directly in the sea.

Other Mollusk Culture

Clams. Clams are cultured in some volume in the United States, although at present only *Mercenaria mercenaria* and *Mya arenaria* are employed. The main requirements seem to be a suitable substrate into which the clams can burrow and protection from predators, particularly sea stars and crabs. The Virginia Institute of Marine Science (VIMS) has determined that spreading shell, gravel, or aggregate over the burrowing (sand) substrate is an effective protection against predators [31]. This method can be applied even before seed clams are added, and so can be used on natural seabeds. Off exposed coasts, however, one would expect wave and current action to transport any but the heaviest surface layer; consequently, the technique probably is not practical for true open sea mariculture on the natural seabed except where unusually stable bottom conditions may exist, as in atoll lagoons.

If an inexpensive large tray device could be designed for association with Texas-tower-type platforms, it might be possible to culture several layers of clams in trays filled with light coral sand 15 to 30 cm deep—assuming, of course, that adequate nutrition would be available for these filter feeders. Might this approach be extended farther into the open sea? Buoyant or bottom-mounted submerged structures containing multiple tiers of such trays seem to be a possibility, although clever design would be required to make them durable and economical. Certainly, if any such grandiose schemes as the seabed system presented in Chapter 12 become practical, the addition of tiered clam trays might be attractive. One possible application of this technique is shown in Figure 10.6. Multiple clam-raising trays are stacked in a frame structure that is slightly buoyant. The entire structure can be raised to, or near, the surface during calm seas to facilitate growing, servicing, and harvesting. But it can be submerged more deeply during heavy weather to escape the scouring action of heavy surge. For a bottom-mounted structure, negative buoyancy and legs would be added. Harvesting would be accomplished with a suction dredge, which would take sand and clams to a surface facility, where the clams would be screened out and the sand returned to the tray for reseeding. The more obvious questions raised by this idea involve the economical design and operation of the structure, the adaptability of clams to both stacked-tray culture and rapid changes in pressure, and the provision of adequate nutrition to a clam population of such high density.

Scallops. Experiments in culturing scallops are underway in Japan and the USSR [4]. The Russians are working with forms that are naturally free living on sand bottoms. The Japanese, on the other hand, are experimenting with means of containing or attaching the animals to ropes, as is done with oysters.

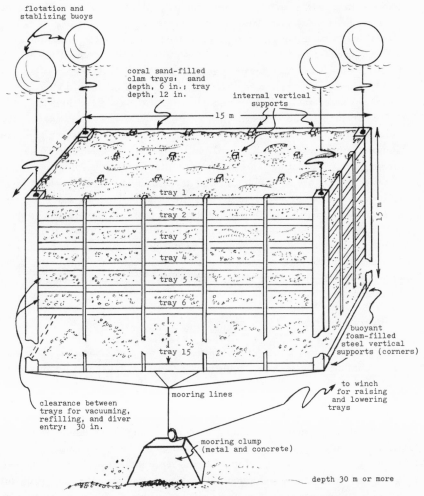

flotation and
stablizing buoys

coral sand-filled
clam trays: sand
depth, 6 in.; tray
depth, 12 in.

internal vertical
supports

15 m

15 m

15 m

tray 1

tray 2

tray 3

tray 4

tray 5

tray 6

tray 15

buoyant
foam-filled
steel vertical
supports (corners)

to winch
for raising
and lowering
trays

mooring lines

clearance between
trays for vacuuming,
refilling, and diver
entry: 30 in.

mooring clump
(metal and concrete)

depth 30 m or more

FIGURE 10.6 *Conceptual drawing of a subsurface clam culture device.*

With modifications, it would appear that the subsurface clam grower might be applied to free living scallop culture as well. If the device will work for clams, it seems that scallop culture could be achieved by adding external netting to prohibit the escape of these erratically swimming mollusks.

Abalone. In California, at the Cal-Marine Abalone Farms, well over 1 million abalone ranging from 3 weeks to 3 years in age are under culture in a small mariculture laboratory. Although they require 6 years to reach their full growth of about 17 cm (weight, 1 kg), a market demand is developing in Japan and Hawaii for the 3-year-old, 6-cm product. Adult stock for controlled monthly spawning comes from the approximately, 1,500 red abalone,

Haliotis rufescens, purchased from commercial divers who work in the adjacent Estero Bay. The laboratory is selling abalone seed stock to experimental sea farms as well as raising stock. The larvae are fed macroscopic algae and the juveniles feed on bull kelp, *Nereocystis,* from the bay. The environment for both is aerated seawater pumped from the bay. At 3 years of age the juveniles are transferred to multiple standing panels in an outdoor concrete basin. The enterprise projects an annual harvest of 1 million abalone [*32*].

The Japanese are also experimenting with abalone (*Haliotis discus*) culture [*4*]. At present, hatchery work is producing juvenile abalone in significant numbers; upon reaching a 15- to 20-mm size, they are removed to natural growing areas for maturation because of limited onshore rearing space. Milne states that there are no problems other than space that would prohibit pond culture of abalone using the same corrugated plastic sheets on which they are now raised to the 15- to 20-mm size. It seems possible, then, that once again modification of the clam device might be applied to abalone rearing in the open sea. The necessary modifications would include not only enclosing the frame with netting to prevent the mobile mollusks from crawling off the growing sheets and falling to the bottom, but also probably changing the horizontal trays to vertical sheets.

Crustacea Culture

Shrimp. Shrimp (particularly *Penaeus* sp. and *Metapenaeus* sp.) are cultured commercially in many countries in shoreside tanks and ponds and in intertidal enclosures of various designs [*4*]. So far, it appears that only in Japan, where wholesale prices are excessively high, is commercial shrimp culture financially vigorous. Problems that are being partially solved in a variety of ways include obtaining juveniles, maintaining temperature and salinity within acceptable limits, oxygenation of the growing medium, and nutrition. Assuming that a solution to the reproduction problem is attained, it may be that some development of the multitray structure suggested for mollusk culture might help solve the other problems of shrimp culture. If net-enclosed trays were filled with an appropriate substrate, they could provide for very high density rearing. Oxygenated and plankton-rich water could be supplied naturally by current action or artificially by pumping, possibly employing a pump driven by a conservative energy source. Placement in the open ocean could provide stabilized salinity and temperature. Furthermore, in nature at least, shrimp are known to feed most actively at night. Might growing rates be increased by shrouding these structures so as to reduce light intensity and increase feeding time per unit time ratios? If rearing in these structures could be achieved economically, harvesting could be accomplished nicely with a combination of pulsed direct current to eject shrimp from the substrate and a fish pump to carry them to receiving bins.

Lobsters and Crabs. Lobsters and commercially important crabs are typically secretive, carnivorous, and aggressive to the point of cannibalism. In nature, cannibalistic behavior is modulated by actually or potentially low population densities, which allow most of the molting animals, in which condition they are the most vulnerable, to avoid potential aggressors. In most lobster culture attempts, high population densities and lack of concealment opportunities have resulted in a high incidence of cannibalism and low harvest. There is some indication that pheromones may be effective in suppressing the cannibalistic tendencies of lobsters [1], and intensive feeding also has a positive effect. Yields might be enhanced further if concealment could be added. It might, therefore, be worthwhile to look into something along the lines of the preposterous looking device shown in Figure 10.7. Neoprene pockets provide a degree of concealment and a defensible niche for the growing and molting creatures. When the buoyancy chamber is water-filled, the device has strong negative buoyancy; when the chamber is air-filled, buoyancy is only slightly negative. After seeding with a suitable number of young, the device is set on a sandy bottom in 10- to 70-m depths in moderate currents. Screens at the upper and lower ends provide for water exchange; feed is supplied through a hose from the surface. For harvesting, air is injected through the feed hose to the buoyancy chamber and the lightened structure is raised to the surface. There it is taken aboard a harvesting barge by crane, the base plate is removed, and the entire structure is set on an

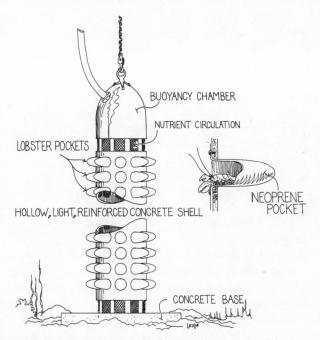

FIGURE 10.7 *Conceptual drawing of a lobster rearing and harvesting device.*

airtight gasket over a partially water filled receiving tank. An airtight collar is fitted around the upper screened area and, since the bottom screen is below the gasket, a light vacuum can then be drawn within the system, causing the neoprene pockets to evert and the lobsters to be ejected into the water-cushioned receiving tanks. The drawing is conceptual only. It is almost certain that the device would not be practicable in the form presented here. For scale economy many more pockets would be necessary. Means for enabling the lobsters to reach the pockets and the food to reach the lobsters also would be needed. Perhaps the rearing position of the device should be horizontal rather than vertical. Nonetheless, it could prove to be a rewarding avenue to explore.

Finfish Containment

There are a number of ideas abroad for employing natural phenomena to concentrate finfish in restricted locales within larger bodies of water, including sound, electrical fields, chemical attractants, bubble fences, and of course physical enclosures. Nevertheless, physical containment appears to be the brightest near-term prospect. The more attractive prospects in both categories are discussed here.

Submersible Cages. Probably the most attractive development avenue at the moment lies in carrying the Taiyo–Niigota submersible-fish-cage concept a few steps farther toward larger enclosures, deeper operational waters, and lowered manufacturing and operational costs.

Certainly, one area worthy of considerable attention is that of long-lasting, fouling-resistant meshes [4]. In his book on fish and shellfish farming, Milne provides a useful discussion of meshes and fouling. Of the commonly available materials, he concludes that galvanized weldmesh appears superior. However, this material tends to succumb to corrosion and would require replacement every 1 to 3 years, depending upon local conditions and type of use. If the demand developed, it seems that copper–nickel alloys could be formulated for enclosure meshes. These alloys are well known for both their antifouling properties and comparatively high cost. Yet, if replacement periods could be lengthened significantly, overall economy could well justify the initial investment.

Another avenue for investigation might lie in the use of antifoulant-impregnated plastic coatings for both plastic and metal meshes. So far as we were able to learn, no work beyond that relating to ship-bottom paints has been done in this realm.

Still another research avenue might lie in the use of the natural predators of fouling organisms. Mullet, for example, will graze on algae and diatoms attached to mesh, as will a variety of other species. Other species might control ascidians and mussels. The required research involves selecting species that could survive in the culture environment and control mesh fouling without disturbing other critical parameters of that environment. With answers to these factors, marine systems design teams could begin to develop designs which could tolerate the degree of fouling that would exist at the point where fouling organisms and predators established ecosystem equilibrium.

Finally, one other solution to the marine-enclosure fouling problem was suggested by Caillouet of the University of Miami [*33*]. The proposed technique involves the use of a slowly, or intermittently, rotating cage; a large portion of the vulnerable area is above the surface repeatedly and for extended time periods. The exposure allows drying, heating, and direct ultraviolet rays in sunlight to kill young fouling organisms before they reach sufficient maturity to become resistant. The price paid by this approach is the loss of usable rearing volume and the power and mechanics required to rotate the cage. Furthermore, repeated wetting and drying in the marine environment is frequently the most corrosive situation to which structural materials can be exposed. Consequently, materials that are highly corrosion resistant would be required for the construction of such a device.

In pondering this idea in the context of large-scale finfish culture in the open sea, we carried the concept somewhat farther, as shown in Figure 10.8. In this conceptual version, the large, drum-shaped cage is rotated slowly by alternately blowing and flooding buoyancy tanks at the periphery of the drum. The rotating force would be applied by blowing the buoyancy chambers after they had passed bottom-dead-center, thus applying a lifting force through a quarter of a revolution. As the tank entered the water on the

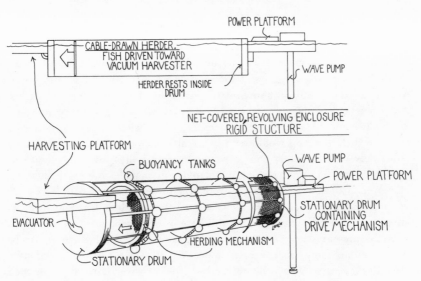

FIGURE 10.8 *Conceptual drawing of a large-scale, open sea, rotatable fish enclosure.*

opposite side, the vent holes would face upward and one-way valves would allow them to fill. Air pressure and volume requirements would be low since depths would not exceed 1 atmosphere pressure and slow blowing of the tanks would be entirely acceptable. Consequently, it seems possible that a low-density, conservative-energy device might provide the power (see Chapter 13). If this concept was to be developed to the point of feasibility, it might be combined with the Japanese idea for submerging cages during heavy weather to achieve a survivable system. The questions left unanswered in this notion are many, but it offers yet one more potential research-and-development avenue.

Bubble Fences. Experiments, and indeed limited commercial applications, with bubble fences offer some encouragement that they might be developed into effective enclosures for selected sites. We already know that they tend to lose effectiveness as the tracks of rising bubbles digress from vertical. This probably indicates an inappropriateness for areas having even sporadic currents of any strength; it would, of course, exclude areas in which occur winds of sufficient strength and duration to create any significant Ekman drift. It is not yet known which species could be contained (or excluded) with the bubble fences or how long the fences would remain effective, that is, how rapidly each species would adapt and learn to ignore the essentially harmless fence. Moreover, all work so far has been done in shallow water. In all probability, operation in deep water would exhibit different characteristics, but we cannot guess which characteristics would change and to what degree.

If one ignores all the unknowns and assumes that bubble fences could be effective in deep water, some exciting possibilities emerge. Assume the water depth to be 600 m; this is potentially sufficient to reach waters rich in inorganic nutrients. Conceivably, an enclosure several hundred meters in diameter could be created costing no more than the price of the bubble-generating ring, its weight, and the equipment that supplies it with air or oxygen at about 800 pounds/square inch (psi). The advantages of such a device would be multifold: it would be survivable in any weather; it would not foul; it would automatically oxygenate the enclosed water column; the enclosed surface waters would tend to be calm since the surfacing bubbles are effective wave absorbers, and it would automatically provide a degree of nutrient upwelling by entraining bottom water with the rising bubbles. Moreover, if open sea mariculture were ever to be associated with large-scale offshore hydrogen production (see Chapter 13), the bubble fence might be an attractive use for the waste oxygen this latter process produces.

At the present stage of our knowledge, deep-water bubble fences can be viewed as no more than a fond hope. Yet it seems that their potential attractiveness is so great that research—sufficient at least to disprove their feasibility, if nothing else—is advisable.

Conditioned Behavior. A notable difference between aquaculture and fishing is that in aquaculture the organisms may be under some human control during all or most of their lives. Therefore, opportunities to condition organisms to engage in desired behavior on cue should not be ignored. It is well demonstrated that fish can be conditioned to congregate at a feeding site [4]. Therefore, it seems that one of the most attractive approaches to achieving harvesting densities in aquaculture systems lies in just this sort of conditioning. Once harvesting densities are achieved by feeding cues, the final step can be accomplished with seines, lift nets, or fish pumps.

CONCLUSIONS

We have reviewed the present state-of-the-art in concentrating and harvesting marine organisms and then proceeded to apply imagination and creative speculation to shape some innovative concepts for application in the open sea that might eventually be developed to a point of practicality. Although these speculative concepts are just that and no more, they may point the way to further creative investigation that will someday prove fruitful. Specific areas that appear to merit on-going research-and-development efforts are offered as a conclusion to this discussion.

Open Sea Enclosures

Most experience with floating enclosures developed to date has shown a complete lack of survivability under rough water conditions and vulnerability to fouling. The recent Japanese approach, using submergence to escape rough water, seems at the moment to offer promise. We suggest, then, that a modest but well-focused research program for the development of open sea enclosure technology is in order now. Primarily, this program should consider survivability, fouling, longevity and maintainability, operational expense, and initial-cost minimization. All construction and protective materials should be catalogued and evaluated, as well as all possible mechanisms for avoiding rough water, practicable construction methods (with emphasis on modular approaches), and non-labor-intensive procedures for maintenance and servicing at sea. The same thing should be done for all possible techniques to achieve biological and chemical control of fouling, and to assure compatibility with cultured crops. Because of the high degree of interaction among the various parameters of this problem, this should be an integrated attack pursued by one or a few well-coordinated institutions rather than a series of small, piecemeal, research activities. Since the likelihood of beneficial information transfer seems high, the pursuit of finfish cages and mollusk and crustacea enclosures by the same team is recommended.

Bubble Fences

Although their practicality is in doubt, the possibility of bubble fences in deep waters appears to hold such a high potential that, as a minimum, a research program sufficient to disprove feasibility is warranted. Generating sizable bubble fences in deep water probably is within today's capabilities, given an adequate applications development program. We suggest that two concurrent initial research thrusts should focus on (1) biological responses to bubble curtains, and (2) the behavior of bubble curtains generated in water depths of at least 600 m. These two avenues could probably be pursued effectively for some time by different institutions.

Fish Pumps

To the technologically oriented mind, harvesting by "batch" methods, such as nets and trawls, appears a temporary necessity to be eventually designed away. Batch harvesting causes queues in the harvest process cycle which necessitate holding, and that in turn necessitates temporary preservation of spoilage-prone products. Continuous processing from the point of harvest to the point of final preservation seems preferable if it can be achieved economically. As a beginning, further development of fish-pumping technology is recommended. This should include high-concentration techniques where needed, means of rendering inactive the larger and more vigorous species, and pumping technology for organisms up to 5 kg. The first two developments might be pursued by a single integrated research team. The last could probably be accomplished best by a competent hydraulic engineering firm.

Biological Factors Pertinent to Containment and Harvesting

Spatfall and the growth characteristics of sessile organisms; the adaptability of mollusks and crustacea to proposed methods of high-density culture; population density tolerances of finfish; species and size variability in responses to electrical fields by finfish and crustacea; chemical communication and its effect at least on finfish homing behavior and crustacean aggressiveness; natural and conditioned responses to auditory, visual, and chemical signals; the growth rates of mollusks and crustacea as a function of light; and probably many more factors are all pertinent to the concentration and harvesting of marine organisms generally. We suggest that each offers a potentially rewarding research program for a competent institution.

REFERENCES

1. Todd, J. H., J. Atema, and D. B. Boylan. 1972. Chemical communication in the seas. *J. Marine Tech. Soc.* 6(4).
2. Fujiya, M. 1970. Oyster farming in Japan. International Helgoland Symposium. *Helgolander wiss. Meeresunters. 20:*475.
3. Figueras, A. 1970. Flat oyster cultivation in Galicia. International Helgoland Symposium. *Helgolander wiss. Meeresunters. 20.*
4. Milne, P. H. 1972. *Fish and shellfish farming coastal waters.* Fishing News (Books) Ltd, London.

5. Saila, S. B., and C. E. Williams. 1972. An electric trawl system for lobsters. *J. Marine Tech. Soc. (Sept.-Oct.).*

6. Anonymous. 1969. Turning day into night. *World Fishing* (Mar.).

7. Klima, E. F. 1973. Personal communication.

8. Wickham, D. L. Collecting coastal pelagic fishes with artificial light and 5-meter lift net. *Com. Fisheries Rev. 898.*

9. Wickham, D. L. Harvesting coastal pelagic fishes with artificial light and purse seine. *Com. Fisheries Rev. 900.*

10. Richard, J. D. 1968. Fish attraction with pulsed low frequency sound. *Can. Fisheries Res. J.* (July).

11. Dijkgraaf, S. 1963. The functioning and significance of the lateral-line organs. *Biol. Rev. 38:*51–105.

12. Harris, G. G., and W. A. van Bergeijk. 1962. Evidence that the lateral-line organ responds to near-field displacements of sound sources in water. *J. Acoust. Soc. Amer. 34*(12):1831–1841.

13. Lawenstein, O. 1957. The acoustico-lateralis system. In M. E. Brown, ed., *The physiology of fishes.* Vol. 2, *Behaviour.* Academic Press, Inc., New York.

14. Carlisle, J. G., Jr. et al. 1964. Artificial habitat in the marine environment. *Calif. Dept. Fish and Game Fishery Bull. 124.*

15. Turner, C. H., et al. 1969. Man-made reef ecology. *Calif. Dept. Fish and Game Fishery Bull. 146.*

16. Anonymous. 1972. Artificial reefs of waste material for habitat improvement. *Marine Pollution Bull 3*(2).

17. Anonymous. 1972. Australia builds artificial havens to attract fish. *Fishing News Internatl. 9*(9).

18. Thompson, R. R. et al. 1972. Cooperative environment projects, High Island Block 24L, Offshore, Texas. *Offshore Tech. Conf. Preprint 1676.*

19. Klima, E. F., and D. A. Wickham. 1971. Attraction of coastal pelagic fish with artificial structures. *Trans. Amer. Fisheries Soc. 100*(1):86–99.

20. Sakai, K. 1972. Fish farming machines from Japan. *Fishing News Internatl. II*(8).

21. Bedell, G. W., and P. D. Flint. 1969. Pumping fish in California. *Prog. Fish-Cult. 31*(4).

22. Grinstead, B. G. 1969. A fish pump as a means of harvesting grizzard shad from tailwaters of TVA reservoirs. *Prog. Fish-Cult. 31*(1).

23. Baldwin, W. J. 1973. Results of tests to investigate the suitability of fish pumps for moving live baitfishes. *Prog. Fish-Cult. 34*(1).

24. Anonymous. 1969. Fishing by suction. *World Fishing 18*(8):22–24.

25. Klima, E. F. 1970. An advanced high seas fishery and processing system. *J. Marine Tech. Soc. 4*(5).

26. Cahn, A. R. 1950. Oyster culture in Japan. *Rept. Nat. Resources Sect. Gen. HQ 134.*

27. Fujiya, M. 1970. Oyster farming in Japan. International Helgoland Symposium. *Helgolander wiss. Meeresunters. 20:*469–472.

28. Landers, W. S. 1971. Some problems in the culture of oyster larvae. *Proc. Second Ann. Workshop, World Mariculture Soc.*

29. Walne, P. R. 1970. Present problems in the culture of the larvae of *Ostrea edulis*. Internatl. Helgolander Symp. *Helgolander wiss Meeresunters.* 20:514–525.

30. Nash, C. E. 1971. Unpublished report (Apr.). Oceanic Inst.

31. Castagna, M. A. 1970. Hard clam culture method developed at VIMS. *Virginia Inst. Marine Sci. Res. Advis. Serv., Ser. 4.*

32. Bailey, J. H. 1973. Test-tube abalone. *Sea Frontiers 19*(3):148–153.

33. Caillouet, C. W. 1972. Rotatable cage for high density aquaculture. *Prog. Fish-Cult. 34*(1).

Additional Readings

Morton, K. E. 1963. A preliminary report on a new multipurpose fish moving machine. *Proc. Northwest Fish Culture Conf.* (Tumwater, Wash.), pp. 20–24.

Novak, P., and W. F. Sheets. 1969. Pumping device used to collect small-mouth bass fry. *Prog. Fish-Cult. 31*(4):240.

Robinson, J. B. 1971. Effects of passing juvenile king salmon through a pump. *Prog. Fish-Cult. 33*(4):219–223.

11

ON-SITE HARVEST PROCESSING

J. M. Collier

Seafoods have been a major protein component of man's diet in every era; as such they are harvested and preserved worldwide, wherever men have access to the sea. The simple, freshly caught or air-dried fish that were the only available seafoods in man's very early history[1] and for many suceeding centuries are still with us, as are the smoked and salted products that have sustained man since the Bronze Age. To these now have been added a tremendous variety of frozen, canned, preserved, dried, concentrated, precooked, and convenience products.

Today, as always, the range of products depends upon three basic factors: available raw material; market demand as determined by the needs and tastes of the consumers; and technology to process any given product efficiently and economically enough to market it at a price that the consumer is willing and able to pay, and that assures the provider an adequate reward for his efforts. The consumer market itself is subject to changes in demand as (1) growing population outstrips the production of protein from land-based resources; (2) rising standards of nutrition create a demand by existing populations for increased quantity and quality of protein in their diets; (3) increasing affluence in developing countries enables their populations to indulge a taste for variety in the diet and for aesthetically appealing foods; and (4) growth in leisure-time activities and the increase in employed women spur the demand for convenient food products with short preparation times. These changes in the market spur the development of techniques to harvest greater quantities of preferred fish species and, in turn, the development of improved processing methods.

BACKGROUND

The flesh of marine organisms is subject to extremely rapid deterioration, beginning to some extent at the point of capture even if life is sustained for a period, and shifting into high gear at the instant of death. A complex breakdown results from a varying combination of bacterial action, chemical decomposition, and autolysis. Exposure to air speeds the action of some

---

[1] Sea fish bones are found far inland in the refuse heaps of the Late Stone Age cave dwellers in the Dordogne, dating from about 40,000 B.C. [1].

bacteria; other [anaerobic] types multiply in organisms immersed in fluids or packed in airtight containers. The enzymes in fish tissues activate an auto-lytic, or self-digestive, process in a variety of environments.

Many factors affect the speed of breakdown and the nutritional quality of fish flesh, such as the condition of the living fish as determined by its position in its life cycle (e.g., capture immediately before or after spawning); the state of fatigue it may have reached in its struggle against capture; the degree of pollution of its environment before capture; disease; species characteristics; the sanitary conditions under which it is handled prior to and during process-ing; and so on. But whatever the given set of modifying circumstances may be, the basic premise remains: fish muscle provides a hospitable medium for many pathological organisms, and this characteristic, in combination with high water content, leads to rapid putrification.

Throughout history man has devised often crude but reasonably effective methods for coping with the problem of storing and transporting reserve stocks of fish protein. He relied principally upon drying, salting, and smoking until the use of ice as a temporary preservative was introduced to Europe from China in 1786, a development that led to the formation by Britain of trawler fleets with swift carrier vessels at the beginning of the 19th century. That same century, which witnessed the flourishing of the industrial revolu-tion, saw the development of the canning process and the beginning of mechanization in fish processing.

Freezing by artificial means rapidly took hold in western Europe, the United States, and Russia as an early industrial use of the newly invented machine compressor. By 1880, freezing machines were towed on barges along the Volga River to collect raw material for the canning industry in Russia. Quick freezing, first of whole fish and then fillets, for direct consumption was pioneered around 1916.

Filleting of white fish at the ports in the mid-1920s to make distant-water cod more attractive in the markets led to an industry utilizing filleting offal for fish meal to provide animal fodder. An increasing demand for this product led to a similar utilization of surplus catches of whole fatty fish, such as herring, usually after extraction of the valuable oil. A fried-fish trade using small, less marketable fish developed in Britain in the mid 1800s—a fore-runner of the modern precooked, convenience-foods industry, and now the major user of British distant-water cod fillets [1].

Thus, the current processing methods that produce a diversity of fish products for the relatively affluent markets had a simple beginning almost two centuries ago, but in the main developed only in the last 100 years or so. Interesting as history may be, however, it is the trends in seafood harvest utilization, as they may be deduced from the relatively recent past and projected into the future, that are of value in planning for and designing a mariculture operation.

Trends in Harvest Utilization

The total world catch of fish and shellfish tripled from 1948 to 1967, going from 19.6 to 60.5 million metric tons. During this period of increase in overall catch, per capita consumption remained approximately stable owing to two primary factors: (1) the growth in world population, and (2) the dramatic rise in the proportion of total catch converted into fish meal and by-products and thereby lost to direct human consumption.

Over the past two decades, the marketing of fresh (iced) fish declined from nearly 50 percent of the catch in 1948 to about 31 percent by 1967. Although this was historically the preferred product, auction marketing of the highly perishable catch, as was customary in many areas, was subject to the vagaries of demand and caused high waste. The growing practice of freezing at sea greatly decreases perishability and so provides a stabilizing influence; today much of the catch now marketed as "fresh" is actually frozen at sea and thawed before sale.

The pattern of use of the total world catch of fish and shellfish from 1938 to 1960 is shown in Table 11.1, extracted from data published by Georg Borgstrom of Michigan State University [2], to which we have added the figures for 1967 [3].

Although the proportion of the total world fishing yield being converted to fish meal and oil was close to 25 percent as early as 1960, many individual countries actually use well over 50 percent in this manner. Peru is the top manufacturer of meal, converting 94 percent of its catch to that purpose.

It might be interesting at this point to note an encouraging movement in the direction of resource husbandry. For example, in 1950 half the waste from USSR processing plants was discarded, and an estimated 40 million kg of fish viscera was being discarded annually by Canadian Atlantic fisheries as late as 1960. Likewise, the offal from the Atlantic salmon industry, although tried successfully as animal feed, was being poorly utilized principally because of the distances between plants and their remoteness from large-scale meal processors. However, during the 1950s and 1960s efforts were launched to better utilize the waste from processing plants. One interesting development in the 1960s was the use of offal meal and by-products in fish culturing ponds in Denmark, Israel, and elsewhere. Moreover, the Soviet fishery found invertebrates, and mussels specifically—which they had customarily discarded—to be particularly valuable for meal. A similar use was found for trawler "refuse," that is, catches of nonmarketable species of marine animals, which formerly had been shoveled back into the sea [2].

Harvest Preservation at Sea

The post-World War II period has seen a strong trend toward transferring the principal harvest preservation onto the seas and into distant waters as increasingly sophisticated machinery has been developed for shipboard instal-

TABLE 11.1 *Utilization of Total World Catch of Fish and Shellfish (%)*

Mode of disposal	1938	1948	1952–1953	1954–1955	1956–1957	1958	1959	1960	1967
Fresh	50	51	44	44	43	44	42	40	30.9
Freezing	5	5	5	6	7	8	8	9	12.0
Curing	24	24	25	25	24	23	20	19	13.2
Canning	8	7	7	9	9	9	9	9	8.6
Reduction[a]	8	8	14	13	14	13	18	20	35.2
Miscellaneous	5	5	5	3	3	3	3	3	

[a] Fish meal, oil, and FPC (includes offal).

lation. Three major factors influenced this move to ocean processing: (1) the extreme perishability of the raw material during storage and transit to shore installations, with the resulting advantage of processing at the source; (2) the growing need to fish distant waters as stocks near home shores became depleted; and (3) advancing technology, which provided the ships and machinery to process at sea. Also, the ability to do all processing in floating factories provides nations with a degree of freedom from political alliances and complications attendant upon maintaining shore facilities in foreign countries adjacent to distant fisheries.

Some modern vessels combine catching and processing operations; others are factory, or "mother," ships devoted only to processing and to servicing the fishing fleets. The latter contain a variety of equipment and facilities to process the huge catches of mixed species and sizes, only part of which are marketable for human consumption. For example, by 1963 one particular 165-m vessel could in 1 day salt 181 metric tons of herring and store it at $-5°C$, process 136 metric tons of fish into meal, fillet and freeze 90 metric tons of bottom fish, and manufacture 4.5 metric tons of fish oil, 18 metric tons of ice, and 90 tons of distilled water. Another, a crab-canning factory ship, has a daily capacity of 200,000 cans [4]. Some factory ships carry large crews, up to 500 or more; other very advanced ships are designed for almost completely automated processing, comprising a range of operations from drying small fish to freezing packaged products. As an example of the relative effectiveness of this approach, some 78 percent of the Soviet catch is produced by 7 percent of their fishing vessels operating in conjunction with highly automated processing ships [5].

CURRENT METHODS OF PROCESSING AND PRODUCT CHARACTERISTICS

A brief review of the principal methods currently in use or under development to prepare seafood products for volume markets shows us that some of these appear readily adaptable to the processing component of a potential mariculture system. Others may well be unsuitable or require substantial modification or development to become economically attractive. This inventory is by no means complete. It can only summarize the operating principles discussed in great variation and detail in the technical literature on freezing, canning, drying, and the production of fish protein concentrate (FPC) for human consumption, and of fish meal, oils, and solubles for animal feed and other uses.

We also touch on smoking and salting, which produce commodities for a more limited market. However, for practical reasons we omit products such as fish marinades, pastes, sauces, and sausages, which are popular primarily in Japan and countries of Southeast Asia and are admittedly an important protein supplement in the diet of the countries where they are favored. These require special and, in some cases, complicated processing, and they are in too limited demand in the world market to be considered commercially feasible in the near future.

The simple, basic operations currently handled by open sea floating factories appear to offer the real near-term potential for supplying high-volume protein for hungry populations as well as for producing high-quality seafoods for the relatively affluent market. Complex processing of speciality products can more practicably be done ashore where labor is readily available—at least until such time as floating cities may become a reality.

Drying

Although drying is the earliest method of preservation known to man and one of the simplest, it has not only remained important but now, through advancing technology, it shows promise of future growth as a major industry. Today many tens of millions of the human race avert protein deficiency through consumption of dried fish from distant waters. Any nonfatty or extracted dried fish represents the most potent protein commodity in world trade, with 2.5 times the protein per equivalent volume of nonfat milk solids [6]. Moreover, it is one of the cheapest.

The drying of whole or split fish was originally achieved by simple exposure to the elements; in time, various arrangements were worked out for adding heat and forced air to speed evaporation. Air-dried fish undergoes changes in texture and flavor, yielding a product very different from fresh fish and with much more limited appeal. Major markets are in the Mediterranean countries and Africa, where its keeping qualities in the absence of refrigeration are valuable, and in Scandinavian countries where it has been a diet staple for centuries.

Most modern methods of dehydration evolved under the impetus provided by war, which required large amounts of easily transported, relatively imperishable foodstuffs. Products were developed by artificially drying cooked minced fish and by vacuum freeze drying fish in flake form. Accelerated freeze drying (AFD) is a relatively costly process that requires filleting or slicing of the raw material to ensure uniform thickness for the freezing and vacuum-drying operations [7]. In this process prefrozen fish are placed in contact with heated plates in an evacuated cabinet. Under low pressure, evaporation occurs so rapidly that the flesh remains frozen while it dries. Two advantages to this method are the virtual elimination of oxidation during drying and the absence of shrinkage, coupled with ease of reconstitution. The resulting improvement in flavor and texture makes this a fairly appealing product. Moreover, the dried foodstuff can be stored for years in inexpensive plastic wrapping at normal temperatures, while light weight and low volume reduce storage and transport costs still further. Initially, crustaceans, particularly shrimp, produced the most acceptable products. Current development of freeze-drying technology is rapidly solving both quality and cost problems for many types of seafood.

Salting

Salting is, in effect, an extension of drying; its main importance as a fish preservative is its ability to withdraw water from the flesh and so to provide a less attractive bacterial environment. Although this is a comparatively simple method of preservation, varying somewhat with the species, a relatively high degree of labor is involved. In a typical example involving nonfatty species, the fish are gutted, headed, and bled, after which the carcasses, either whole or split, are packed in layers alternating with layers of salt. At frequent intervals the fish are rotated and restacked to ensure even penetration of the salt. At this "green cure" stage the fish are said to "be lying in wet stack," where they may remain for months. Eventually they must be dried, an operation carried out on shore.

For fatty species, oxidation of the fat that occurs in this technique renders the fish unpalatable; hence, salt curing or pickling was developed. For this process, salting down is followed by packing in barrels, where the fish shrink and a brine accumulates. Some of this is drained off and additional fish are added in a series of manual operations until the product has "cured" and the barrels are sealed for storage. The market for salt fish has been declining and this method appears to hold little promise for future mariculture operations [7].

Smoking

In applying a smoking process to seafood, bactericidal substances from the smoke are deposited on the flesh, but preservation is only partly due to the properties of the smoke; additional preservation is supplied by the brining that precedes smoking and by the drying that occurs during smoking. The final product still is perishable and must be kept cool. Overall consumption of dried smoked fish is declining, although there is a growing gourmet trade in delicacies such as smoked oysters, salmon, sturgeon, herring, and the like. Addition of the smokey flavor to these products might, however, more properly be considered an extention of other processes, since the smoking (either actual or by addition of a smoke-flavoring ingredient) is purely for flavoring, and preservation itself depends upon canning, freezing, or vacuum packaging [5, 7, 8]. We are led to conclude then that smoking as a process affords insufficient preservation to accommodate extended storage or extensive transportation, but that the addition of a smokey flavor to many seafoods may increase their market potential.

Canning

Although canning technology developed relatively early, a comparatively small proportion of the world fish harvest is preserved by this method. One reason for this is that canning is suitable for only a limited number of species; another is the alteration of the product characteristics that occurs during the heat processing. Canning is an eight-stage operation that consists of gutting and trimming the carcasses, washing and descaling, brining, packing, "exhausting" the cans to obtain a partial vacuum, sealing, heat processing, and labeling. Extreme care is needed to prevent spoilage by contamination, underprocessing, or from faulty can seams or flawed can lining. Precooking is an additional step involved in the canning of some species, such as tuna [7].

Crustaceans were among the first seafood products canned, and they still remain a popular product, although essentially in the luxury category. The high costs of canned crustaceans may be attributable in part to labor costs associated with removal of the exoskeleton, in most cases a more complicated process than the trimming pelagic fish carcasses.

As a food commodity, canned seafood is a ready-cooked food with the advantages of ease in handling and long-term keeping qualities. The nutritive values are retained by this processing, and under normal storage conditions canned food shows an excellent degree of resistance to vitamin loss. For maximum retention, canned food should be stored under cool conditions; elevated temperatures (such as may be encountered in tropical climates under primitive storage conditions) have some detrimental effect, but little research on long-term storage has been done on canned fish [9].

The advantages of canning are offset, on the one hand, by consumer preference for fresh fish, a taste the relatively affluent nations can afford to indulge because of the growth in their freezing industries. On the other hand, the relatively high cost of canned fish places it beyond the reach of low-income populations in developing areas where the consumer is less finicky but where the real need is for a high volume of cheap protein foods.

Freezing

The technology of freezing fish at sea has undergone more rapid and probably more diversified development than the technology for any other seafood preservation method. Some reasons for this are that (1) a variety of products ranging from untrimmed whole fish to fillets and crustaceans, in consumer-ready packages, can be processed at the peak of freshness; (2) freezing at sea offers the possibility for preserving raw material in its best condition for reprocessing elsewhere, with a resulting versatility of the product for canning, filleting, smoking, marketing fresh, or any of a wide variety of uses; and (3) there is a high market demand for the fresh product, which has been preferred by the consumer throughout history. Entire books have been written on freezing techniques. The processes are far too involved to cover in any detail here, but the principal methods may be summarized very briefly from the excellent presentations to the United Nations Food and Agricultural Organization (FAO) Technical Conference in Madrid in 1969 [10].

Freezing in Air. Freezing of seafoods by exposure to superchilled air is accomplished principally by the following techniques.

Frozen Fish Hold. The frozen fish hold is a primitive method in which fish and crustaceans, whole or packaged, are packed into a cold chamber that freezes and stores them as a single operation. This technique is used successfully for large fish too bulky for other freezers and for operations by small shrimp vessels. A refinement of this method is the shelf freezer, in which the fish are placed on shelves or trays in intimate contact with cold pipe grids arranged below them. Further improvement is possible by fitting small fans to circulate cold air along the shelves; this method has been used for lobsters.

Freezing Tunnels. Freezing tunnels are used by the great majority of freezing vessels. In this technique, air-blast freezing tunnels circulate cold air at high velocity over fish, either individually or in a variety of packages or trays. Three basic subtypes of freezing tunnel are used: batch freezers, continuous tunnel freezers, and fluidizing tunnel freezers.

Batch freezers are usually used in pairs, loaded alternately, and the fish are packed in racks or trays (Figure 11.1). With a typical maximum capacity (using two tunnels) of 27 metric tons/day, each tunnel is loaded every 6 h. Precooling is difficult and not generally done. Several disadvantages include waiting periods before loading each tunnel for freezing, thereby allowing deterioration of the raw material, and periods of high peak load on the electrical system as freezers are rapidly chilled after a warm-up period during loading. The use of six to eight tunnels eases the problems somewhat.

Continuous tunnel freezers handle individual carcasses, blocks of fish, or trays. They are highly automated and have evolved into some very sophisticated conveyor systems using minimum personnel, continuous throughput, and a more constant power load to maintain freezing temperatures. As an example, one design with a capacity of 21 metric tons/day (in 10-kg blocks) requires only two operators (Figure 11.2).

In fluidizing tunnel freezers the product is individually suspended in an upward-flowing cold air stream. Suitable for shrimp, small fish, and fillets of sufficiently light weight in relation to the area presented to the air stream, this method is very fast and produces individually frozen fish of fine quality and appearance. Figure 11.3 shows a simple diagram of the basic operation. Typical sizes used on land can freeze from 43 to 130 metric tons/day. So far as is known, fluidizing freezers have not yet been used on board ship, but they appear to offer an attractive method to apply to several products. Moreover, they are much more compact for a given output and require substantially less fan power and compressor capacity than other tunnel

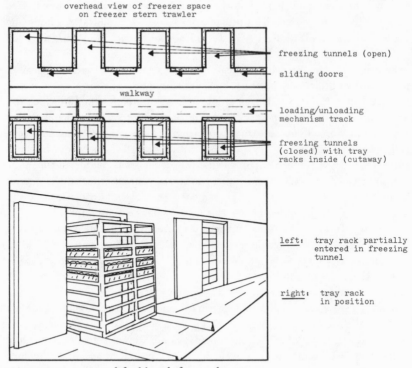

overhead view of freezer space
on freezer stern trawler

freezing tunnels (open)

sliding doors

walkway

loading/unloading
mechanism track

freezing tunnels
(closed) with tray
racks inside (cutaway)

left: tray rack partially
       entered in freezing
       tunnel

right: tray rack
       in position

FIGURE 11.1 *Simplified batch freezer layout.*

freezers. It would be simple to combine the fluidizing freezer with a conventional freezing tunnel arranged on the opposite side of the air coolers. This is an example of how the modular nature of cold-processing equipment can achieve economies by allowing more than one process to share in the same processing element.

Freezing with Liquified Gases. The use of liquid nitrogen (LN) for flash freezing is a recent development given impetus, interestingly enough, by the need to find uses for enormous quantities of LN by-product resulting from the huge production of liquid oxygen in the U.S. space program and, to a lesser extent, from increased steel production in many countries. The technique involves the application of cryogenic gases to freezing, especially for high-value products such as shrimp. Initially accomplished by direct immersion in the liquified gas, the method has evolved to freezing in the vaporized gas and in a spray, on somewhat similar lines to a conventional air-blast freezing tunnel.

The product conveyor is contained in a double-walled, vacuum-insulated, open-ended cylinder with high-velocity nitrogen gas curtains at the ends to reduce warm-air infiltration. As indicated in Figure 11.4, the raw material

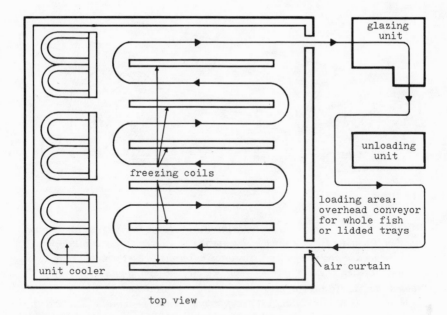

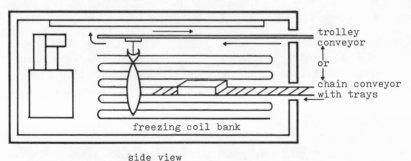

FIGURE 11.2 *Simplified typical continuous conveyor freezer.*

first enters a precooling section where cold gaseous nitrogen exhausted from the main tunnel is circulated over it to freeze the outside. This process continues in the first section of the main tunnel; in the second section, liquid nitrogen is sprayed around the product; and last, it enters an equilibrium section, where the temperature has a chance to equalize throughout the product.

A typical nitrogen freezing tunnel handles about 22 metric tons/day. At atmospheric pressure LN has a temperature of $-196°C$, which permits the design of unusually compact freezers. Moreover, the inert nature of this refrigerant makes it possible to spray it directly on the product, thereby eliminating heat exchangers. Conveyor speed is adjustable to suit the product; a fish fillet remains in the freezer only about 3 min, for example, and the

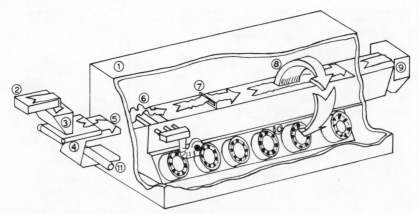

FIGURE 11.3 *Simplified typical fluidizing freezer: 1, freezing tunnel chamber (insulated); 2, primary dewatering shaker; 3, metering hopper; 4, variable-speed conveyor belt; 5, product entering freezer chamber; 6, "suspension booster": extra jets of high-velocity air, "low-phase" freezing; 7, "reload bar": assures perfect distribution of product; 8, "dense phase" suspension: minimum fluidization for maximum freezing efficiency, no product damage or small-particle loss; 9, collecting hopper; 10, adjustable recirculation fans; 11, automatic belt washer and dryer mechanism.*

time must be accurately gauged to avoid excessive speed or degree of cooling. The approximate evaporation rate is 1 kg of LN/kg of product frozen. Overall costs currently run three to four times conventional processing costs. Economy depends very much on conserving the liquid gas, which is recirculated through the tunnel by pump from a tank. The practicability of this method for open sea processing depends on the ease of replenishment, the efficiency of liquid storage to limit evaporation losses, and the value of the product to be handled, as well as on overall capital and operating costs.

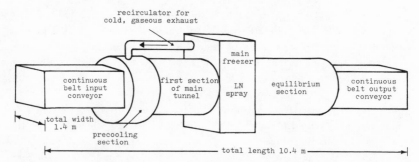

FIGURE 11.4 *Simplified flash freezer using the liquid nitrogen (LN) process.*

Contact Freezers. Probably the most efficient freezers in current use at sea are the varieties of contact freezers. There are four types: horizontal, vertical, and radial plates, and rotary drums. The first two are in most general use.

Horizontal Plate Freezers. In horizontal plate contact freezers, whole fish or fillets are frozen in aluminum or steel trays with lids, which are placed between adjacent horizontal freezer plates and maintained in contact with them by light hydraulic pressure. The nest of freezer plates, with associated structure and hydraulic equipment, usually is contained in an insulated cabinet with heated double doors on front and rear to permit through-loading. Each cabinet (see Figure 11.5) is known as a station; freezer sizes range from 6 to 7 up to 20 stations.

Automatic horizontal plate freezers are available for freezing consumer packs, and designs for the mechanized loading of tray racks holding whole fish have been developed; these have not yet been installed on ships, where this method of freezing is used mainly for fillets and small fish like herring. Considerable development is still needed to mechanize the operation and eliminate the trays.

Vertical Plate Freezers. Using a design adapted from the principle of the horizontal plate version, the vertical plate freezer offers greater versatility in handling the frozen product. All vertical plate freezers are designed for loading from above, but unloading can be arranged in any direction to suit the layout of the freezing and storage system, as indicated in Figure 11.6.

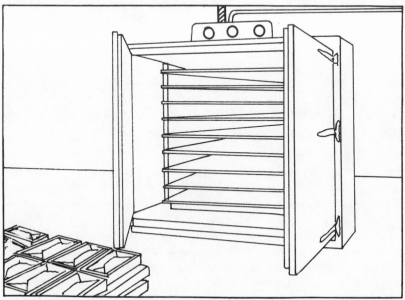

FIGURE 11.5 *Horizontal plate freezer station.*

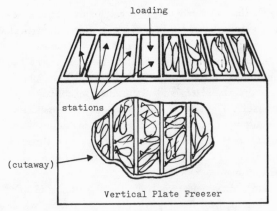

typical block sizes: 800-1000 X 400-1000 X 80-150 mm

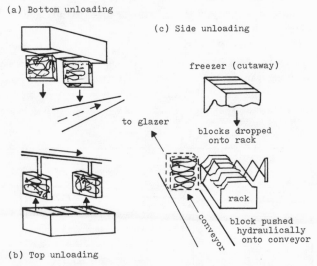

FIGURE 11.6 *Typical vertical plate freezer types.*

Each direction for unloading has its peculiar advantages. A design that discharges the blocks upward hydraulically is used for freezing fillets, with the options of inserting cartons between the plates before freezing and for producing blocks of varying thickness.

The substantial advantages of the vertical plate design include the simplicity of operation, elimination of trays, a considerable degree of mechanization, less space requirement for installation and operation per given output than any other freezer presently used at sea, and a product quality equal or superior to that frozen by any other means. The principal disadvantage is irregularity in the shapes of the frozen blocks and arrangements of fish. However, the large blocks are intended for the wholesaler only, where proper thawing and handling can produce attractive products for the retail trade.

Radial Plate Freezer. A novel design for a radial freezer has been developed in the USSR. The machine has a capacity to freeze 7 metric tons/day (comprised of 6-kg blocks) with automatic loading and unloading and a freezing time of 2 h, 18 min. No data on the extent of its use or the quality of the product were available to this study.

Rotary Drum Freezer. The rotary drum freezer, adapted from the revolving drum flake ice maker, is a recent development for individual freezing of shrimps, fillets, and the like. Fed by a conveyor onto a revolving stainless steel drum held at low temperatures, the products immediately adhere by the freezing of water. At the end of one revolution the frozen product is scraped off onto a conveyor and passed through an automatic glazer for packaging, as shown in Figure 11.7. With no air movement and very rapid freezing, the makers claim that loss of weight from the raw material is eliminated. No data are available as yet on the output capacity for this process.

Immersion Freezing. Freezing by immersion is less commonly used and generally less satisfactory than the techniques already described. Of four

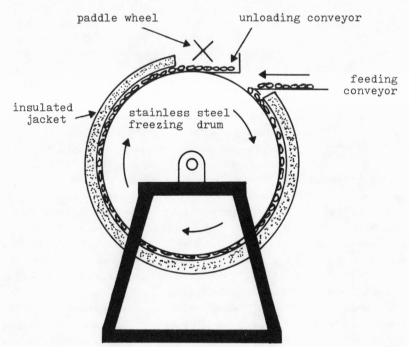

FIGURE 11.7 *Wentworth rotary drum freezer.*

methods developed for this technique, only brine freezing has had significant application. Freezing in brine spray and by direct immersion in liquid nitrogen have proved unsatisfactory; and a recent development in the United States that entails freezing high-value products by immersion in Refrigerant 12, although it shows considerable promise, has yet to be cleared on health grounds for commercial use.

Brine freezing has been employed almost universally in the U.S. West Coast tuna purse seiners, and also for sardines as well as shrimp and other crustaceans, and has been adopted by a number of other countries. Since salt is absorbed from the brine, its use for pelagic fish has been principally for tuna bait and for species that will ultimately be canned; white fish varieties have not been commercially processed by this method. Brine freezing normally used on the tuna vessels involves stowing the freshly caught tuna in tanks of 27- to 55-metric-ton capacity, where the fish are first cooled in seawater and then chilled in a sodium chloride brine to about −8°C or lower. The brine is then drained and the fish are cooled further and stored dry. Each tank is lined on all surfaces with ammonia direct-expansion cooling coils, requiring up to 884 m of pipe coil for a 55-metric-ton tank. Installation is bulky and corrosion is a constant threat.

In immersion freezing, practical problems include space considerations for both freezing and storage, difficulty in handling the frozen product, a need for more satisfactory freezing mediums, and corrosion protection. Although some products, such as brine-frozen shrimp, for example, are comparable to those produced by air-blast freezing, this technique may not be sufficiently versatile and economical for an open sea mariculture operation.

Product Variety and Processing Issues. A review of the technical literature reveals diverse viewpoints on the degree of processing that can most practicably be done at sea. Should the processing be minimal, that is, just enough to preserve the raw material for later processing ashore, or should the sea-based operation attempt to produce finished market products? At most, fish frozen whole are only headed and gutted, and these operations are readily automated. Carrying the process further to, say, filleting involves problems of proper bleeding, removal of pin bones, suitable packaging, and other material- and labor-intensive operations. These more expensive operations usually are transferred to shore factories, where space and labor are more economical. The machinery that may in time answer the objections to full processing at sea is now under development. For the present, the advantages and disadvantages of each approach from the point of view of the fishing industry are summed up in Table 11.2.

We must emphasize here that these pro and con considerations are oriented toward fishing and not mariculture. Many arguments for whole-fish freezing and against fillet freezing would not apply in a mariculture operation where

underutilization of machines, aggravated increasingly by declining catching rates[b]

Productivity no higher except in much larger vessel[b]

Difficulty of maintaining quality and hygiene, especially with heavy catches[b]

Complicated equipment[b]

High capital cost

High maintenance costs

Larger space occupied by machines[b]

Much larger crew and hence more space on board for accommodations, amenities, stores, etc.[b]

Much higher operating costs of vessel[b]

Usually longer voyages, resulting in difficulty in obtaining and retaining good crew[b]

No-better price for product

[a] Substitute "mariculture stock feed" for "fish meal."
[b] Considerations that may be of little or no concern, or that may not apply, in a mariculture operation.
[c] Substitute "FPC and/or stock feed" for "fish meal."

279

harvest types and rates would be predictable, and more energy, space, and facilities very likely would be available.

From this review it is apparent that the overall technology for freezing fish harvests at sea serves two broad and distinguishable purposes. The first is the initial or primary preservation of the raw material in the best possible condition during periods of storage and transport prior to secondary processing ashore to produce varied products for the consumer market and by-products such as meal, oil, vitamins, and so on. The second purpose is to combine preservation and processing to produce products suitable for direct marketing to the consumer, either as fresh (thawed) seafood in bulk or as frozen packaged fillets, shellfish, or whole fish. In either case, the freezing process yields a high-quality, high-cost product, in popular demand by the affluent consumer market but impractical for solving serious nutritional deficiencies in underdeveloped countries.

A complicating factor in this latter instance is the necessity for maintaining a *cold chain* from the source to the consumer. The term means, simply, that refrigerated storage spaces, containers, and vehicles must be available to maintain the cold temperatures of frozen products at every step of their progress from the instant of freezing at sea to the point of preparation for the table in the home, restaurant, or institution. Obviously, the capital investment would be utterly impractical were only one type of commodity to be handled; consequently, frozen seafoods can be considered a substantial protein resource only in areas where a highly developed frozen-food industry and distribution system already exist. This eliminates currently underdeveloped areas from consideration.

There is, however, a significant exception to this situation. If a vacuum-drying process is added to whatever freezing system is designed for an open sea operation, the capability for producing accelerated freeze dried (AFD) fish products, as discussed previously, can be achieved with minimal additional investment. The prospect for addressing protein needs in underdeveloped areas through open sea mariculture would then appear to offer some real possibilities.

Fish Meal

Used principally for animal feed, fish meal has been an important protein supplement in the diet of agricultural stock for several decades. The raw material for this product comes from three sources: (1) residual offal from various processing operations, such as freezing, filleting, and canning; (2) trawler catches of small fishes and miscellaneous species unsuitable or unprofitable for the market; and (3) surplus from peak catches of desirable food fish that overload the normal processing facilities or market demand.

In the simplest method of fish-meal production [which is suitable for use only with nonfatty species such as stockfish] the flesh, bones, and offal are minced, dried, and ground into meal. Fatty species require precooking and pressing to extract their oil, which itself is a commercial by-product. The fluid other than oil that is extracted by the pressing operation is known as *stickwater*. Originally, it was discarded; but once it was realized that it contains 20 percent of the original solids in the raw material, it proved more economical to condense it by evaporation and either add it back to the meal or market it as condensed fish solubles. In general, the processing chain follows the steps shown in Figure 11.8. Many variations in the several steps have been developed—such as dry rendering, solvent extraction, or the use of supersonics or ultrasound to extract oil. By and large, these have not yet proved feasible on a commercial scale [12].

The use of crab and shrimp meals is increasing as the processing of crustaceans grows and the efficient utilization of offal and shells gains importance. The production of meal from crayfish and dried mussels also is growing. The protein content of the latter is rated as good as, or superior to, fish meals [13]. In general, it may be said that fish-meal production requires a constant and high-volume source of cheap raw material to offset the costs of processing in order to produce a low-cost meal profitably.

In terms of food resources per se, the protein reserves of the sea and fresh water can be utilized more quickly and efficiently by eliminating the extra step in the food chain now represented by using fish as stock feed for the secondary production of animal protein. The large proportion of total fishing yields processed into meal and oil is lost for direct consumption, and most of it ends up in the markets of well-fed nations. There appears to be some strong feeling among world nutritionists that fish yields should be utilized to the maximum for human consumption, including the production of fish protein concentrate, and that production of meal and oil eventually should be restricted to offal [9].

Fish Flour, or Fish Protein Concentrate (FPC)

Fish flour is essentially a fish meal prepared for use as human food; it is more properly designated a fish protein concentrate, or FPC, since it is not a substitute for grain flour. Manufacturing methods range from simple washing and thorough drying of the raw fish, followed by pulverizing in a ball mill to produce a crude but efficient flour, to relatively sophisticated processes that produce deodorized and highly palatable products. A number of such processes have been developed and patented in the United States by General Foods, Lever Brothers, and others. One of the better known, the VioBin process, is in large-scale operation in Massachusetts. It involves the extraction and low-temperature azeotropic dehydration of degutted, pulverized fish flesh, using ethylene dichloride. This is followed by grinding and repeated washings with alcohol to remove traces of solvents and to deodorize the flour [9].

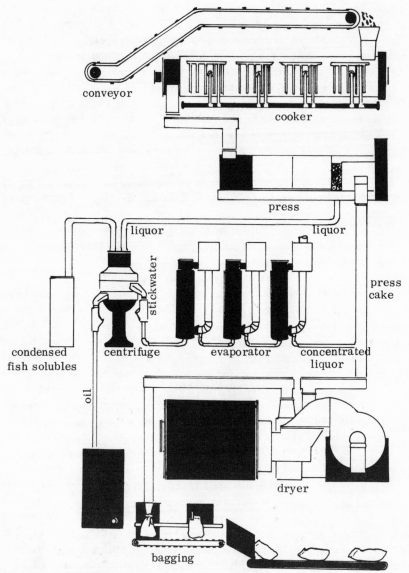

FIGURE 11.8 *Sequence of operations in fish-meal production.*

As another example, the National Marine Fisheries Service has developed a process to make FPC out of otherwise unmarketable red hake. In this instance, whole fish are minced and treated three times with separate batches of isopropyl alcohol, once cold and twice hot. The alcohol extracts both water and oil, and the fish residue is dried, ground, and packaged. For this process, the ratio of raw material used to product volume produced is about 6 to 1. At a raw fish cost of $0.01/lb, a plant with a daily capacity of 50 tons could produce FPC at $0.139/lb and sell it profitably at $0.20/lb. Should the market accept a faint fishy taste, the selling price could be reduced to about $0.135/lb.

As an excellent and versatile protein-rich, low-fat diet supplement, FPC holds great promise for meeting worldwide nutritional deficiencies, especially where low bulk, easy transportability, and keeping qualities independent of refrigeration are essential. All the fish now used to make fish meal for animal feeds could theoretically be made into fish flour [4]. Thus, we may conclude that, given a source of cheap raw material, FPC (and dried fish) represents the cheapest form of fish protein available for human consumption. This process would seem to be unsuitable for a mariculture operation concentrating on selective harvesting of quality species, but it might prove to be practicable for low-grade species that may be particularly amenable to culturing.

Irradiation

Work in a number of laboratories in the United States and European countries during the decade of the 1960s confirmed the effectiveness of cold-pasteurizing doses of low-level radiation in prolonging the shelf life of fishery products held under refrigeration. The degree of shelf-life extension is dependent on the original condition of the raw material and the level of radiation dose. Some experimenters recommend a primary low-level radiation dose at the point of harvest, followed by a stronger secondary dose onshore approximately 7 days later, as the most effective procedure [14]. Continuing investigations of the techniques and products determined that irradiation processing of fish can be carried out in a number of different ways, depending on the objectives. Examples include decontamination or sterilization of fish meal or flour; irradiation of fresh fish to extend shelf life; and irradiation of frozen, canned, or prepackaged products to reduce or eliminate the presence of bacteria. Irradiation does not, however, prevent autolytic reactions, which in time affect the quality of stored fish and make it quite distinguishable from the freshly caught product.

Research on irradiation of fish (haddock, smelt, and surf clams) at sea has been accomplished by the National Marine Fisheries Service Technological Laboratory in Gloucester, Massachusetts, under contract with the U.S. Atomic Energy Commission, following an earlier determination that shipboard installation of irradiators is feasible. Figures 11.9 and 11.10 show the operating principle and layout of a small experimental cobalt 60 irradiator designed at the Brookhaven National Laboratory on Long Island, New York, and installed aboard the service's research vessel, M.V. *Delaware* [15].

Descriptions of experimentation in this field in the several countries involved, and discussion of problems yet to be resolved, are treated in detail in

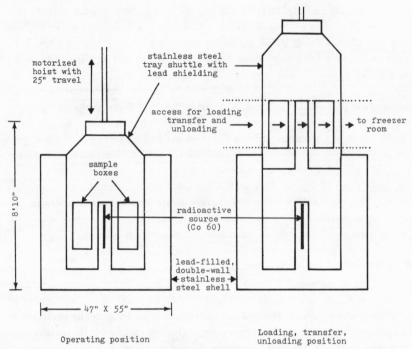

FIGURE 11.9 *General operating principle of small experimental irradiator (pilot model) for ship installation (cutaway view from side).*

the papers presented at the FAO Technical Conference on the Freezing and Irradiation of Fish in 1969. The *World Fisheries Abstracts* of the FAO contain many additional accounts. Work in the 1960s indicated the effectiveness and low cost, and established that little real question of safety existed. However, the commercial use of irradiation is dependent upon the feasibility of fitting the process into existing fish-handling procedures and upon verification of the capital costs and operating expenses of commercial-scale equipment. Commercialization of radiation-treated seafoods also must wait on clearance for general sale by the U.S. Food and Drug Administration (FDA). Clearance has already been granted for some irradiated foods, such as bacon, wheat, and potatoes, and scientific data are being mobilized to support clearance for fish, clams, crabmeat, oysters, and shrimp. Evolution in this field bears watching.

Other Technology

The development of fish-processing technology is accelerating rapidly in the face of rising pressure to increase production and upgrade quality. Brief examples include the use of antibiotics in the water used for washing fish

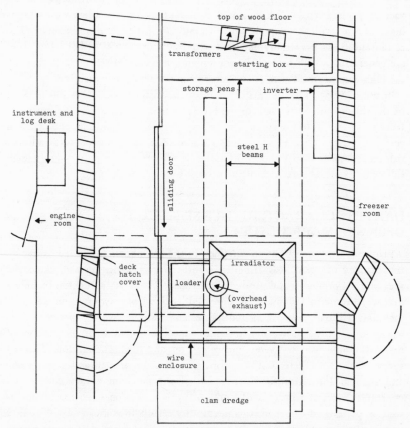

FIGURE 11.10 *Layout of irradiator hold (pilot model) aboard M. V. DELA-WARE (top view).*

carcasses before processing and in water frozen into ice for storage, to retard spoilage, and the use of microwave radiation for thawing frozen fish and for opening the shells of live mollusks such as oysters and clams. Another example is the development by the National Marine Fisheries Service of automatic gutting—or vacuum evisceration—to keep fish intact and reduce external bacterial contamination. This system is now commercially manufactured for shipboard installation.

PRODUCT MARKETS

The two broad markets for consumption of fish products in the United States are the chain-store supermarkets or retail stores and the institutional or restaurant trade. Fresh and frozen seafoods and precooked convenience products such as fish sticks are the most popular in the retail markets;

vacuum-packed frozen fish in plastic cooking pouches, complete with seasonings, are a typical example of ongoing new-product development. The restaurant trade preferably and traditionally buys fresh fish, but labor costs are generating a high interest in convenience products here, too, especially in the high-volume, moderate-priced establishments and the institutions.

In the United States the major share of shellfish consumption (60 percent of the shrimp and 81 percent of the lobster versus 37 percent of the fresh and frozen fish [9]) is in restaurants, which also are considered to be potentially a very large user of radiation-treated seafoods. The retail market for irradiated fish may depend in part upon promotional efforts to gain public acceptance, following eventual clearance of the process by the USDA [16].

HARVEST CHARACTERISTICS SPECIFIC TO OPEN SEA MARICULTURE

Fish processing methods to date have in the main been developed to handle the catches of open seas fisheries whose characteristics have dictated the spectrum of processes and products developed. These characteristics include a massive volume of raw material generated by fleets of trawlers, a mixed catch including a broad range of desired species and trash or "weed" species, a wide variation of sizes between and within species, and a high proportion of nonmarketable fish to convert into meal, oil, and FPC. The random mixes of species and sizes create problems for mechanization as well as for packaging and storing. Furthermore, the volume of the catch may vary widely and constantly, creating periods of underutilization of processing machinery as well as periods of overload; this necessarily results in some waste of quality species. By contrast, a mariculture operation would ideally be characterized by a uniform and relatively continuous harvest of one or more desired species, with the harvest containing individuals of nearly uniform size. Consequently, the entire yield would be marketable, and mechanization would be greatly facilitated. The only "waste" material would be the offal from the processing operations. Since this probably could be ground up and utilized as stock feed in various stages of the fish farming itself, it is likely that a mariculture station would have a much less complex processing chain than, say, a fishery factory ship. A flow diagram comparing fisheries and mariculture operations is shown in Figure 11.11.

OPEN OCEAN SEAFOOD PROCESSING FOR MARICULTURE

As we have seen, factory ships currently are equipped to perform most or all of the major conventional processing operations. An open ocean mariculture station possesses many physical characteristics common to floating factory ships, but it offers far greater freedom for the design of the total

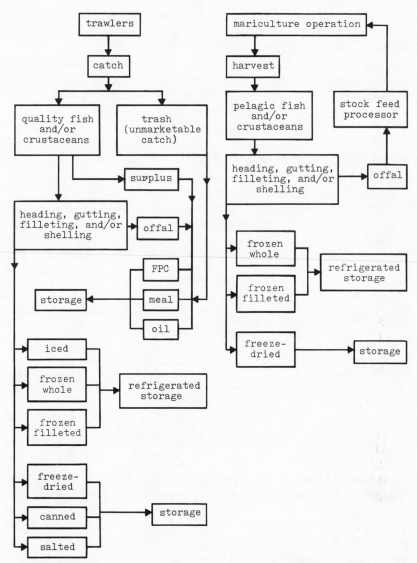

FIGURE 11.11 *Comparative processing chains for fisheries and mariculture.*

system. First, not only is it free from the constraints of conventional hull design, but it may take the form of a floating base, an undersea installation, an atoll system, a bottom-mounted platform, or any of the other configurations proposed in this study or yet to be conceived. Second, the mariculture operation may not be subject to the uncertainties of a random harvest as is a fishery. The unique features of culturing stations may impose limitations upon the choice of practicable processing methods or they very well may open entirely new possibilities. For any given combinations of species selected for culture, desired outputs will be identified on the basis of market value and demand, and of flexibility of product use in direct marketing or secondary processing ashore. The ultimate selection of product lines will then be determined by capital investment requirements and operating costs.

Preliminary Design Considerations

The achievement of a reasonable profit return on investment will require that efficient and economical use be made of the limited and relatively high cost space available for on-site processing operations. The personnel needed to operate processing equipment will require support facilities, and they may be difficult and/or expensive to retain if the station lies beyond commuting distance. Since most material resources will, in the main, have to be transported to the station, economical use of these also is indicated. We can, then, identify many of the criteria for the design of a processing operation within the cost and space constraints:

- Maximum throughput in minimum space
- Economical use of energy unless or until local, low-cost conservative energy extraction is achieved
- Minimum labor requirements
- Simplicity of processing
- Multipurpose processing modules adaptable to a variety of species and products
- Ease of product storage and handling
- Minimum waste
- Protection of the station environment from pollution
- Moderate capital investment

It is apparent that processing in an open sea mariculture station would be limited, in most cases, to the simplest possible operations that can take the raw material at the source, before deterioration can begin, and preserve it in forms which best retain its fresh qualities and offer the greatest versatility of use and subsequent processing. Space and labor costs suggest that only primary processing be done at sea, unless or until highly automated and fast processing becomes a reality. Secondary processing to produce variety products such as smoked, canned, or precooked convenience foods should be accomplished ashore where lower-cost space and labor are readily available. An exception to this generality might occur in the case of atoll-based mariculture stations, where space would not be a constraint and indigenous labor might be inexpensive and readily trained to perform secondary processing operations.

Mariculture operations most likely will concentrate, at least in the beginning, upon species having a relatively high market value to offset initial capital investment. Since the economic production of FPC, fish meals, and oils depends upon a volume source of cheap raw material, it seems apparent that this will, at least for the foreseeable future, continue to be restricted to the large and heterogeneous fisheries catches. This prediction could change if lower-grade species prove to be particularly amenable to culturing and economical to produce in high volume.

Elementary Design Scheme

If we look more closely at the very elementary scheme for a mariculture processing chain in Figure 11.11 to determine ways in which the requirements for personnel might be minimized and the unique features of a mariculture operation might be incorporated, we might come up with something like Figure 11.12. An operation of this type might use electrical power as its predominant energy form. For efficient use of energy, the refrigeration machinery and insulated spaces required to generate and maintain the freezing process might be utilized also for a parallel operation to vacuum freeze dry some portion of the output. This is still an oversimplified version, which does not explore the mechanics of the operations, power generation, physical equipment, and the like. Were we to attempt detail, we would have to acknowledge, for instance, that not all the offal may be suited for stock feed; the shells of lobsters probably would be otherwise utilized—perhaps reduced to derive chitin and chitosan for use elsewhere in the system. This scheme may assist us, nevertheless, in forming a generalized mental picture of the potential operation. We are not ready to get down to specifics until we know what species will be cultured and what the desired products are, but we can keep in mind the principles of maximum automation to reduce manpower and flexibility of design to enable modular addition of processing subsystems.

Potential Liquid Nitrogen Generation System

As another example of a potential mariculture subsystem, we might explore an interesting design for a seagoing fish processing and liquid-nitrogen (LN) generation system, using an alternative form of power. This design was developed by Integral Process Systems, Incorporated, and was presented at the FAO Technical Conference in 1969 [17]. It was intended for a factory ship operating with small vessels, but it appears to be quite feasible for a mariculture station, with some adaptation. The design provides flexibility of operation without high capital cost, and could possibly be economically attractive. All the major components are standard production items. Figure 11.13 is a flow diagram of this system.

Although the diagram is reasonably self-explanatory, we might add these comments. The power unit is a gas turbine that develops 1,100 horsepower (hp) and weighs 590 kg. This "total-energy" system would generate the

290

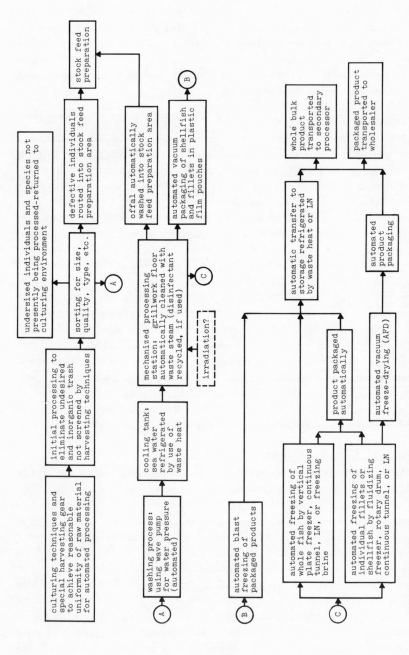

FIGURE 11.12 *Preliminary design for a mariculture processing subsystem.*

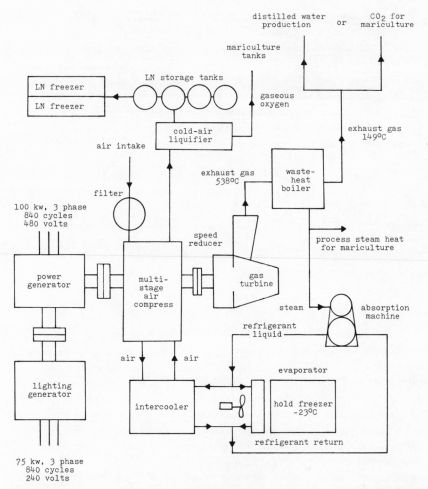

FIGURE 11.13 *"Total-energy" liquid nitrogen (LN) generation and utilization system.*

electricity for lighting and power for the station, in addition to operating the processing equipment. Overall thermal efficiency is high, and space requirements are minimal. We have modified the original to indicate use of the exhaust gas to produce distilled water. The process steam, suggested as input to a fish-meal plant, could instead be used for additional processing to vacuum freeze dry, or for whatever was needed. In fact, a compressor could be coupled to the turbine to add modules for tunnel freezing or other processing. Actually, the three outputs, gaseous oxygen, exhaust gas (carbon dioxide), and steam, could be utilized totally in the culturing operation—in part to modify the environment of the immature species. The designers

292   ON-SITE HARVEST PROCESSING

developed cost figures based on diesel oil as the fuel, although they suggest the possibility of using liquified natural gas. Current and projected costs of possible fuels would determine the choice at the time of construction. According to the estimated cost of operation and of LN to run a plant capable of freezing 22 metric tons of fish/day, the product cost per pound frozen would run $0.0166. This, we hasten to add, does not include cost of space, raw material, preparation and/or filleting equipment, labor, or storage room. The turbine and nitrogen plant are automated, and a minimum of labor is required to load and unload the freezers, each of which is rated at 907 kg/h.

A more specific description of the design with a detailed cost breakdown may be found in the source document, as indicated. The design possesses some interesting possibilities in the context of a mariculture system, and appears to offer the necessary flexibility to add other processing modules as needed. Further consideration of this system will require study to determine more specifically the relative costs of such a process, using LN, versus the cost of conventional freezing methods.

CONCLUSIONS

In the course of the foregoing review, it has become evident that the technology for open sea fish processing provides many possibilities for system design. We have also seen that the field is undergoing rapid development and that new techniques are continually being generated and/or refined for commercial use. It becomes apparent, too, in looking at the environment of the industry, that the consumer has a great deal to say in determining the selection of products to be produced for profit. Other pressures are generated by the availability (or scarcity) of raw material of given species and by the sheer weight of the world's need for protein foods.

We may conclude, then, that the choice of processing methods to be incorporated into a mariculture operation will be determined at least in part by the choice of species to be cultured and the analysis of market demand. Other criteria, such as location of a given station, available energy sources, size, and crew, also will have to be considered. Given these parameters and the rapidly developing technology in the processing field, it seems reasonable to assume that modular processing units will be commercially available, or can be designed and fabricated, to assemble any required combination of automated processes to produce desired products with a reasonable profit margin. In any case, should the demands of open sea mariculture create a significant need for such units, it seems quite clear that the industrial complex is equipped now to satisfy that need.

In terms of future trends, it appears, in the words of G. Meseck of Germany, speaking to the FAO in 1961, that

In the fish processing industry deep freezing and the production of protein concentrates should become increasingly important. The greatest attention should be given to freeze drying which, in many respects, may cause a truly revolutionary change in the fishing industry. By means of the freeze-drying method, the food resources of the seas could be utilized more quickly, and to a greater extent for the human population, than may seem possible from the present point of view [9].

REFERENCES

1. Cutting, G. L. 1962. Historical aspects of fish. In G. Borgstrom, ed., *Fish as food,* Vol. II. Academic Press, Inc., New York.
2. Borgstrom, G. 1962. Trends in utilization of fish and shellfish. In G. Borgstrom, ed., *Fish as food,* Vol. II. Academic Press, Inc., New York.
3. California Agricultural Experimentation Station. 1970. The California fresh and frozen fishery trade. *Calif. Agr. Expt. Sta. Bull. 850* (Oct.).
4. Idyll, C. P. 1970. *The sea against hunger.* Thomas Y. Crowell Company, New York.
5. Borgstrom, G., and C. D. Paris. 1965. The regional development of fisheries and fish processing. In G. Borgstrom, ed., *Fish as food,* Vol. III. Academic Press, Inc., New York.
6. Borgstrom, G. 1962. Fish in world nutrition. In G. Borgstrom, ed., *Fish as food,* Vol. II. Academic Press, Inc., New York.
7. Burgess, G. H. O., C. L. Cutting, J. A. Lovern, and J. J. Waterman, eds. 1965. *Fish handling and processing.* Ministry of Tech., Torrey Res. Sta. H.M. Stationery Office, London.
8. Burgess, G. H. O. 1965. *Developments in handling and processing fish.* Fishing News (Books) Ltd, London.
9. Heen, E., and R. Kreuzer, eds. 1962. *Fish in nutrition.* Fishing News (Books) Ltd., London. 450 pp.
10. Kreuzer, R., ed. 1969. *Freezing and irradiation of fish.* Fishing News (Books) Ltd., London. 548 pp.
11. Ranken, M. B. F. 1969. Evaluation of modern techniques and equipment for freezing whole fish at sea. In R. Kreuzer, ed., *Freezing and irradiation of fish.* Fishing News (Books) Ltd., London.
12. Sparre, T. 1965. Fish meal: manufacture, properties and utilization. In G. Borgstrom, ed., *Fish as food,* Vol. III. Academic Press, Inc., New York.
13. Borgstrom, G. 1962. Shellfish protein—nutritive aspects. In G. Borgstrom, ed., *Fish as food,* Vol. II. Academic Press, Inc., New York.
14. Liston, J., A. M. Dollar, and J. R. Matches. 1969. Effects of multiple dose irradiation on the bacterial flora of seafoods. In R. Kreuzer, ed., *Freezing and irradiation of fish,* Pt. VI. Fishing News (Books) Ltd, London.

15. Carver, J. H., T. J. Connors, and J. W. Slavin. 1969. Irradiation of fish at sea. In R. Kreuzer, ed., *Freezing and irradiation of fish,* Pt. VI. Fishing News (Books) Ltd., London.
16. Ronsivalli, L. J., J. D. Kaylor, and J. W. Slavin. 1969. Status of research on irradiated fish and shellfish in the United States. In R. Kreuzer, ed., *Freezing and irradiation of fish,* Pt. VI. Fishing News (Books) Ltd., London.
17. Breyer, F., R. C. Wagner, and J. Seliber. 1969. A seagoing fish processing and liquid nitrogen generation system. In R. Kreuzer, ed., *Freezing and irradiation of fish.* Fishing News (Books) Ltd, Londond.

12

PLATFORMS AND HOUSING FOR OPEN SEA MARICULTURE

S. B. Ribakoff, G. N. Rothwell, and J. A. Hanson

Like any other farming or industrial enterprise, open sea mariculture must have a base of operations. It is dissimilar to conventional enterprises, however, in the potential design flexibility of its operational bases. An open sea facility may be located on the sea surface or on the ocean floor or somewhere in between. On the other hand, the foundation might be a naturally occurring land formation such as an atoll. Likewise, the structure may be firmly and permanently affixed to its location, floating but moored, or mobile. In form it may be a permanent structure, massive or lightweight, or a temporary structure shaped from any of a variety of materials, ranging from steel or concrete to rubberized fabric or netting—or, most likely, some combination of these. In each case it is likely to represent some sort of platform, or platform surrogate, to offer physical support to culturing and harvesting activities. In some instances, existing designs and construction technology developed originally to serve other functions in the marine environment may be adapted. In other cases, entirely new facility designs may be called for, developed primarily and specifically for mariculture. The determination of the most practicable path to follow will most logically rest upon the requirements of each given type of mariculture operation as shaped by the species to be cultivated, culture and harvest methods, processing needs, geographic location, and other appropriate factors.

In considering this support aspect of mariculture, we review the present state-of-the-art in offshore platform design and evaluate how each specific type of platform might serve open sea mariculture. It would be illogical to overlook the obvious—those platforms which occur naturally in the open sea environment, atolls. Enclosed basins or lagoons are always associated with atolls, and their morphology suggests a number of open sea mariculture applications that might require a minimum of modification and capital outlay where suitable configurations can be found. Accordingly, these are dealth with first, and then followed by two broad categories into which man-made offshore platforms can be divided on the basis of their mode of installation:

- Bottom-supported structures
- Free-floating platforms

Next, provisions needed for housing the equipment and crews of mariculture operations are examined. Then, having looked at what is possible today, we propose some entirely unconventional approaches to offshore platform design—approaches that are possible with today's technology but for which the questions of feasibility and practicality are still to be answered.

ATOLLS AS NATURALLY OCCURRING PLATFORMS

General Characteristics

Several hundred coral atolls display interesting possibilities for the purposes of this study. All share some basic characteristics of potential use to mariculture: (1) Atolls rim lagoons of varying depths and diameters, which in many cases may be nearly enclosed naturally, or may be fairly easily enclosed artificially; (2) because of a volcanic foundation, atolls are usually very close to deep water where phytoplankton nutrients are available; (3) indications are that subterranean wells within the base structure of many atolls may be a source of nutrient-rich water; (4) many atolls lie within human-inhabited island complexes where some form of economic development is sought; (5) the open sea platforms that atolls represent are highly stable and survivable while requiring little maintenance. Additionally, many atolls are so situated and configured that free-energy fluxes in their environs are quite high. These free, or *conservative,* energy fluxes are contained in winds, waves, ocean currents, solar radiation, and tidal currents flowing through lagoon passages to and from the open sea. Their potential value is discussed in Chapter 13. The detailed characteristics of atolls are well covered in standard texts and so will not be examined in depth here. An excellent study published by Wiens of Yale University [1] provided the information summarized briefly in the next section.

Distribution and Morphology of Atolls

The distribution of atolls appears to be an integral function of tectonic and associated volcanic activity, already described in Chapter 4, and of the patterns of ocean currents and marine climate. Most atolls in the area of interest lie in the northern and central Pacific basin, with only a few in the Caribbean. The Pacific basin atolls can be grouped into six major chains: the Marshalls, Gilberts, Carolines, Marianas, Line Islands, and the Hawaiian Archipelago, as may be seen in Figure 12.1. (It should be mentioned that the majority of submerged atolls are loosely associated with atoll chains, as are the seamounts and guyots that approach the surface, and are therefore possible mariculture sites. This is clear in Figure 4.5.)

In shape, atolls depart freely from the oval or circular outlines commonly associated with them. Virtually the only generalization that might be made is that each has its own peculiar shape, reflecting most likely the particular

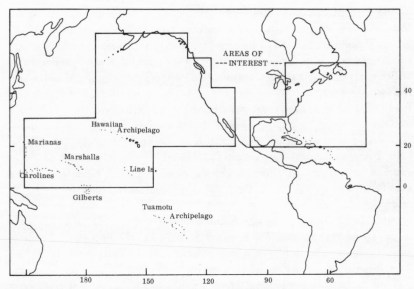

FIGURE 12.1 *Coral atoll distribution in the Pacific and Caribbean.*

submarine contours of its volcanic foundation. Figure 12.2 shows typical examples of the variety of configuration. In a chart analysis of 162 atolls and table reefs (reef islands without lagoons), Wiens found no correlation between their axis alignment and the prevailing winds and currents. Rather, alignment seemed to coincide generally with the submarine tectonic and volcanic topography at the foundation.

The land area of any given atoll is rarely, if ever, continuous; it usually is broken up into small fragments or islets. Of 125 atolls examined by Wiens, 55 had fewer than 10 islets of differing sizes, 27 had more than 10 but less than 20, and 43 had more than 20. Some of the latter group had more than 100, as

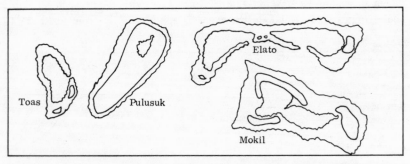

FIGURE 12.2 *Examples of atoll shapes.*

in the case of one with 280 islets. However, only the large atolls of the Marshalls appeared to have very numerous land fragments; in most of the groups (Carolines, Gilberts, Ellice, and others) the majority of atolls had land area concentrated in a few large islets. In a survey of Navy Oceanographic Office charts of these same 125 atolls to determine the proportion of the reef rim occupied by land, as measured by length irrespective of width, it was found that approximately 72 percent of the atolls had less than half their outlines occupied by exposed land, 11 percent had one half to two thirds occupied, and six, or nearly 5 percent, were essentially encircled by exposed land. These percentages were extrapolated to all Pacific atolls. The location of exposed land is generally toward the windward side, with elevations limited usually to under 15 to 20 ft. Figure 12.3 is a generalized diagram of an atoll cross section.

Dimensional magnitude is significant in several ways, some of which apply to geological time frames. For mariculture purposes the important aspect of atoll size is its relationship to lagoon structure. Large atolls generally enclose large lagoons and zones of living reefs, both of which contribute to greater reef and lagoon fish production. The Marshall atolls are, in general, the largest in the Pacific; in this group, Kwajelein (122 by 28 km) is the world's largest. In the Carolines the largest is 83 by 50 km. On the other hand, the smallest one studied (only 4.2 by 1.7 km) is also in the Carolines. Of 56 charted atolls in the Marshalls and Carolines, it was found that atoll size related to lagoon depth generally as follows:

Atoll Width (km)	Lagoon Depth (m)
Less than 9	Less than 30
9 to 18	30 to 45
18 or more	45 to 90 or more

With exceptions noted, it was concluded generally that although lagoons of small atolls may be deep or shallow, the larger atolls all have relatively deep lagoons. Many atolls have two or more lagoons of varying sizes. An example is shown in Figure 12.4. Water temperatures vary with depth as in all stable bodies of water, in this case the range is approximately from 29 to 35°C.

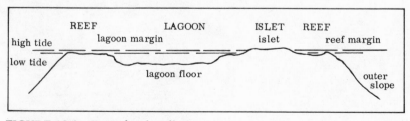

FIGURE 12.3   *Generalized atoll cross section.*

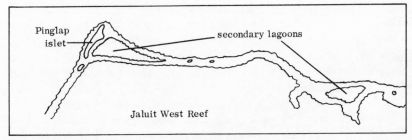

FIGURE 12.4 *Atoll lagoon with secondary lagoons.*

Most lagoons are joined to the open sea by passages, which in many cases are either actually or potentially navigable by ocean-going vessels. Most of these deep passes are found in the Marshalls, the Carolines, and the Tuamotu Archipelago.

Atolls as Open Sea Mariculture Platforms

The use of atolls as open sea mariculture sites already has been suggested by several authors, among them Pinchot [2] and Pryor [3]. For this purpose, the exposed land area could provide the platform while the lagoons served as rearing ponds or enclosures. The steep slope associated with the volcanic foundation of atolls places nutrient-rich deep water resources close to the reef perimeter, usually within 2 km. It is also conceivable that pelagic species could be attracted in numbers to the periphery of atolls by providing fish tents (shelter) and a heightened availability of food fish. Most atoll-based schemes envision the initiation of a complete marine food chain, beginning with either deep nutrient water or nutrient-enriched water pumped from subterranean wells within the atoll's base structures. The chain might then end with mollusk production and harvesting or might be continued on to form a polyculture system of some complexity. If free swimming forms are included, lagoons of course would require naturally or artifically completed enclosures. Yet another possibility involves employing lagoons as hatchery stations for pelagic forms. We explore the atoll possibilities further in Chapter 14.

MAN-MADE OFFSHORE PLATFORMS— STATE-OF-THE-ART

In this section we examine a variety of man-made offshore platforms. For convenience and clarity, they are grouped into two general categories: (1) bottom-supported platforms, and (2) floating platforms.

Bottom-Supported Platforms

Since 1940 the United States has erected over 2,300 offshore platforms in the Gulf of Mexico alone. The bulk of all gulf platforms are devoted to petroleum production, but a few are used for other purposes. Over this same period, the fishing industry in the gulf area has experienced a sixfold increase in catch, going from 113 million kg in 1940 to 0.63 billion kg in 1970 [4], as shown by the curve in Figure 12.5. Can this dramatic increase be attributed to the introduction of the platforms alone? Probably not. Over the same period of time, the number of fishing vessels has increased and fishing technology has advanced. Nevertheless, empirical work leaves little doubt that, other factors being equal, the introduction of structures such as these into a marine environment does tend to increase the local biomass significantly [5, 6]. It seems reasonable, then, that these structures may either form, or initiate the formation of, ecological niches that serve to relax certain limiting factors in some marine populations. This warrants investigation of and by itself; but the central question involves the potential applicability of this type of platform to open sea mariculture. We divide man-made bottom-supported structures into those which pierce the surface and those which do not.

Bottom-Supported, Surface-Piercing Platforms. This subcategory of man-made offshore platforms is represented by five groups having design differ-

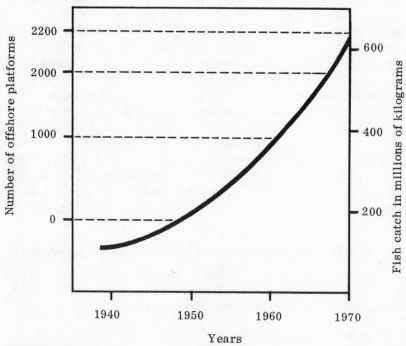

FIGURE 12.5 *Fish-catch history for the Gulf of Mexico since introduction of offshore oil platforms.*

ences of significance to mariculture interests. The groups are (1) the classic Texas tower, (2) jack-up rigs, (3) monopods, (4) petroleum storage facilities, and (5) perforated wall structures.

Texas Towers. The basic design of the Texas tower type of platform now in common use consists of a flat deck, a supporting "jacket" framework, and pilings that act as rigid anchors when driven through the hollow legs of the jacket into the sea floor and secured inside the legs by concrete filling. The Texas-tower-type radar platform shown in Figure 12.6 is an example [7]. The platforms typically are constructed on land, and the jackets transported horizontally to the marine installation site. Once on site they are rotated to a vertical position by differential ballasting, set in place, and secured by the pilings. Once the jacket is secure, the deck and auxiliary equipment are installed. Being permanently moored and rigid, no motions other than vibrations resulting from wave impact are usually discernible on the platform's

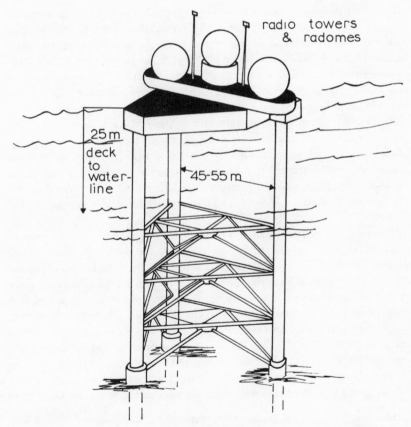

FIGURE 12.6 *Texas-tower type radar platform.*

deck. Hence, these structures represent highly stable bases. With very few exceptions, they repeatedly survive hurricane-force winds and seas.

At present most applications of Texas towers are limited to water depths of 92 m or less; they do, however, appear to have potential for modular expansion. Sections or entire new platforms could be constructed, assembled, and deployed directly from an existing platform and joined together as modules to form a much larger base of operations with added payload capacity.

Jack-up Rigs. Jack-up platforms are movable like surface craft and have the capability of raising (or lowering) their deck structure by "jacking" up (or down) on supporting legs. To achieve this capability, the legs cannot be the simple, but comparatively heavy, cylindrical legs used on permanent structures. Instead, the legs are a triangular or rectangular lattice-type framework, which is supported by the deck structure when under tow. When the platform is in position, the legs are jacked down and embedded in the bottom, and the deck is raised above the surface.

Two configurations currently are employed. One consists of four vertical columns arranged in a rectangular configuration; the other is the three-legged platform with columns arranged in the triangular configuration shown in Figure 12.7 [7]. When this latter platform is on station, the columns are embedded in the bottom at some angle less than vertical. The design gives the unbraced columns greater stability through increased resistance to horizontal forces induced by waves, wind, and current.

The largest jack-up platform currently in use is 70 m^2, with an in-place payload capacity of approximately 2,540 metric tons. As with permanent platforms, the jack-up platforms have a theoretical potential for modular joining of several platforms to form a larger structure. Owing to flexure in the supporting columns when in place, and to other limitations related to the floating mobile configuration, these platforms are limited to depths of 91 m or less.

Monopods.[1] Another example of a bottom-supported, surface-piercing platform is the fixed oil-production platform for the Cook Inlet, off the Alaskan Coast. A 31- by 34-m deck is supported by a single, large-diameter column that pierces the water surface. The platform's base has two cross-braced pontoon sections connected to the center column by two sets of four diagonal braces (see Figure 12.8) [7]. The single, large-diameter column was designed to give both lower resistance to waves and greater strength to resist the concentrated forces of ice loading at the water line. The large pontoons serve as buoyancy chambers in towing and as oil storage tanks once the platform is in position on the bottom. The monopod platform is secured by pilings driven into the bottom through the ends of the pontoons. Once fixed

---

[1] Monopods are classed as "submersible" platforms by the American Bureau of Shipping. The classification employed here suits our purposes.

FIGURE 12.7 *Jack-up type oil-drilling platform. (Adapted from Marks and Kim [7].)*

to the bottom, this type of platform is subject to the same constraints as any offshore structure mentioned previously. The Cook Inlet monopod is presently operating in 45 m of water. It is quite possible that the dimensions could be increased to support a larger deck or deeper operating limits, but construction techniques and costs may be prohibitive. A more feasible scheme might

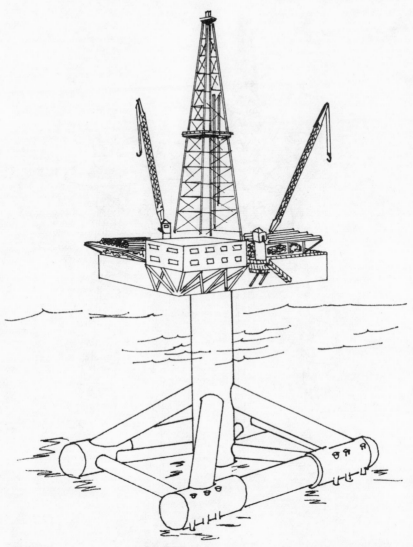

FIGURE 12.8 *Monopod-type drilling platform. (Adapted from Marks and Kim [7].)*

be to join several such structures together to provide a larger deck area and increased storage capacity.

Undersea Storage Facilities. The development of submarine oil fields has instigated the use of large-capacity, submerged oil storage tanks to solve crude-oil storage problems. A few, of relatively conventional design, are already in operation. These are conical or teardrop in shape, with the apex of the cone piercing the water surface. Two tanks in operation at the Fateh field in the Sheikdom of Dubai have 500,000-barrel storage capacities, are 82 m in diameter at the base, and stand 64 m high, as shown in Figure 12.9 [8]. A third facility has been planned and should be in operation in the near future.

Other designs for oil storage containers consist of cylindrical or spherical tanks that are moored to the bottom and have a storage capacity of 250,000 to 500,000 barrels. All submerged storage tanks are linked to the production wells by underwater pipelines and have ship terminal and loading facilities at the surface. As such, of course, they are not entirely submerged, but the bulk of their volume and surface area is.

The potential usefulness of submerged storage tanks to open sea mariculture is somewhat nebulous; large surface area, regular shape, and their increasing numbers are the main attractions. Given minor modifications in exterior configurations, they might be attractive habitats for a number of sessile organisms and might even act as bases for fish cages of significant magnitude. Association with the offshore oil industry appears to offer some labor and transportation advantages, too. In short, as the number of these tanks increases, it might be worthwhile to investigate inexpensive modifications to their exterior designs that could allow them to serve as mariculture bases as well as serving their primary purpose. Clearly, such an idea presupposes that the aquatic environment in their immediate vicinity can be maintained in acceptable conditions.

Perforated Wall Structures. The oil industry also has been dealing with the

FIGURE 12.9 *Submerged oil storage tanks at the Fateh field.*

problem of offshore storage of crude oil with some interesting new developments in bottom-mounted structures. Representative of a surface-piercing design is the Ekofisk oil-storage and tanker-loading caisson constructed in Norway and installed as a component of Ekofisk Center, a multistructure drilling, storage, and loading complex in the North Sea. The 8,800-m concrete "island" rests on the sea bed in nearly 70 m of water. The design comprises a central core (a nine-compartment tank) nearly 91 m tall and with a 1-million-barrel storage capacity, surrounded by a perforated concrete outer shell 99 m in diameter [9]. Figure 12.10 shows a cutaway view of the structure. The perforated outer shell, or breakwater, acts as a wave damper between the central core and the heavy seas prevalent in the area, while allowing free circulation of water around the caisson. Connecting the outer shell and the caisson are four radially positioned concrete reinforcing walls, which divide the enclosed water into four segments. These walls are also perforated to allow for circulation and wave damping.

The structure is designed for operation at a depth of 70 m with a survivability of sea state 7 on the Beaufort scale. It was built on land, towed to position, and sunk to rest on a 6-m-thick concrete pad; once positioned on the seabed, the caisson is anchored by its own weight plus sand ballast and the weight of its liquid contents, either oil or seawater. Total deadweight is over 272 million kg (300,000 tons). Its designers envision the principle being applied to other purposes than oil or gas storage, such as drilling platforms, gas liquefication plants, and the construction of various plants, such as nuclear power stations, on the continental shelf [10]. Looking at it from the mariculture perspective, the structure provides an underwater surface, vertical and horizontal, totaling more than 40,470 m^2 (10 acres) and an enclosed volume of circulating seawater exceeding 1.46 million m^3. Although no cost data are available yet, it seems safe to conclude that structures of this type will be too expensive to be used solely for the support of mariculture operations. However, there does seem to be a distinct possibility that mariculture operations could be associated with their originally intended functions in instances where the structures are appropriately located for culturing purposes.

Bottom-Supported, Submerged Platforms. At this point the term platform begins to be used a good deal more loosely. This subcategory includes artificial reefs, immobile habitats, and even mobile habitats. All may find application in offshore mariculture at some point in the future.

Artificial Reefs. One of the more successful attempts to increase marine productivity has been the placement of artificial reefs. We know that fish tend to congregate around reefs, floating objects, isolated rocks, and, indeed, any object whatsoever. This behavior is most often attributed to two factors: (1) the object may provide shelter, and (2) it provides visual reference in the reference-poor marine environment. Since 1958, over 100 artificial reefs have been placed in various coastal localities throughout the United States to

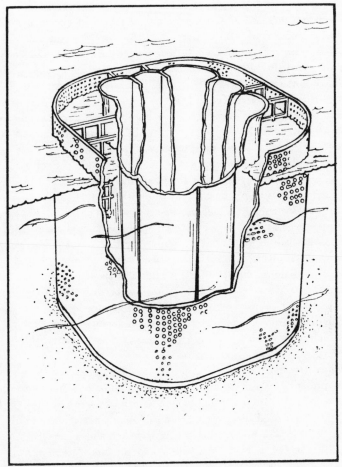

FIGURE 12.10 *Ekofisk offshore oil storage facility. (Courtesy of the Phillips Petroleum Co.)*

enhance fish abundance. Although to varying degrees, all such projects can be considered successful. Some areas showed less productivity than others, but all projects exhibited higher fish populations with the artificial reef than without it.

One artificial reef off the coast of South Carolina had drawn a resident population of hundreds of black sea bass, porgies, grunts, sheepshead, and bluefish schools within 2 months of placement [11]. In May, 1958, 20 old automobile bodies were placed in 15 m of water at Paradise Cove, near Malibu Beach, California. By September, 1960, approximately 24,000 fish were counted in the immediate area [12].

Hawaii dumped abandoned automobiles in three sites offshore during a

cleanup campaign in November of 1972. Seven months later divers reported a
fish population increase to 435 lb/acre on the reef off Waianae, compared to
the previous count of 100 lb/acre in 1959. In Maunalua Bay the fish
population jumped from 40 to 300 lb/acre over the same period. The junked
cars at the third site, off Kaaawa, could not be located; they apparently were
swept farther offshore by the ocean currents. No visible evidence of pollution
from the artificial reefs was found [13].

Several materials have been used; car bodies work exceptionally well as
habitats, but they only last 3 to 6 years in the open ocean and are also
expensive to replace. Street cars have also been used, but these too are
expensive to place and the supply is somewhat limited. The most successful
material in terms of expense and longevity has been rubber car tires. The
capital cost of securing old tires is virtually nothing, deployment is extremely
easy, and they have shown a longevity in salt water in excess of 20 years.

Most reef-dwelling nekton are of minor economic importance today. But
some species, such as the jacks, certainly might become commercial attractions if made available in sufficient numbers. Sessile forms, such as oysters,
mussels, and abalone, might also be encouraged to inhabit artificial reefs in
commercial quantities if adequate nutrition were provided. These reefs can
usually be placed below the zone of major turbulence in the open seas, and
thus remain little disturbed by extreme environmental conditions. So it might
be that such reefs, when combined with other factors, could provide bases for
commercially significant productivity.

Immobile Undersea Habitats. Conshelf, Sealab, Aegir, Tektite, and a host
of other present and past experiments in extended underwater human habitation leave little doubt concerning human abilities to live and work in the sea
under hyperbaric conditions. Aegir, with a crew of six, remained for 5 days at
a depth of over 152 m, and the others have remained for weeks at lesser
depths. But, so far, these have all been experiments. Although little doubt
remains that man is capable of inhabiting the sea floor to continental-shelf
depths, all his real experience has been under tightly controlled experimental
conditions. Long-term, economically feasible habitation of the sea floor with
practical results has yet to be demonstrated.

Nevertheless, the offshore petroleum industry, leading again, is developing
practical habitats appropriate for limited bottom excursions in over 100-m
depths. These developments are for uses related to the installation and
maintenance of underwater wellheads and pipelines. All habitats to date are
designed for occupation by no more than six persons, and more usually for
two or three. They are expensive to deploy and to maintain on station;
probably it will be quite some time before they find practical application in
open sea mariculture.

Mobile Undersea Platforms. Eugene Allmendinger has offered conceptual
designs for mobile undersea habitats operating on networks of fixed mono-

rails [14] to be used in installing and servicing the larger wellhead—pipeline—storage tank complexes on the ocean floor. They could, of course, also find application in servicing sea bed mariculture should it be feasible to associate the two.

In the area of unmanned equipment, the Japanese have developed an undersea, remote-control bulldozer that has been demonstrated in shallow water. Modifications of this design have clear implications for offshore shellfish farming. So far these two concepts are all that appear in the literature dealing with mobile undersea platforms. Speculations on the subject are offered in a later section of this chapter.

Bottom-Mounted Platforms and Open Sea Mariculture. The bottom-mounted platforms we have discussed range in cost from a few to several tens of millions of dollars. But if mariculture operations are to be based at sea, some sort of platform will be required. One solution to the prohibitively high costs lies in associating mariculture with other offshore activities in order to share platform costs. It might also be born in mind that relative platform costs are likely to diminish as (and if) designs are standardized for continuous multiunit production and applications of such platforms are broadened and intensified.

Another valid generality appears to be that all structures of this type will attract marine life to some extent, although it is not entirely clear that a platform will, solely by its existence, increase nektonic productivity. It can be stated, however, that the new surface area it introduces almost certainly increases sessile biomass and, therefore, seems likely to increase nektonic biomass. Numerous experiments have demonstrated that thermal enrichment (within limits) of a marine ecosystem tends to increase biproductivity quite markedly [15–19]. So it seems that a combination of platform structure and thermal enrichment is quite likely to result in dramatic bioproductivity increases in many instances. That this will actually be the case in real-life applications, that it can be done with marketable species, and that its effects can produce protein on commercially attractive scales remain to be demonstrated.

Floating Platforms

The second category of man-made offshore platforms includes conventional ships and barges and semisubmersible platforms. The platforms comprise structural approaches to motion stability, ranging from large, manned spar buoys through a variety of moderately large to very large multilegged or multihulled configurations. Submerged floating platforms, such as research submersibles and submarines, have been omitted here since their characteristics are well known and their utility as mariculture platforms is doubtful.

Conventional Ships and Barges. It does not really seem appropriate, here, to discuss conventional ships and barges in any depth. They are common and familiar figures and their great variety generates a mass of relevant information impossible to compile, much less to present in a brief review. It will suffice to say that these structures in some forms may be appropriate to some open sea mariculture operations. Since ships are surface followers, they are inherently sea-state limited, and economic concerns will very likely act to limit their size, which further limits sea-state capability. Even if constructed of corrosion-resistant material, they all require periodic haul-out for maintenance. For these reasons, we find our thoughts drawn toward retired conventional surface craft that might be obtained at little cost, employed for a time, and then converted to submerged structures that would form artificial attractants for marine life—or constitute ocean dumping of solid waste—depending on one's point of view. If such be the practical role of conventional craft as open sea mariculture platforms, their application will be limited to moored configurations in fairly shallow waters seldom subjected to severe environmental conditions.

Spar Buoys. Spar buoys come in a variety of shapes and sizes, as shown in Figure 12.11 [7], but all are long, tubular configurations designed to be towed to sea horizontally and subsequently translated to a vertical position on station. The buoys may be moored or free-drifting, depending upon the particular mission. As an approximate statement, it can be said that in the vertical position their motions with respect to most seaways are small because of their very high ratio of total mass to water plane area, and because their natural periods of oscillation typically are significantly longer than the periods of surface-wave input forces. The most successful and most publicized spar buoy is the Floating Instrument Platform (FLIP), which has been operated since 1962 by the Scripps Institution of Oceanography [20, 21]. Because of experience with FLIP and several other large examples, the wave-induced motions of spar buoys are fairly well understood by now, and several computer programs have evolved for predicting their motions in open seaways. These theoretically derived predictions and actual data obtained in open sea conditions are in fair agreement.

Without disturbing the hydrodynamic characteristics appreciably, spar buoys can within limits vary their draft by adding or reducing water in the ballast compartments. This affords a buoy the ability to increase the distance from the water line to the deck during heavy seas and to reduce it to a minimum for work at the sea surface in calm weather. Spar-buoy payloads are usually quite limited, as is working space. Natural heave motion periods in large spar buoys are usually well in excess of 20 seconds, and seas of any size near their resonant periods are highly unlikely.

The Manned Open Sea Experimentation Station (MOSES) spar buoy has been developed by the Oceanic Institute in Hawaii through preliminary engineering specifications only [22]; it is the sole example specifically designed for biological research (see Figure 12.12). MOSES was designed with

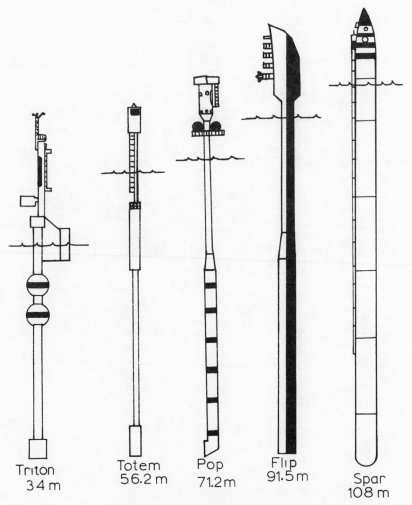

FIGURE 12.11 *Spar-buoy configurations. (Redrawn from Marks and Kim [7].)*

open sea mariculture experiments in mind, and buoys of this type might be employed fruitfully in the development of culture data and techniques and in assessing the potential productivity of proposed sites.

Simple spar buoys produced in numbers could most likely be manufactured quite inexpensively. If constructed of reinforced concrete or ferrocement, they could also be quite durable. Some versions might find application as floating "posts" for "marine corrals," as in Figure 12.13. As such, they could serve double duty as work stations, too. Aside from these uses, it is not

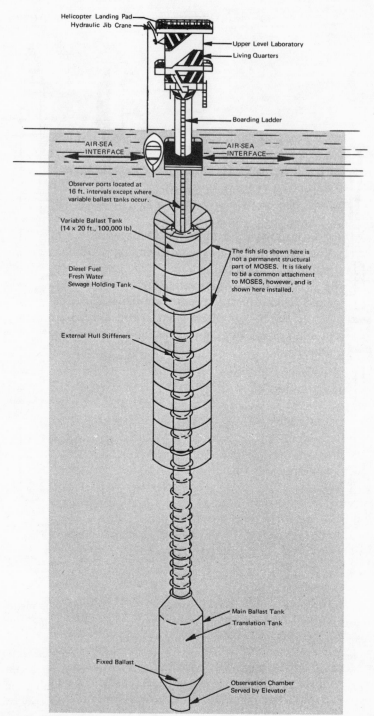

Helicopter Landing Pad
Hydraulic Jib Crane

Upper Level Laboratory

Living Quarters

Boarding Ladder

AIR-SEA INTERFACE

AIR-SEA INTERFACE

Observer ports located at 16 ft. intervals except where variable ballast tanks occur.

Variable Ballast Tank (14 x 20 ft., 100,000 lb)

The fish silo shown here is not a permanent structural part of MOSES. It is likely to be a common attachment to MOSES, however, and is shown here installed.

Diesel Fuel
Fresh Water
Sewage Holding Tank

External Hull Stiffeners

Main Ballast Tank
Translation Tank

Fixed Ballast

Observation Chamber Served by Elevator

FIGURE 12.12 *Spar-buoy configuration for the MOSES (Manned Open Sea Experimentation Station).*

312

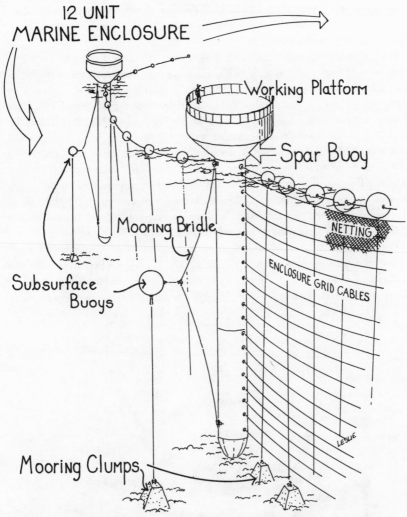

12 UNIT MARINE ENCLOSURE

Working Platform

Spar Buoy

Mooring Bridle

Subsurface Buoys

NETTING

ENCLOSURE GRID CABLES

Mooring Clumps

LESLIE

FIGURE 12.13 *Marine corral: hypothetical employment of spar buoys as floating posts and work stations for a net-enclosed marine corral.*

obvious how spar buoys per se would be employed effectively in open sea mariculture. But the column-stabilized, semisubmersible platform discussed next is in truth a rigid spar-buoy cluster. So the basic concept is important.

Semisubmersible Platforms. Virtually all operational semisubmersible platforms and those presently under construction (numbering now about 100) are designed for offshore oil drilling. Because of their potential versatility, this is probably a transient phase; the next few decades very likely may see exponential broadening of semisubmersible platform applications for urban and industrial purposes. These platforms are quite expensive at present; large versions may cost tens of millions of dollars. Extending their application, however, should result in the eventual reduction of costs, particularly if designs can be standardized.

The spectrum of semisubmersible design seems to range from a hull orientation on one end to a column orientation on the other, with essentially all presently operational examples falling somewhere in between and combining elements of both. The hull-oriented platform concept is represented by Zapata Corporation's SS-3000, shown in Figure 12.14 [23]. But it may be noted that the SS-3000 and similar examples have large, surface-piercing columns. Column orientation is seen in the Pentagone 82 in Figure 12.15 [24]. An even more extensive employment of column orientation is seen in Hawaii's floating-city conceptual design. Preliminary engineering on this latter project has proceeded through the development of computer simulations of the hydrodynamic characteristics and the construction and testing of the 100-ton, 1-to-20 scale model shown in Figure 12.16 [25].

The floating-city design, and Scripps Institution of Oceanography's four-legged platform model shown in Figure 12.17 [7], are the first to employ the concept of modularity in the sense of joining large semisubmersibles at sea to form still larger platforms. The Scripps model has two, two-legged mod-

FIGURE 12.14 *Zapata semisubmersible SS-3000.*

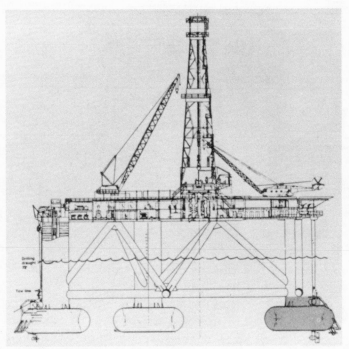

FIGURE 12.15 *Semisubmersible Pentagone 82.*

ules joined to produce one four-legged platform. It can be towed long distances in two parts horizontally and then erected and joined on station. The floating-city concept, on the other hand, involves multiple, three-legged modules that can be joined in concentric circles (see Figure 12.18) to develop modular complexes so large that entire urban systems could be supported [25].

The chief advantages of the semisubmersible platform are a high level of motion stability and high survivability in extreme environmental conditions. The major technical disadvantages stem from the fact that the same designs which produce motion stability reduce water-plane area, and therefore give relatively little buoyancy force per unit of submersion. Thus, overturning moments and added live loads must be compensated for by ballast adjustments, in contrast to the spontaneous compensation occurring in conventional surface hulls. Another disadvantage is their presently high cost.

Articulated Columns. One new approach to motion stability has been proposed by Offshore Technology Corporation in the form of an *articulated-column* semisubmersible [26]. In this approach (see Figure 12.19) the inherent buoyancy of the semisubmersible is provided by surface-piercing buoyancy chambers. These, however, are connected to the submerged hulls in

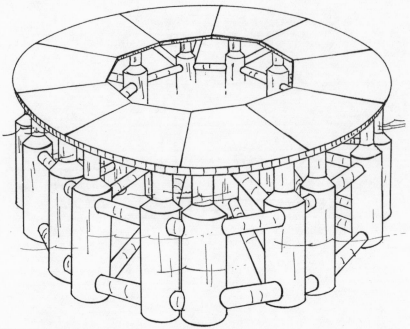

FIGURE 12.16 *Hawaii floating-city conceptual design—1-to-20 scale model.*

such a way that they are free to rotate in a hemisphere about their point of attachment. This, of course, allows wave-induced pitch forces to be largely absorbed rather than transmitted to the main structure. Although scale-model tests seem to support the validity of this approach, no full-scale versions have been constructed as yet.

Floating Platforms and Open Sea Mariculture. Spar buoys may have some limited applications to mariculture, but their principal contribution will be their basic design concept as it applies to stable, semisubmersible platforms and their value as research tools. The implications of semisubmersible-platform technology for open sea mariculture seem quite clear and indisputable. Simply put, open sea mariculture on an extensive scale in deep water will necessarily be based on this type of platform; it alone can provide the stability, size, durability, and survivability for a deep-water operation of commercial magnitude. Furthermore, it seems likely that concrete construction will be advisable, unless the plastics industry is able to provide a noncorrodable material having a sufficient strength-to-weight ratio at an acceptable manufacturing price. At present, concrete is the only economical material offering long-term durability with a minimum of maintenance in the marine environment. Thus, prospects for extensive open sea mariculture would be brightened considerably by practical designs for large semisubmersible platforms to be constructed inexpensively from concrete.

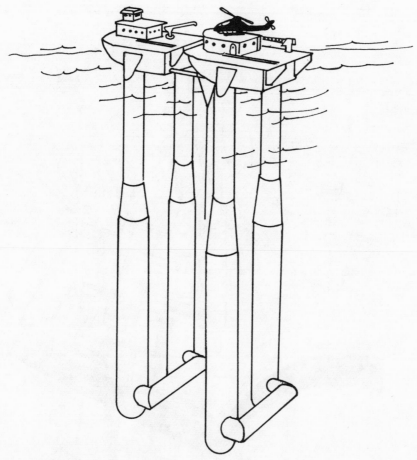

FIGURE 12.17 *Sea Legs platform of the Institution of Oceanography. (Redrawn from Marks and Kim [7].)*

HOUSING FACILITIES FOR OPEN SEA PLATFORMS

The term *housing* is employed here to signify all shelter requirements for both equipment and personnel. Among the fundamental design criteria will be survivability, that is, strength and flexibility to withstand the heaviest sea and wind conditions that may possibly be encountered in any selected area; ease of emplacement; durability under normal wear and tear; and, of course, cost minimization. With machinery and equipment only, these criteria would be the principal guides. However, the requirements of the resident operating crew necessitate the incorporation of human-factors engineering to create a physically and psychologically confortable environment. The provision for adequate space to properly position equipment and to design work flow,

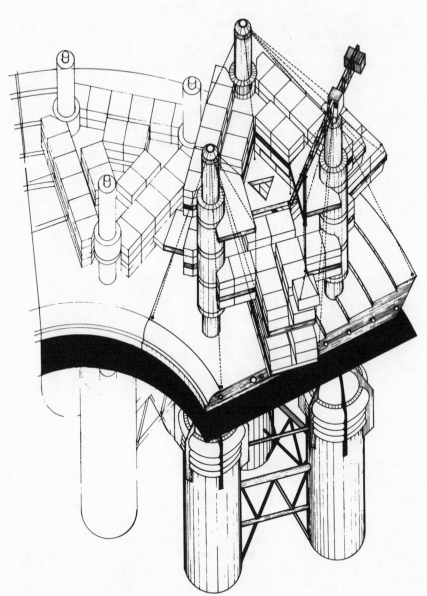

FIGURE 12.18 *Floating-city module perspective.*

FIGURE 12.19 *Articulated-column floating platform. (Adapted from* Ocean Industry *[26].)*

moreover, may contribute to efficiency and to minimizing costs and personnel requirements.

It is not yet possible, of course, to identify with even the broadest approximation the number of personnel or types of individuals that will be needed to staff any given type of station; this projection must await decisions on what operations will be selected and how and where they will be accomplished. It is possible, nevertheless, to look at the problems to be solved in providing acceptable housing, sufficiently attractive to enable retention of trained personnel.

The characteristics of the marine environment itself, exclusive of the on-going activity, impose definite constraints on design and construction materials. The following discussion covers an admittedly conjectural view of the general problems to be faced and some design principles that can be incorporated in the light of current knowledge about marine construction.

Physical Characteristics of an Open Sea Mariculture Facility

An open sea mariculture operation will differ sharply from the typical industrial activity in that the processing plant, with its machinery, noise, and odors, will be located in tight proximity to living quarters and recreational facilities for station personnel, together with such other components as storage for feeds, waste-management processes, crop-culturing activities, hatcheries, rearing enclosures, laboratories, cold storage for intermediate and finishes products, and materials-handling equipment for loading and unloading the station. In fact, except for warships and offshore petroleum rigs, it would be hard to find a mixture of activities more unfavorable for human physical and psychological comfort.

In an onshore industrial activity the classical solution to this problem is simply to put distance between the workers' dwellings and recreation areas and their place of work. The present practice in the offshore oil industry is to provide soundproofed, air-conditioned dormitories for the crew of a production or drilling platform, and to rotate crews at frequent intervals. Even so, crewmen willing and able to tolerate the rigors of offshore platform life can command premium wages.

Design Criteria for Mariculture Installations

Separation of Working and Living Areas. The physical characteristics of a mariculture operation make it apparent that the prime criterion for the design of housing facilities is the separation of living and recreational areas from the work area. For an atoll, or for a mariculture station located close to a hospitable shore, this may be most easily achieved by distance. If, however, a man-made station is located far enough at sea that daily transportation of the station's work force to and from shore is impractical, separation may have to be achieved by the station design. This will require a decision early in the design process as to whether the station will be the principal domicile of its crew for extended periods of time, or whether some intermediate period longer than 1 day will be chosen. In shipboard practice, where vessels touch port at scheduled intervals, the opportunity for crew members to go ashore periodically has undoubted psychological value, allowing provisions for crew comfort to be minimized. This approach (in slightly modified form) will certainly be useful with open sea mariculture, too.

Another consideration to be examined early in the design process is the degree of skill, training, aptitude, and experience the crew members must possess. The desired crew-retention rate will be a function of the operator's investment in training, whereas the turnover rate actually achieved will be strongly influenced by what provisions the owner has made for the comfort and satisfaction of his crew members.

Environment Control. Once the period of residence, the degree of comfort, and the extent of the living and recreational facilities are determined, it remains to examine the environment suitable for equipment and work areas, and the factors from which the living and recreational facilities must be protected or isolated.

The important components of an open sea mariculture station's environment may be classified under two general headings: (1) weather or external climate, over which the designer has no control, and (2) environmental components emanating from the activities of the station itself, over which he may or may not be able to exercise control. For an example of the latter, the noxious or unpleasant aspects of some processes may be eliminated only at prohibitive expense, whereas the proper selection and location of the station's power plant may easily result in an acceptable abatement of noise and exhaust gases near the living area. At any rate, the designer must be aware of all the important environmental stresses, ameliorate them where practical, and provide as much isolation as possible from those he cannot change.

Interior Climate. Air conditioning, heating, ventilating an open sea station generally should conform to accepted practice for coastal areas in similar latitudes, with due attention to the filtering of makeup air and the location of intakes so as to avoid noxious odors, fumes, and sea spray. Obviously, corrosion-resistant materials should be used in all air ducts and plenum, grills and registers, fan-coil units, filter frames, and machinery housings.

Mobility and Flexibility. Another criterion for good design of housing is portability. Plumbing, wiring, carpentry, painting, plastering, and the like, require a surprising amount of labor in the assembly and finishing of a structural unit, and any work at sea is always significantly more expensive. If all or most of such work can be done ashore by prefabricating units or modules, work at sea can consist merely of delivering the units or modules and putting them in place—resulting in savings of time and cost. Moreover, units of the prefab type could easily be removed and replaced when worn out, or as requirements change.

Construction Materials for Open Sea Facilities

Especially for metals, but in general for all materials, the ocean environment is severe. In addition to durability, weight may be an important consideration if the housing units are located in the superstructure of floating platforms.

Exterior Structure. Aluminum is an excellent material for the exterior skins of buildings. For most exposures the alloys normally recommended for exterior use on shore should be adequate. Pure aluminum cladding over high-strength alloys may be specified if maximum corrosion resistance is desired. Aluminum- or cadmium-plated steel fasteners should be used. Where aluminum comes in contact with other metals, some protection against electrolysis is required. Where aluminum contacts steel, hot-dip galvanizing of the steel is preferred. In all cases, the faying surfaces should be insulated from one another by asphaltic paints or gasketing. Aluminum may be installed bare, anodized, painted, or finished with a vitreous enamel. In warmer climates, exterior surfaces should probably be bright to reflect solar radiation. This, of course, conflicts with the need to minimize glare on deck; it might be handled with wide overhangs on roofs. In colder climates, dark, more absorptive colors seem in order.

Glass-reinforced polyester is a very suitable material for light building exteriors also, especially if formulated with ultraviolet light inhibitors and fade-resistant pigments. It is already being employed in the manufacture of preformed kitchen and bathroom sections; in fact, complete prefabricated bathrooms are now available in this material, with fixtures, plumbing, and wiring already installed. This approach could be translated into designs for laboratory, seafood-processing, and larval-rearing facilities as well.

Steel standing hardware should be hot-dip galvanized wherever possible. Operable hardware, such as window sash and door hardware, should certainly be at least galvanized steel or, better, bronze or passivated stainless steel.

Interior Structure. Interior partitions of wood, composition materials, or standard dry-wall construction seem acceptable, as is wood trim. Lath and plaster, because of weight and labor, seem unsuitable for most marine applications. Light metal partitions appear acceptable for most interior applications, and would be indicated for sanitary facilities such as seafood processing and laboratories.

Nonbearing wall framing for small buildings may be of wood, depending on fire-resistance requirements. For most buildings, light-gauge steel with open-web steel joists and cold-formed steel studs is preferable.

Thermal and Acoustic Insulation

Thermal insulation for marine structures should be inorganic. Spun-glass batting with integral vapor barrier is one type that would be suitable for all but the most severe climates.

Acoustical isolation is a more complex problem. Sound and vibration may be transmitted through air or through structural members. In a lightweight structure, the transmission of high-frequency sound may be reduced somewhat by establishing a complete airtightness of the building skin, and by providing for energy dissipation within the exterior walls. For most sound, however, the fundamental method of reducing transmission is to increase the mass of the partition, floor, or ceiling; this clearly is in conflict with any weight-minimization requirements. Yet lightweight thermal insulation and acoustical tile are relatively ineffective in inhibiting noise transmission unless they are combined with mass and resilient mountings. Acoustical treatment of interior surfaces does reduce sound reflection, however, and so leads to reduced noise levels within an enclosed space in which noise is produced. The transmission of the vibration of the structural frame of the platform into housing units may be reduced by mounting units on resilient supports and taking special care to avoid transmission paths along pipes, conduits, cables, and the like. For an excellent text on this subject, see Harris, *Handbook of Noise Control* (listed in the Additional Readings).

Natural Lighting

Windows of various size, type, and location form an important part of the character of almost any building. The distribution and control of daylight, views of outdoors, privacy, ventilation control, heat gain or loss, and weather resistance, all can be achieved to varying degrees that are determined by the choice and placement of windows.

A good general rule for window area is that it should equal 20 percent of the floor area served. On some faces of an open sea structure, wave or wind hazard may limit size or require the omission of glass entirely. For sound or thermal insulation, double- or triple-glazed windows may be used. Heat-absorbing glass in the outer panel can provide substantial reductions in air-conditioning requirements. In all cases, permanent (nonmovable) glass should be set in resilient, airtight seals or gaskets. Metal window frames of either galvanized steel or aluminum are extensively used and are available in a wide variety of strengths and thicknesses, including obscure, patterned, wire, and tinted. Of particular interest are wire glass for its fire resistance, and the various high-strength glasses for their resistance to breakage in high winds. For open sea stations in warmer climates, open window design should usually be explored as a less expensive alternative to air conditioning. At higher latitudes proper window designs can, of course, reduce heating requirements.

Construction Materials for Atoll-Based Installations

For housing units to be placed on atolls, an entirely different construction method may be indicated: for instance, if coral suitable for concrete aggregate can be obtained locally, building elements could be made readily and inexpensively of precast concrete slabs. Concrete has good thermal insulation value and long, maintenance-free life. As in all marine exposures, special attention should be paid to corrosion resistance in all metal items: aluminum should not be used in contact with concrete; and galvanized or bronze door and window hardware are most suitable. For roofs, some slope and broad overhang are desirable; in low latitudes additional thermal insulation would be needed. Large, fully shaded wall openings would provide copious cross ventilation as an alternative to full-time air conditioning. In the western Pacific, where typhoons are severe, roofs must be secured against uplift, and all openings should have removable, bolt-on hurricane shutters.

Conclusions

From the foregoing discussion we may conclude that safety and comfort will be major issues in designing open sea mariculture housing facilities, for a high turnover of trained staff could well prove more costly in the long run than good initial design and construction. Durability is also an important consideration, as is conservation of energy, and superimposed on all these functional concerns is the question of capital and maintenance costs. With all these objectives in mind, we have looked at the practicability of designing for transportability, ease of assembly at sea, and flexibility. These factors, taken together, give strong support for a modular approach.

In general, the use of exotic space-age materials and techniques does not seem indicated, since it appears that adequate designs can be achieved with the much less expensive methods and materials now in wide use in the shoreside and marine construction industry.

SPECULATIVE OPEN SEA PLATFORM CONCEPTS

To this point the subject has been an overview of open sea facilities and structures currently in use or design, all of which (with the halfway exception of artificial reefs) have, as their *raison d'être,* functions essentially unrelated to mariculture. The principal concerns have been to evaluate (1) the impact these structures may have on the marine organism populations for which they represent an environmental modification; (2) the potential utility and adaptability of the structures to the requirements of an open sea mariculture operation; and (3) the possibility of adding mariculture operations to the ongoing functions for which the structures were built. So for the most part we have been talking about adapting mariculture requirements to predetermined facility designs. If open sea structures were to be designed specifically and solely to support mariculture operations, what forms might these take?

Permanent Ocean-Bottom Habitats

Despite the promising results of experiments with manned undersea habitats such as Aegir, Sealab, and others, no man has yet remained under extreme hyperbaric conditions for time periods that would allow extensive physiological adaptation to occur. Might there be problems in readapting to 1 atmosphere of surface pressure after several months under hyperbaric conditions? There is no present-day technology directly applicable to on-site, human-attended farming of the seabed at depths in the continental-shelf ranges. But, if such technology is postulated, how might it appear?

Many schemes have been offered for large-scale, permanent, ocean-bottom habitats; some are farfetched, and others within or close to the present state-of-the-art. Although it is hard to imagine a commercially viable enterprise that could afford such a facility, some combination of sea conditions, intensive culture methods, and world protein need may someday prove supportive to an ocean-bottom installation.

To indicate the type of facility that may be feasible, we can postulate a set of cultured food organisms that might include a bottom-dweller such as lobster or shrimp, a sessile filter feeder such as the oyster, and a free-swimming herbivore, or a combination of herbivores and detritus feeders that would serve as food for a marketable carnivore; for such groups we can postulate a set of design requirements.

The facility might be required to house breeding and rearing operations for each species, to provide enclosures and substrates in the sea in close proximity to the facility, and to provide means for feeding, harvesting, and processing, and maintaining the facility itself.

Site-selection criteria would include temperature, solar radiation, mean current velocity, richness and variability of planktonic organisms, shape and character of the bottom, wind and wave climate, and the distance from supporting facilities and markets ashore.

After carefully weighing these factors, analyzing the feeding habits and other behavior of the selected species, and determining the most economic set of breeding, hatching, harvesting, and processing alternatives, we would arrive at a plan that states, for example, the number of hectares of bottom to be

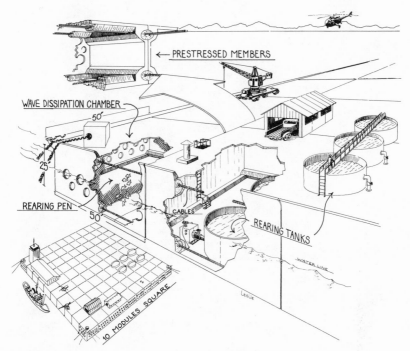

FIGURE 12.21 *Concept for a modular, rigid floating structure.*

this shape is the ability to provide support for very large loads that can be lowered through a center well. Figure 12.22 is a sketch of this concept.

Ice Floes. In the past decade or so, a few people have advanced the idea of separating icebergs or large natural ice floes from polar sea ice and towing them away to provide drinking water for thirsty coastal cities. These projects have all implied the solution to problems of towing, protection from solar radiation, and mooring these extremely large objects (1 km² or more) [29]. If man ever resorts to ice floes for drinking water, perhaps he can also use them as platforms from which to raise seafood.

A similar notion, which has received some study, is that of manufacturing an ice floe. If a suitable insulating jacket were prepared in advance, an ice floe could be constructed within by freezing the impounded seawater. The structure could be hollow with internal bulkheads and columns, with the chilling pipes acting as reinforcements. Although the power cost for freezing such a structure would be high, the maintenance cost would be rather nominal.

Flexible Structures. A variety of flexible structures can be envisioned, which, although perhaps appearing especially exotic at first glance, may in

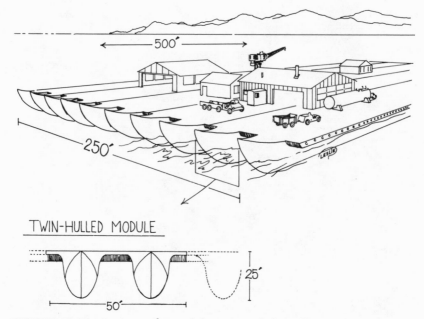

← 500′ →

250′

TWIN-HULLED MODULE

50′

25′

FIGURE 12.22 *Concept of a modular, oriented-shape floating structure.*

some cases offer attractive economy and mobility. Several of these have, in fact, already undergone some preliminary design and testing.

Wave-driven Raft. The wave-driven raft is one such idea. Several years ago, researchers at Scripps Institution of Oceanography discovered that a train of rafts, hinged together end to end, each equipped with a movable flap on its underside, would proceed forward by wave power, independently of the incident wave direction. This rig, if equipped with rudders and means to control the driving flaps, might use wave power to maintain stations, or even to follow drifting fish enclosures.

Surface-following Fish Enclosures. A flexible, surface-following fish enclosure could be assembled from an air-filled flexible tube arranged in the form of a toroid float with a large ratio of major to minor diameter, from which a net or curtain would be suspended. The net could be either open or closed at the bottom, depending on the behavior of the impounded fish; in shallow water the bottom edge of the net could lie in contact with the bottom. An extension of the ring-shaped float would be a cellular square or hexagonal array of enclosures, with small rigid decks affixed at the nodes to support air compressors or net-handling machinery.

Mooring or towing such arrays would present some interesting problems, but a free-floating version might survive at sea for extended periods, and would provide for very large enclosures at relatively low cost.

Semisubmerged Breakwaters. Experiments have been conducted with several types of semisubmerged breakwaters [30]. Among these, the flexible spherical and cylindrical types appear interesting for deployment at sea to support nets, enclosures, and the like, while at the same time affording a measure of wave attenuation. Inflated mostly with water, plus enough air to remain at the surface, they present a low profile to advancing waves and derive part of their effect from wave overtopping. The effect is heightened when they are arrayed in several rows. A properly designed deck structure could be placed on short vertical struts and saddles to lie across a row of such tubes. This deck would carry the machinery, air and water pumps, and so on, for controlling a section of breakwater, as well as for raising and lowering fish enclosures.

A variation of the preceding type would consist of an air-inflated ring made from a tube of rubberized fabric supporting a fish enclosure made from flexible netting. Another variation would consist of a hexagonal arrangement of straight tubing sections joined together at nodes, which would have additional buoyancy to support machinery on rigid platforms. Access to such an array could best be gained by air-cushion vehicles, which could pass over the rubberized tubing floats without damage to either structure.

Goodyear Inflatable-Column Semisubmersible Platform. The Goodyear Aerospace Corporation under contract to the Federal Advanced Research Projects Agency has proposed and investigated a column-stabilized semisubmersible platform to be constructed mainly of inflated rubberized material. The vertical buoyancy columns would be of the inflatable rubberized material while the above-the-surface decking would be either a rigid or an inflated structure composed of hexagonal plates [31] (see Figure 12.23). The design objectives for this structure were centered about transportability and survivability; long-term durability was not a major criterion. Still, such a structure might well be as durable as steel in the marine environment, and its light weight and compactness when deflated might make it attractive for temporary additional work loads, such as harvesting and processing.

SUMMARY AND CONCLUSIONS

If mariculture extends into the open seas on a commercial scale, it will need bases of operation on, in, and possibly under the oceans. From this condensed review of open sea platforms we see that, aside from conventional ships and a few spar buoys, the petroleum industry so far stands alone, for all practical purposes, as a developer of new offshore platforms. Thus, the only all-new approaches to offshore platforms that have been taken past the experimental stages have been those which were economically attractive to the offshore petroleum industry.

There are two major implications in this. First, it seems that new ap-

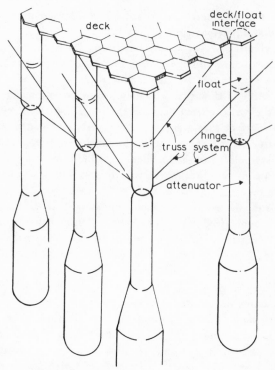

FIGURE 12.23 *Goodyear inflatable-column semi-submersible platform. (Adapted from Goodyear report, [31].)*

proaches to offshore platforms can and most likely will be developed when the economic attraction is sufficiently strong and the economic power of whatever segment of society may be attracted is sufficiently great. It is important to recognize that oil companies emerging near the turn of the 20th century probably could not have afforded to develop offshore platforms; it was necessary that they grow and fatten on less expensive terrestrial oil reserves first.

Second, because offshore platforms represent large investments on the part of the oil industry, even the major companies would likely be attracted to schemes for increasing the return on their investments. Hence, because these platforms are already in existence, they may very likely become the first seeds of open ocean mariculture. In fact, some experiments have already begun [6]. In addition to these petroleum-fertilized seeds, current interest in such possibilities as offshore airports and offshore primary industry, if developed into reality, might serve as additional rooting sites for open sea mariculture.

But what might lie beyond these slightly parasitically oriented prospects?

Further prospects can be identified only through engineering studies that will first define at some level of detail the platform support requirements of open sea mariculture, and then systematically explore all alternative engineering approaches. The end result of this initial work would be preliminary specifications for discrete types of platforms. Such platforms would be modular; they would be as inexpensive to build, operate, and maintain as possible; and they would serve specified types of open sea mariculture operations. In all probability they would be of concrete construction; so research into new types of concrete and new methods of nearshore or at-sea concrete fabrication also seems indicated.

These engineering studies could be conducted most effectively if they were directed toward well-defined types of mariculture operations. Since offshore oyster and mussel production appear to be, at the moment, the most economically attractive open sea mariculture potential, one such project might explore platforms of this possibility. Looking farther into the future, artificial upwelling of deep, nutrient-rich water is likely to be the fulcrum for the development of extensive open sea mariculture. Therefore, preliminary design specifications for platforms that could support large-scale culture of sessile organisms based on artificial upwelling would be an attractive place to begin.

REFERENCES

1. Wiens, H. J. 1962. *Atoll environment and ecology.* Yale University Press, New Haven, Conn.
2. Pinchot, G. B. 1970. Marine farming. *Scientific American 223*(6):15–21.
3. Pryor, T. A. 1971. The development of total atoll production system pilot project in Rarotonga, Cooks Islands. Proposal submitted to the Rockefeller Foundation, May.
4. Treybig, D. L. 1971. How offshore platforms help fishing. *Ocean Industry 6*(4):64.
5. Klima, E. F., and D. A. Wickham. 1971. Attraction of coastal pelagic fishes with artificial structures. *Trans. Amer. Fisheries Soc. 100*(1):86–99.
6. Thompson, R. R., B. D. Honeycutt, and J. C. Parker. 1972. Cooperative environment projects, High Island Block 24L, Offshore, Texas. *Proc. Offshore Tech. Conf. 1676.*
7. Marks, W., and W. Kim. 1971. The stable platform in marine enterprise. *Poseidon Scientific Corp. Tech. Rept. 1011-TR-1.*
8. Anonymous. 1971. *Ocean Industry 6*(3):42.
9. Anonymous. 1973. Ekofisk One becomes an island in the North Sea. *Ocean Industry 8*(8):21–24.
10. Anonymous. 1972. Ekofisk—prestressed concrete oil storage caisson. C. G. Doris, Paris.
11. Anonymous. 1972. Artificial reefs of waste material for habitat improvement. *Marine Pollution Bull. 3*(2).

12. Carlisle, J. et al. 1964. Artificial habitat in the marine environment. *Calif. Dept. Fish and Game Fishery Bull. 124.*

13. Honolulu Advertiser, Aug. 8, 1973.

14. Almendinger, E. E. 1971. The conceptual design of an advanced undersea habitat. *Pacif. Symp. on Hydrodynamically Loaded Shells,* Pt. 1, *Paper 2–1.*

15. Anonymous. 1969. Priming a marine food chain by artificial upwelling. *Underwater Sci. Tech. J. 1*(3):110–112.

16. Anonymous. Heat pollution—or enrichment? *Industrial Res. 10*(7):31.

17. Nash, C. E. 1966–1969. Plaice rearing at Hunterston and Carmarthen Bay electrical generating stations, 1966–1968; Dover sole rearing at Hunterston nuclear generating station, 1968 and 1969; Plaice rearing at Hunterston, 1969; Turbot fattening at Hunterston, 1969; The influence of water temperature on the growth of Plaice and Dover sole at Port Erin, Isle of Man. British White Fish Authority.

18. Nash, C. E. 1972. Power stations and fish farming. In E. Dennis, ed., Everyman's nature reserve. David & Charles, Newton Abbott, England, Chapt. IV.

19. Nash, C. E., and J. E. Shelbourne. 1967. Power station effluent as an environment for flatfish culture. *I.C.E.S. Fisheries Imp. Comm, CM E*:10.

20. Bronson, E. D., and L. R. Glosten. 1962. FLIP floating instrument platform. *Scripps Inst. Oceanogr. Rept. S10-62-64.*

21. Glosten, L. R. 1962. FLIP—some remarks on certain design considerations. *SNAME Pac. Nw. Sect. Paper* (Oct.).

22. Hanson, J. A., ed. 1971. Manned open sea experimental station (MOSES) A feasibility study. *Oceanic Inst. Rept. OI–70-28-1.*

23. Anonymous. 1972. *Ocean Industry* (Feb.), p. 41.

24. Anonymous. 1972. *Ocean Industry* (Sept.), p. 106.

25. Hanson, J. A., and J. P. Craven. 1972. First annual report, Hawaii's floating city development program—fiscal year 1972. *Oceanic Inst.*

26. Anonymous. 1971. Floating ocean platforms. *Ocean Industry 6*(11):24.

27. Picard, J., and J. Manson. 1972. Retrieval of the ELF-ERAP platform. *Ocean Industry* (Apr.), pp. 124–125.

28. Hromadik, J. J. et al. 1971. Mobile ocean basing systems—a concrete concept. *Nav. Civ. Eng. Lab. Tech. Note N-1144.*

29. Weeks, W. F., and W. J. Campbell. 1973. Towing icebergs to irrigate arid lands. *Bull. Atomic Scient. 29*(5):35–39.

30. Griffin, O. M. 1972. Recent designs for transportable wave barriers and breakwaters. *J. Marine Tech. Soc. 6*(2).

31. Anonymous. 1972. Final technical report: expandable floating bases. *Goodyear Aerospace Corp.* (Akron, O.) *GER 15491.*

Additional Readings

Bader, J. 1970. Ocean platforms—state of the art. *Offshore Tech. Conf. Paper 1282.*

Harris, C. M. 1957. Handbook of noise control. McGraw-Hill Book Company, New York.

Isaacs, J. D., and W. R. Schmitt. 1969. Stimulation of marine productivity with waste heat and mechanical power. *J. du Conseil 33*(1):20–29.

13

ENERGY FOR AND WITH OPEN SEA MARICULTURE

J. A. Hanson

The total electrical production capacity of the United States in 1970 was 340,000 megawatts (MW). This is expected to increase nearly fourfold by 1990 to 1,260,000 MW. In 1970, nuclear power comprised less than 2 percent of the total capacity, with fossil-fuel power plants accounting for nearly 80 percent. By 1990, the percentages are expected to be nearly equal, with nuclear power accounting for slightly less than 40 percent and fossil-fuel power slightly more [1].

What relevance do these facts hold for open sea mariculture? In Chapter 2 we noted in passing our interest in exploring ways to work in harmony with the open sea environment, rather than competing against it in an effort to exploit its resources without regard to consequences, as has too often been the pattern of the past. Within this concept, we noted that any energy conversion and storage system employed in an open sea mariculture station ideally would rely on conservative energy and also would be nondegrading to its environment. In the practical world, however, we must realize that compromises with technological and economic inertia will be necessary. Let us take a closer look, then, at this practical world within which we are projecting the development of mariculture.

AVAILABLE ENERGY

Going beyond present trends only, we may view energy availability on the planet earth as falling into four general classes: (1) stored solar energy (fossil fuel), (2) energy stored in elements (nuclear energy), (3) geothermal energy, and (4) ambient solar energy flux, or *conservative energy*. Each source imposes its own requirements for storage and handling. Since we may expect to be dealing with most, if not all, of them in designing mariculture systems, and since, for reasons that we hope to make clear, it may occur that associating open sea mariculture with offshore production of electrical power will offer one of the early stepping-stones for mariculture's development, it might be useful at this point to briefly summarize some of the characteristics of each of these prime energy sources.

334

Fossil Fuels

Solar energy stored in organic matter over millions of years by the photo-synthetic process is currently available in the form of coals, natural gas (methane), and crude oils. We have learned to release it through exothermic oxidation reactions. Although we do not know exactly how fossil fuels were formed, conservative paleontological estimates place the beginning of this formation in the Carboniferous period of the Paleozoic era, about 300 million years ago. At the same time, even optimistic estimates of our current fossil-fuel reserves and projected usage rates indicate that civilization's dependence on fossil fuels cannot last more than about 300 years from beginning to end—until about the middle to latter half of the 21st century A.D. [2]. Thus, it appears we are managing to use fossil fuels at possibly one million times their rate of production. It seems very likely that civilization's fossil-fuel energy epoch will exhibit the brevity of a flash bulb if viewed against the human-history time scale. Nevertheless, in our considerations of open sea mariculture systems we cannot ignore the fact that we are deep into the fossil-fuel epoch now, even though the apparent environmental conse-quences of fossil-fuel use may conflict with our idealized objectives.

Nuclear Energy

Although many problems are yet to be solved, we have made considerable progress in learning to release energy stored in atomic elements through controlled nuclear fission and fusion reactions. There are basically three reaction types of interest here: fission burner, fission breeder, and fusion reactions. In burner reactors the quantity of fissile material at the end of a given reaction cycle is less than at the beginning of the cycle. In breeder reactors, "fertile" material is converted to fissile material during the reaction, so the quantity of fissile material increases during any given cycle. Fusion reactions occur when two heavy isotopes of certain elements combine to form a new element with a higher atomic weight than the original. Currently, the most common fusion reaction is the combination (fusion) of hydrogen isotopes having atomic weights of 2 (deuterium) or 3 (tritium) to form helium (atomic weight 4) with a resultant release of energy in very large quantities. It appears that materials practicably available in economically recoverable concentrations for fission burner reactions are even more limited than fossil fuels. But controlled fusion and breeder reactions, if their com-mercial development is attained, seem to offer civilization the potential of high levels of energy for at least tens of thousands of years into the future, and probably much longer [3]. However, not only have we yet to bring these promising reactions under control for purposes of useful energy production, we have not yet learned how to manage the biologically destructive radio-active wastes produced, particularly if and when these wastes are produced in quantity. Furthermore, the overall efficiencies of these energy converters impose lower scale limits: minimum practical sizes may be too large to use solely for powering open sea mariculture stations.

Geothermal Energy

Geothermal energy is energy available in the heat of the earth's mantle. It derives from sensible heating of underground regions by volcanic, magmatic, or metamorphic processes, sometimes accompanied by hot water or steam. The escape of water or steam through natural fissures produces hot springs or geysers.

Geophysicists do not seem to be in complete agreement concerning the mantle's mineral composition nor even concerning its state—solid or liquid. However, it is clear that the mantle temperature is sufficient to produce vast quantities of steam if water is available. Its total mass is so large that, given its average temperature (at least $1200°C$ and very likely much higher as the core is approached), huge quantities of energy might be extracted as steam power without significantly reducing the mantle's overall temperature. Liquid magma ejected from the mantle to, and sometimes through, the earth's thin crust, as in volcanoes, suggests a high likelihood that this source of energy could be tapped. The crust itself is only 3 to 5 km thick over large areas underlying the oceans; hence, access to the mantle is well within current deep-ocean drilling technology [4]. Thus, there seems to be a fair likelihood man might in the not-too-distant future find himself tapping the high temperature mantle of his earth from large ocean platforms to derive energy for his myriad technological activities.

Conservative Energy

Solar energy flux that permeates and affects the earth and its atmosphere in one way or another is manifested in several forms. As we saw in Chapter 4, these include (1) radiant energy derived directly from the sun, (2) a solar-induced temperature differential between warm oceanic surface waters and cold deep waters, (3) winds induced by differential surface and atmospheric heating and cooling, and (4) ocean waves induced in turn by these winds. Moreover, there are kinetic energy sources in the gyroscopically induced winds and major ocean currents, and in the tides induced by cyclic variations in interplanetary attractions.

This classification is coming to be known as *conservative energy*. The term arises from the fact that these sources of energy will be available as long as the sun remains active and the earth remains in solar orbit; furthermore, none of them is expected to produce any serious environmental degradation. This contrasts favorably with the preceding three categories, all of which must be considered nonconservative since the energy resources are finite, however large. Furthermore, fossil fuel and nuclear energy have obvious environmental disadvantages; geothermal energy also has the potential for harm to the environment if tapped excessively or ineptly.

Energy Storage and Transmission

Energy must be converted from the prime sources into forms in which it can be stored, transported, and used by whatever devices it is to power. Fossil fuels such as coal or natural gas are in usable forms when extracted from the earth, whereas crude oil is usually fractionated into a variety of products appropriate for use in machinery. Nuclear energy is most frequently employed to produce steam, which, in turn, drives steam turbines to produce kinetic energy to drive generators that produce electrical energy. Driving turbines of one sort or another to produce electrical energy is typical of concepts for converting geothermal and conservative energy sources, too. There is, however, an alternative energy possibility on the horizon—hydrogen. Although the experts are not now in accord, it seems this gas may offer a variety of economic and environmental advantages over either electricity or fossil fuels.

Having looked at the prime sources of energy, let us now consider sea-based technology for converting, storing, and transporting energy to be used for human activities, and specifically mariculture. For if societies begin to move their energy-conversion activities offshore, the offshore sites might well find a mutually beneficial symbiosis with open sea mariculture.

OCEAN SITING OF FOSSIL-FUEL POWER PLANTS

Disadvantages

There are at least three economically sound reasons *not* to base large, fossil-fueled, electrical-generation plants at sea. First, they are large and heavy; therefore, the platforms required to support them will be very expensive. Second, such facilities are typically designed to burn whatever type of fuel (oil, natural gas, or coal) is closest to the plant and therefore least expensive. A remote location at sea is likely to increase fuel transport and storage problems and costs, unless power plants are associated with offshore oil recovery and offshore refining, since the latter seldom burn crude oil. Third, technology for transmitting high-voltage electricity through or across water is not yet well advanced. Such transmission is likely to be expensive for some time to come. The second and third factors are relatively self-explanatory; let us examine the first one a bit more closely.

Plant Size. The trend in fossil-fuel power generation is toward larger and larger units. Between 1930 and 1950 the largest units in service were about 200-MW capacity; today they are exceeding 1,300 MW. These units are frequently multiplied to achieve total plant capacities totaling several thousand megawatts. The reasons for this trend revolve around typical economies of scale—certain fixed costs, costs not directly relating to plant size, and dependent costs that increase at a ratio lower than capacity.

Yet, in the terrestrial environment there are factors that tend to limit plant size. Among these are land availability, water-coolant supply, public reaction, and reliability. Relative to the last factor, a 3,000-MW outage is more serious than a 500-MW outage. The problems related to cooling water and public reaction would no doubt be lessened by offshore siting. Space and reliability problems would probably be somewhat intensified.

Advantages

Essentially, there is only one good reason to locate any electrical power generation facility offshore. But this one attraction might soon overwhelm the disadvantages; it is environmental protection.

Water-Coolant Requirements. At the present time power generation accounts for more than 80 percent of all water used for cooling purposes by industry in the United States [1]. The fact that industrial use of cooling water is reaching the carrying-capacity limits of our inland waters is evident from widespread concern over *thermal pollution* and from the billions of dollars being spent on the development and construction of cooling towers.

Since we appear to be approaching, if not exceeding, the limits of our terrestrial, cooling-water capacity in the United States, cooling water must increase in real cost. This cost is manifested in the taxes required to protect or restore "public" waters degraded by thermal pollution, in depressed property values, and in the costs of cooling systems that restore water to its original intake temperature before returning it to its source. Regardless of which share of these costs our society decides to pay, the sum is a function of the electrical energy generated. The term "function," rather than "constant," was carefully chosen: the more power we generate in a finite space, the more limiting will cooling ability become, and therefore the more precious. It seems safe to predict that a theoretical unit of cooling will be significantly more expensive in 1990 than it is today.

Water Cooling in an Offshore Environment. Figure 13.1 shows the estimated patterns of electrical power production in the United States for 1990 [1]. It is clear that a large proportion of all U.S. power production in that era will be concentrated on the East, West, and Gulf coasts. We suggest that these projections, along with advancing offshore platform technology, allow a distinct possibility that many future fossil-fuel and nuclear power generation facilities might be located offshore and thus offer opportunities for associated mariculture operations.

It is not our intention to review here the design and operation of a hypothetical offshore electrical power generation facility. Information on conventional power-plant design is available in a number of standard texts.

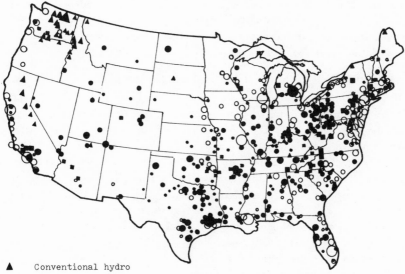

▲   Conventional hydro
■   Pumped storage hydro
●   Fossil steam
○   Nuclear steam
·   0.5-1 GW (gigawatt = 1,000,000 kilowatts)
•   1-3 GW
●   3-9 GW
●   9-20 GW

FIGURE 13.1 *Generating centers in 1990: estimated patterns of electrical power production in the United States. (Adapted from* **The 1970 National Power Survey**—Part I *p. I-18 [1].)* ·

Our basic interest in this area lies in the cooling-water question. By our calculations, a 500-MW unit designed for an $8°C$ temperature increase from intake to outlet of the cooling water could require over 28 m^3 of such water each second, which would mean around 2,000 acre-ft/day. In temperate latitudes the enhancing effect of such temperature increases on bioproductivity is well known. If an offshore facility of this modest size and type were designed to fit the hydrodynamic environment in such a way that the majority of the heated water would spread near the surface (as it would tend naturally to do), where light for photosynthesis is at its peak, and if sufficient nutrients were also made available, the facility by itself could potentially initiate a highly productive marine food chain covering several thousand acres. We do not suggest that the problem is that simple, of course, but thermal energy, light, and nutrients are certainly among the most significant limiters of bioproductivity in mid- and northern latitudes.

If we add to the cooling problem the question of regional atmospheric pollution and the question of aesthetics, both of which incur still further costs for their solutions, the arithmetic answer may point to offshore power generation in the not-too-distant future. This well may involve generating systems using fossil fuel and nuclear fuel.

Fossil-Fuel Power Plants for Mariculture Stations

If we now examine the other side of the coin and look at fossil-fuel systems as power sources for mariculture stations, rather than considering large systems as potential hosts, we find that a wide variety of fossil-fueled power-generating equipment is available off the shelf, in any capacity desired.

Gas-turbine-driven power units are available up to about 5,000 hp; these are compact, efficient, and reliable. For smaller demand, diesel-driven generator sets would be the preferred choice. For installations larger than 5,000 hp, oil-fired steam turbines would be more efficient. In this kind of application, current practice for shipboard power-plant design could be utilized in toto.

In general, for these small power-plant sizes, the most bothersome side effect would be noise, because the relatively small quantities of heat and exhaust gases produced could be dissipated easily in an open ocean environment.

OCEAN-BASED NUCLEAR POWER PLANTS

Current Status of Floating Nuclear-Fission Plants

Large nuclear-fission power plants at sea are already in the making. Westinghouse Electric Corporation and Tenneco have recently joined forces to form Offshore Power Systems (OPS), an enterprise that has designed a floating, standard, 1,000-MW nuclear-fission electrical power plant that can be produced in quantity and seemingly at competitive prices. However, the design requires the presence of a special breakwater, the cost of which may vary widely with location and environment. This is somewhat offset by the fact that two or more identical plants can be combined within a single breakwater to make up a larger facility.

Characteristics of the Design. As designed by Offshore Power Systems, each 1,000-MW unit will be supported by a welded-steel floating platform 121.92 by 115.82 by 12.19 m (400 by 380 by 40 ft high). The platform will displace 147,000 metric tons (145,000 tons) and draw 9.14 m (30 ft) when fully loaded. The principal structures above deck are the nuclear-reactor containment, the turbine enclosure, and housing for auxiliary services. Most above-deck construction is of welded steel, with concrete used where radiation shielding is required. With a maximum overall structural height of 54.5 m above the surface, the design is essentially a conventional shore-based configuration moved to sea and integrated on a single, floating "foundation" (see Figure 13.2).

FIGURE 13.2   *Floating, nuclear-fission power plant design. (Adapted from Schmidt [6].)*

In the floating nuclear power plant, cooling-water discharge is near the surface. We could find no figures for thermal rejection into the surrounding water; but if we assume 8,000 Btu/kilowatt hour (kWh) [5], a 5.5°C rise in the 15.5°C cooling water, and a 24-h period for heated water to cool to ambient temperature, our calculations indicate that each of these plants may be associated with approximately 400 km² (100 acres) of surface water having temperatures 1.5 to 5.5°C above the surrounding water.

Advantages. Offshore Power Systems notes a number of substantial advantages inherent in this type of nuclear power plant. For instance, the concept opens up new areas for power-plant siting. In many cases plants can be placed closer to load centers; moreover, the structures are insulated from seismic shock. There is no requirement for the use of coastal lands or disturbance of the coastal-zone ecosystem, and, since the ocean offers an extremely large area for waste-heat dissipation, environmental impact in the form of thermal pollution is minimized. Another significant advantage is that standardization of design, accompanied by production-line technology, can reduce production costs, shorten construction time, and lead to improved quality control [6].

Radioactive Pollution Hazard to Mariculture. Our literature reviews indicate that the maximum nontritium radioactive discharge annually of a pressurized water reactor of 1,000 MW would be in the range of 10 to 20 curies (Ci), roughly equivalent to 10 to 20 g of radium in equilibrium with its decay products. This would be mostly in the form of the heavier and larger half-life radionuclides, many of which could be concentrated by marine organisms, as discussed in Chapter 5. This appears to be a rather small increase in overall radioactivity for a marine area of the size involved; yet it very definitely is a measurable one and could be biologically significant. Tritium discharge might be expected to be on the order of 500 to 2,000 Ci/year. Tritium has a half-life of 12.5 years but is noted for a very low radiotoxicity.

Given the combination of these two categories of radionuclides associated with discharges from fission reactors, one seems to have no choice at the moment but to conclude that (1) there seems to be a definite probability of unknown magnitude that the volume of radionuclides contained in today's fission-reactor discharges is sufficient to produce deleterious effects in some marine ecosystems, and/or their end products, if the discharges and ecosystems are closely associated in an intensive mariculture system; and (2) the hazard is intensified as reactors become larger or are more closely associated in space. We do not contend that nuclear power production and mariculture *cannot* be associated; only that the hazard appears real at this juncture and that a great deal more evidence will be required before one will be on safe ground in stating that any mariculture activities *can* safely be associated with today's nuclear power production. The lifting of current legal restrictions in the United States will no doubt await development of the evidence.

There are, of course, some nuclear power plants of a size probably suitable to mariculture stations already operating at sea—in ships and submarines. These are special-purpose systems, however, designed for propulsion and operated in the interest of national defense, and they are not subject to the same cost-effectiveness constraints that would control the designs of commercially oriented mariculture power systems. Therefore, it is not clear that they provide even a conceptual base that might apply to our interests.

We know of no other sizable sea-based nuclear power systems in existence or even undergoing detailed design investigations in the noncommunist world at present.

Radioisotope Power Potential for Mariculture. One potential nuclear-power application for mariculture might be the isotope power source. Devices in the 5- to 25-kW range have been developed with very high reliability and inherently very long operating lifetimes between refuelings, that is, greater than 10,000 h. For submerged and bottom-mounted applications, these devices might be justified on the basis of low maintenance and infrequent retrieval. The economics of isotope power sources would have to be compared with the economics of a battery- or fuel-cell-powered alternative. For an example of a radioisotope power supply, the RIPE concept developed by the Electric Boat

Company provides for 25-kW(e) average output and is a complete electrical power system, including shielding and controls, suitable for unattended operation in deep water. It can be housed in a 2.7-m-diameter steel sphere.

Fuel is the outstanding cost item. Cobalt 60, with a half-life of 5.3 years, has been estimated to cost between $7,500 and $33,000/kW(e) [7]. However, the long period between refuelings should be taken into account in assessing the total cost of this type of power, especially when frequent retrieval for the refueling of competing power sources is a factor.

Current Status of Fusion and Fission Breeder Reactors

Controlled fusion and breeder reactors are as yet far from being a commercial reality, even though the United States is pursuing their development on a near "crash program" basis. The intention is to have large commercial breeder reactors in operation by the 1985 to 1990 time period. No serious time estimates are being made yet for commercial-scale controlled fusion reactors within this century, so far as we are aware. Until these two development avenues have been pursued further, we cannot estimate probable reactor sizes, weights, operational requirements, costs, and minimum efficient power outputs. Therefore, we cannot project the probable applicability to, or symbiosis with, open sea mariculture. We might, however, chance the assumption that thermal efficiencies will be roughly equivalent to today's fission reactors or lower, which would lead us to expect that the thermal rejection rate will be similar or higher.

GEOTHERMAL POWER PRODUCTION IN THE OCEANS

Land-Based Technology

Geothermal energy has been successfully harnessed and transformed into electrical power in a number of terrestrial locations, notably New Zealand, Japan, Italy, and the United States. In most cases the output of steam wells is used to spin turbines. The two largest geothermal plants in the world today are the Geysers, 145 km north of San Francisco, with a 192,000-kW capacity, and the Larderello field in Italy, with a 380,000-kW output. Both are natural steam fields. Joseph Barnea, writing in *Scientific American,* has said:

> Interest in this source of energy has quickened in the past few years. Recent explorations have revealed that the resource is larger and more extensive than has been supposed. A generation ago, the hot springs and steam fields that had long been known in a few localities around the world were believed to be merely local freaks of nature. There is evidence now that these reservoirs of steam and hot water are actually widespread in the earth's crust. Signs of their presence have been detected on most of the continents and on a number of islands. It seems possible that such fields will be found under the seas [8].

Present-day thinking among geophysicists extends beyond the notion of tapping natural-steam reservoirs to the concept of creating them. What is needed, they say, is first a source of heat (not excluding even liquid lava) associated with a region of rock that either is porous or can be made so by hydraulic cracking, or even by nuclear detonation. Water may then be injected under high pressure and the resulting steam tapped off by means of wells.

Site Research

The University of Hawaii recently has received a grant from the National Science Foundation (NSF) to initiate a laboratory to study the development of geothermal power in volcanically active parts of the Hawaiian Archipelago. After describing the geology of the region in relation to geothermal activity, the proposal for this grant states:

> It would seem that the most likely geothermal resource areas for power production in Hawaii would be those that are below sea level, as close to stabilized magma bodies, or recently reheated intrusion areas, as possible.
>
> As a matter of likely fact, some of the best locations may be offshore along extensions of recently active rift zones at depths where the hydrostatic pressure of the sea would prevent flashing of hot water steam. Deep drilling from stable floating platforms would be required to explore these possibilities, after sensitive seismic and thermal survey work [9].

Potential Deep-Sea Geothermal Power Production

One seemingly attractive approach to geothermal power would involve deep-sea drilling into a submarine geothermal region, and creating a geotherm by hydraulic cracking of the rock and injection of water. In one possible scheme, large well bores would allow a concentric arrangement of the production steam, with cold water going down and steam or hot water coming up. The drilling techniques practiced today allow for control and valving at the wellhead, in addition to hole reentry, which leads logically to the idea of the formation and completion of a geotherm by a drill platform that would then leave the site, to be replaced by a production platform. Connection and start-up, and if necessary shutdown and departure from the site, appear to represent extensions of current deep-sea drilling technology and stable-platform design. The immediate presence of $5°C$ water in huge quantities under the platform would satisfy heat-sink requirements at minimal cost, and "packaging" the generated power as cryogenic hydrogen–oxygen would make the market into which this form of power could be introduced essentially unlimited.

Geothermal Ocean Sites and Mariculture

Although commercial-scale oil-drilling operations today do not extend far past the 200-m depth, research and development to achieve 500-m capabilities are already in progress. Both these depths reach well into the realms of nutrient-rich water in many locations in both oceans. In addition, the GLOMAR CHALLENGER is routinely drilling small shallow holes in more than 6,000-m depths. Therefore, it seems likely that future technology could allow for geothermal-power-production sites in water of sufficient depth that the rising heated water which would be associated with them would also be rich in phytoplankton nutrients.

CONSERVATIVE ENERGY IN THE OFFSHORE ENVIRONMENT

Harnessing Wind Energy at Sea

Wind energy, an indirect form of solar energy, is one of the oldest natural sources of power harnessed by man. Wind-driven sailing ships and windmills to grind grains and pump water are indeed ancient applications. Against a historical background of inexpensive fossil-fuel-based electrical energy, serious work on the utilization of wind energy for large-scale applications has been a low-level, sporadically supported effort.

Current Status of Land-Based Technology. The Smith–Putnam 1.5-MW wind turbine (64-m diameter with two blades) located in the mountains of New England the the John Brown 100-kW Aeroturbine (15.24-m diameter with two blades) in England are among the few major devices actually built and tested. Estimates of the design details and costs of wind turbines have been carried out since at least as early as 1945 by Palmer C. Putnam [10]. Cost figures for a complete wind-to-electricity plant range from $68 to $323/kW, depending upon the amount of "productionization" assumed. A range from 1 to 1,000 units was included in these studies. Much larger units (50-MW) were studied in Germany.

What William Heronemus at the University of Massachusetts has referred to as three major misconceptions, or "myths," have contributed to the obvious lack of support for the development of wind-energy utilization in any consistent, modern-day manner. These are

1. Total wind power available in the United States is insignificant.

2. Annual average wind speeds of 48 knots/h are required for economic power generation.

3. Wind-power generating capacity must be backed up one-for-one by other forms of generating capacity.

In his paper treating wind-power and sea thermal gradient energy as "Some Proposed Gentle Solutions (to the U.S. Energy Crisis)," Heronemus counters these myths and attempts to show that this source of energy generation is entirely practical and should be pursued.

One approach that he advocates to provide energy storability and transportability suitable to the variable source (and optimum location) of wind-energy is the electrolytic production of hydrogen fuel from water (with oxygen as a by-product). Heronemus examines the costs of generating electricity, presumably near load centers, from wind-derived gaseous and liquefied hydrogen, and notes costs to the consumer of $675 and $720/kW generated, respectively. These are significantly higher numbers than present-day nuclear plant costs, which are in the range of $400/kW and climbing. Noting this disparity, he asks: "How much per kWh will absolutely pollution-free electricity be worth by 1980?"

Offshore Wind Power Plants. Moving to a sea-based scheme, the same author presented a paper to the Marine Technology Society in which he examined the potential of a large number of floating and tethered wind-turbine-equipped towers in the Nantucket Shoals and Georges Bank areas [11]. Hydrogen once again is envisioned as the means of energy storage and transportation.

Generally, wind turbines should be placed on towers to be exposed to higher wind speeds and more consistent winds. The use of towers or other elevated structures over the water will therefore be needed. Hopefully, an integrated structure for supporting wind turbines, solar collectors, and so on, can be engineered from the floating-city type of open sea supporting base.

Energy from Sea Thermal Gradients

In Chapter 4 we examined the differences in temperature that usually exist between Surface and Intermediate waters in the oceans. In France, prior to the 1930s, George Claude actively experimented with the development of electrical power from this difference. The governing Carnot efficiency for this very low density energy source is quite low, less than 8.2 percent for a perfect system operating between a $21°C$ surface source and a $5°C$ deep sink, and is about what can be anticipated in many areas in the open sea. Nonetheless, such sea thermal gradients provide huge and invariant sources of indirect solar energy that conceivably can be tapped for energy conversion.

Most French and American investigators have tended to dismiss this approach as not being of interest to nations whose shorelines are completely in the temperate region, an attitude indicating that open ocean basing for the production of some form of stored energy has apparently not been much considered. However, in the United States, Anderson and Anderson [12] pointed out the Gulf Stream as being a potential source of 182×10^{12} kW/year. Heronemus et al. currently are working under an NSF grant on the development of thermal-gradient power systems.

As opposed to the Claude concept, in which seawater is used directly to operate the power cycle, a process that he proved operable, but one doomed to "economic failure" [13], more recent approaches propose heat exchange with propane, ammonia, or other liquids as the Rankine cycle working fluid. The problem appears to be one of optimizing between heat-exchanger sizes and pumping-power requirements, as well as turbine size.

Thermal-Gradient Power Plants Related to Open Sea Mariculture. The Andersons advocate an ocean-based floating plant, pointing out that fresh water can be manufactured very economically using the same thermal-gradient energy, and they estimate production costs at $0.03 to $0.04/1,000 gallons (gal). They also cite the open sea mariculture possibility and the opportunity to capitalize on the minerals in seawater, particularly as concentrates result from water purification. However, to the extent researched, they do not note the technique of producing hydrogen and oxygen for storability and transportability. Rather, they suggest the use of superconducting power transmission to shore and into the existing power network.

Energy from Ocean Currents

Another path being explored at the present time is the possibility that kinetic energy in major ocean currents, particularly eastern boundary currents such as the Gulf Stream and the Florida Current, might be extracted by bottom-tethered water turbines. Within a core of water, say 15 km wide by 130 m deep by several hundred kilometers long, large amounts of electrical power could be generated using rotor-type machines or free-stream, propeller-type machines [14].

In proposals to the NSF, the University of Massachusetts suggests that 73-m diameter, four-disc, propeller-type machines operating in a current of 2+ m/s might generate 24 MW. If 12 of these are mounted abreast along the core of such a current and then this assemblage repeated every kilometer for 550 km, some 100,000 MW of power could be generated, representing about one fourth the present U.S. installed capacity.

Energy from Ocean Waves

Wave energy reaches significant proportions where wave periods are 5 to 7 s or longer and heights are 1 m or more; these conditions prevail in extensive areas of the oceans much of the time. But any attempt to extract this energy mechanically is faced with two major problems: first, the low water head usually represented by each wave's height; and second, the oscillatory nature of waves, in which the oscillations are complex and also variable in both frequency and amplitude over a broad range. Thus, any mechanism for extracting this energy must be so designed that it is survivable in storm waves while its mechanical efficiency remains useful in relatively small waves. It also must be designed in such a way that it can build up a sufficient store of energy from the oscillatory motions to perform useful work.

Wave Pump. A device under development at Scripps Institution of Ocean-ography proposes a simple solution to this problem [15]. It is a pump which responds directly to wave motions and exploits the fact that the natural heaving frequency of the buoyant system as a whole (pumping mechanism plus water column) is always higher than the natural period of the water column it encloses. This concept is shown in Figure 13.3. The valve is open as the system descends and closes as it ascends. It thus acts as a half-wave rectifier and forces the water head in the stand pipe ever higher. This effect is amplified by the upward momentum of the water as the pump system begins to descend. The process continues until a pressure head sufficient for electri-cal power generation is reached. The wave train then maintains the flow at this pressure. A small working model of this concept has already been demonstrated, and a larger one reportedly is under construction.

Wave Momentum Generator. Another example, potentially useful with atoll-based mariculture systems or possibly with very large floating platforms, is shown in Figure 13.4. This mechanism employs the net momentum transport of shoaling or breaking waves to maintain a water head by concen-trating the waves with a converging channel. With a few possible exceptions,

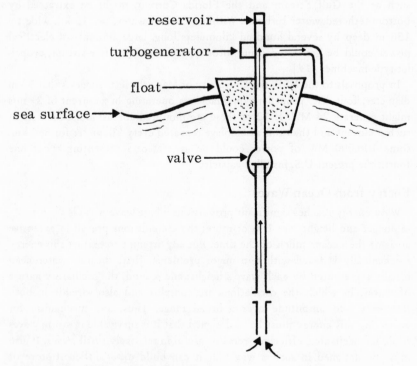

FIGURE 13.3   *Wave-powered generator.*

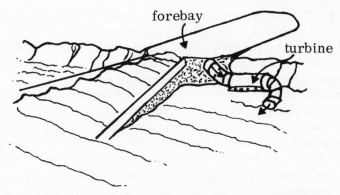

FIGURE 13.4 *Power generator utilizing momentum transport of shoaling waves.*

as in the trade-wind belts, wave energy is not sufficiently reliable to produce electrical current continuously. Consequently, some form of energy storage is needed. Hydrogen and oxygen are suggested here again. Ample pure water for electrolysis could be produced with solar stills and by recapturing the hydrogen–oxygen combustion product produced as electrical power is generated. As will be seen in the next few pages, it appears also that the production and storage of hydrogen and oxygen under pressure could be simplified by placing production units at some distance below the surface.

Solar Power at Sea

Helios–Poseidon is a concept advanced by Escher [16] for an ocean-based, solar-energy-powered production facility for the hydrogen–oxygen energy form. This concept departs from several other solar-energy utilization concepts in that it is open ocean based. The important technical benefits of a sea-based, solar-energy conversion system are: (1) proximity to an excellent thermal sink and source of working mass, that is, the ocean, and particularly the depths, (2) mobility of rotation and translation, (3) space availability for large solar collector areas, and (4) logistical ease in initial construction, servicing, and distribution of products on a worldwide basis.

Storage and Transmission. Proceeding from the fourth point, the energy form to be produced must be both storable and transportable over significant duration and distance, if it is to be delivered to the ultimate consumer. Therefore, it is proposed that solar energy be used to convert water (purified seawater) into *cryogenic liquid hydrogen and oxygen.* In this form the stored energy of the sun can be readily shipped to points of use on a worldwide basis via "cryotankers." Once unloaded at port, the cryogenic liquids can be stored and eventually transported by rail, over-the-road trailers, or pipelines. Alternatively, the hydrogen and oxygen can be gasified and piped in the manner of natural gas. The energy carrier can be finally consumed in the process of heat release (e.g., residential heating) or it can be converted into an electrical form by fuel cells or by hydrogen-powered turbines driving generators. Clearly, a smaller version of this concept might be someday employed to produce the energy needed for an open sea mariculture station.

Production. In this concept a solar still is employed to produce pure water from seawater, as in Figure 13.5. The distilled water is then decomposed to hydrogen and oxygen in the electrolytic cells. Electrolysis and high-pressure gas storage of both elements take place in deep water. Some portion of the products is then combustively combined to produce the power required to liquefy the remaining products. The prime mover for this system is a steam turbine operating in a Rankine cycle. Solar reflectors are used to generate high-temperature steam in the cycle, and the cold-end condenser utilizes the ocean as its heat sink.

There is little doubt that a system such as this could work. Whether it could be made practical for either large- or small-scale power production remains to be proved. In the future, and if operated to produce pressurized hydrogen—oxygen gas only, it might be one candidate for conservative-energy powering of open sea mariculture operations.

A FUTURE HYDROGEN ECONOMY?

It readily may be noted that most of the new-energy-source research is either based upon or, as a minimum, recognizes the potential of hydrogen as a prime energy storage and transport medium of the future. Should this come about, the symbiotic potentials between at-sea hydrogen production stations and open sea mariculture may be quite high. Therefore, a brief look at a possible future hydrogen economy seems warranted.

Characteristics of Hydrogen as a Fuel

If we consider hydrogen as a potential energy storage and transmission medium, we find some rather encouraging facts. First, hydrogen is, of course, one of the most abundant elements in the universe, occurring principally in organic compounds and in water. Present-day technology is adequate to produce hydrogen in large volumes from fossil fuels or water, and its combustive combination with oxygen produces nothing more noxious than distilled water. When hydrogen is burned in air, oxides of nitrogen are produced, but their concentration is extremely low.

A short comparison of the hydrogen cycle with the fossil-fuel cycle might prove illuminating. The fossil-fuel cycle is diagrammed at a very general level in Figure 13.6, and the hydrogen cycle (assuming electrolytic hydrogen extraction rather than production from fossil fuels) is diagrammed at the same level in Figure 13.7. A comparison of these reveals a number of interesting things and moves one to conjecture.

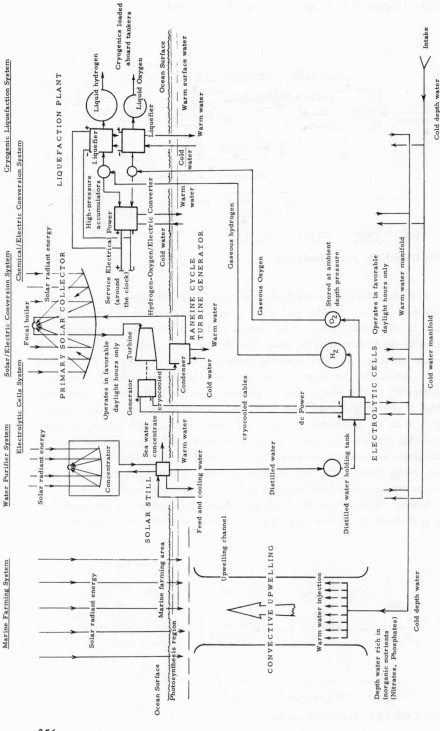

FIGURE 13.5 *The Helios–Poseidon system. (Courtesy of W. J. D. Escher [16].)*

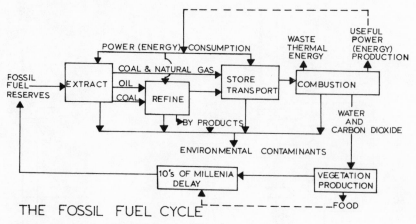

THE FOSSIL FUEL CYCLE

FIGURE 13.6 *Fossil fuel cycle.*

First, it is clear that the hydrogen cycle, if technically and economically feasible, is much simpler and much more direct in terms of both processes and time. Second, there are no obvious sources of environmental contamination in the hydrogen cycle other than thermal waste, whereas the fossil-fuel cycle contains a plethora of all-too-familiar problems of this sort. Finally, and less obvious in the diagrams, we see that the energy yield of fossil fuels compared to the energy cost to prepare them for utilization is sufficiently high that energy derived from fossil fuels can be employed for their preparation with an economic product left over. Not so with hydrogen; some external prime energy source must be employed and this, of course, is where nuclear- and conservative-energy sources come to mind. Hydrogen is merely an energy storage and transmission medium. However, it may be a very attractive one when fossil fuels are no longer obtainable economically.

Mobile Applications for Hydrogen Power. Hydrogen can be carried in cryogenic form abroad any type of mobile platform. Hydrogen-powered

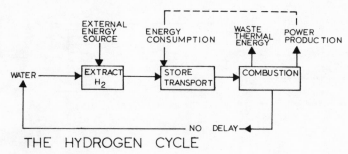

THE HYDROGEN CYCLE

FIGURE 13.7 *Hydrogen cycle.*

spacecraft boosters are common; hydrogen aircraft have been flown experimentally; experimental automobiles are on the roads today; and ships and submarines seem to offer no technical problems since large gas turbines run well on hydrogen. Thus, most mobile applications appear more amenable to hydrogen power than to stored electricity.

Stationary Applications for Hydrogen Power. But what of stationary applications? Is it not more efficient to use electricity? The answer seems to be "possibly not." First, gaseous hydrogen can be piped under pressure through the same piping networks that now supply natural gas. For a variety of reasons this would be more expensive than it now is with methane, but a good deal less expensive than transmitting electricity. Hydrogen can be used in its original form much the same as natural gas and, because of its low ignition temperatures, in new and more flexible ways, too. As Figure 13.8 shows, hydrogen could also be employed to overcome the high cost of electrical transmission by being used to supply power for local electrical power stations. Figure 13.9 shows the computed comparative costs of transmitting hydrogen in underground pipes versus the cost of transmitting electricity underground and through environmentally objectionable overhead wires. Hydrogen appears here to have the economic advantage over both.

Parenthetically, large proportions of hydrogen can be mixed with methane to increase the visibility of the hydrogen flame and the probability of its detection by smell. Or it can be combined with nitrogen to form liquid hydrozine (N_2H_4), which is highly combustible with high energy yields.

Technical Problems. The technical problems associated with hydrogen

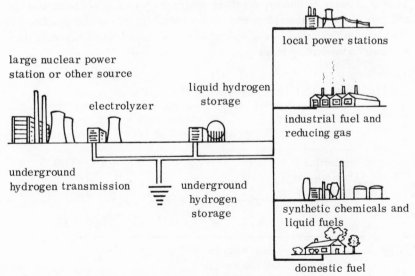

FIGURE 13.8   *Complete hydrogen-energy delivery system.*

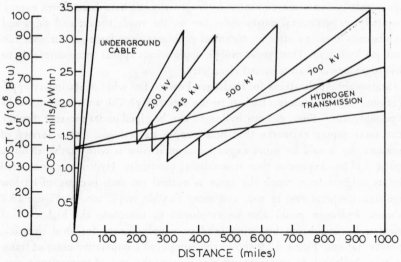

FIGURE 13.9 *Relative costs of transmitting hydrogen and electricity.*

include the following: (1) it is not easily detected by smell or other human senses and so can accumulate in dangerous quantities while remaining undetected; (2) the flame of pure hydrogen is nearly invisible in daylight; (3) the ratio of the energy yielded over the energy required to compress hydrogen is lower than it is for methane; and (4) at present the costs of liquefication and of storage and transport in cryogenic form are much higher than for such gases as propane or butane. All these difficulties are amenable to engineering and procedural solutions, however. It has been pointed out that gasoline and electricity are extremely dangerous also; yet we have learned to handle them routinely in modern societies.

TABLE 13.1 *Relative Prices of Delivered Energy*

	Electricity	Natural gas ($/million Btu)	Electrolytic hydrogen
Production	2.52[a]	0.16	2.81–3.09
Transmission	0.62	0.18	0.22
Distribution	1.61	0.27	0.34
	4.75	0.61	3.37–3.65

Source: 1969 data from the FPC and AGA [17].
[a] At 8.6 mills/kilowatthour.

Prospects for the Future

Might there be a hydrogen economy in our future? The Institute of Gas Technology, which considers it a distinct possibility, has completed recently a voluminous study on the subject [17]. Much of the material in this discussion is derived therefrom. The study concludes that hydrogen can be produced at generally competitive prices, as shown in Table 13.1. Should the hydrogen economy emerge, the utility of ocean-based, conservative-energy-powered hydrogen production systems would almost certainly be increased. If such systems do come into being, they would appear to offer excellent opportunities for associating open sea mariculture systems with them.

REFERENCES

1. *The 1970 National Power Survey—Part I.* Federal Power Commission.
2. Elliot, M. A., and N. C. Turner. 1972. Estimating the future rate of production of the world's fossil fuels. *Amer. Chem. Soc. Meeting,* Boston, Apr. 9–14.
3. Committee on Resources and Man, Natl. Acad. of Sci.–Nat. Res. Council. 1969. *Resources and man.* W. H. Freeman & Co., San Francisco.
4. Macdonald, G. A., and A. T. Abbott. 1970. *Volcanoes in the sea.* University of Hawaii Press, Honolulu. 433 pp.
5. Environmental effects of producing electrical power. P. I. Hearings before the Joint Committee on Atomic Energy, 91st Congress of the United States. Govt. Printing Office, Washington, D.C.
6. Schmidt, P. W. 1972. Floating nuclear power. *8th Ann. Conf. and Exposition, Marine Tech. Soc.,* pp. 483–490.
7. Conceptual study of manned underwater station. 1967. Nav. Civ. Eng. Lab., Port Hueneme, Calif.
8. Barnia, J. 1972. Geothermal power. *Scientific American 126*(1):70–77.
9. University of Hawaii. 1972. Pele Energy Lab. Exp. Proposal to NSF-RANN.
10. Putnam, P. C. 1948. Power from the wind. Van Nostrand Reinhold Company, New York.
11. Heronemus, W. E. 1972. Power from the offshore winds. *8th Ann. Conf. and Exposition, Marine Tech. Soc.,* pp. 435–466.
12. Anderson, J. H., Jr. 1972. Economic power and water from solar energy. Winter Ann. Meeting ASME. *Amer. Soc. Mech. Eng. Paper 72-WA/Sol-2.*
13. Claude, G. 1930. Power from the tropical sea. *Mech. Eng. 52*(12):1039.
14. Green, J. 1970. Concept for a self-contained oceanic resources base. *J. Marine Tech. Soc. 4*(5).
15. Isaacs, J. D., and R. J. Seymour. 1972. The ocean as a power resource. *Internatl. J. Environ. Studies 1.*
16. Escher, W. J. D. 1972. *Nonfossil synthetic fuels.* Rept. to the Synthetic Fuels Subgroup of the Comm. on Energy and Res. and Develop. Goals, Fed. Council on Sci. and Technol.
17. Gregory, D. P. 1972. A hydrogen-energy system. AGA Project IV-4-6, Inst. Gas Tech. *Amer. Gas Assoc. Rept. L 21173.*

PART IV
PROSPECTS
FOR THE FUTURE

14

QUO VADIS?

J. A. Hanson

If mariculture is to move to the open seas, it seems most likely to do so aboard a vehicle that is, or will be, there for other purposes. There appear to be two such vehicles in the offing: atolls and offshore industry.

In Chapter 12 we supplied the rationale for including atolls in an examination of open sea mariculture possibilities: they are natural mariculture platforms of a size sufficiently small that their environment is entirely marine in nature. Later in this chapter we examine atoll mariculture possibilities.

The larger potential lies in associations between mariculture and offshore industry, and here there are marked differences between the terrestrial and the oceanic environments. The essence of these differences lies in the often unnoticed fact that, for practical purposes, the land environment is two-dimensional, whereas the oceanic realm is four-dimensional. By this we mean that, because of its solid state, only the surface area of land is employed for most purposes. Moreover, land changes slowly and, since materials deposited in or on it do not diffuse through it rapidly, it does not offer an effective recycling sink. The oceanic realm, on the other hand, is geometrically three-dimensional, and because of its fluidity it is also three-dimensional as a practical matter. Furthermore, because of fluidity and natural dynamics, the water content of any defined area within the oceanic realm changes significantly and is, in effect, renewed over brief time spans; so it is logical, for many practical purposes, to consider the sea as four-dimensional. This practical four-dimensionality, when contrasted with the two-dimensionality of the land environment, holds important implications for offshore industry and for the association of mariculture and offshore industry. We can see this through the examination of (1) the distinction between waste and pollution, (2) the present incompatabilities between land-based industry and agriculture, (3) the attractions of ocean siting for some offshore industries, and (4) the potentials offered by the association of offshore industry and open sea mariculture.

Waste Versus Pollution

Waste is a parochial word in the sense that it is applied only to by-products that offer no returns to the producer. By-products for which the producer finds beneficial uses are not wasted, from the producer's point of view, and so are not considered wastes. If one's perspective is constricted to include only a single industrial activity, wastes are included in that perspective. If, now, the perspective is broadened to encompass not only the industrial activity but also the environment in which it exists, one is likely to find that some waste products find beneficial uses in the environment, while others can be assimilated by the environment—whether or not beneficially—so long as the rates at which these wastes are introduced do not exceed the environment's capacity to assimilate them. The introduction of wastes at rates greater than this will cause waste buildups, or will cause the environment to change its characteristics, or both. To any of these latter conditions we assign the term *pollution*.

This view of waste versus pollution brings to light several things that are important to offshore industry and to considerations of associating mariculture and offshore industry: (1) the polluting potential of any given industry can be reduced by reducing the variety of wastes it produces and by reducing the rates at which wastes are produced; (2) the polluting potential of a given industry also can be reduced by placing it in a new environment with a higher waste assimilation capacity; (3) the polluting potential of any given industry can be reduced by combining that industry with another in such a manner that each of the two industries beneficially assimilates significant portions of the other's wastes, thereby reducing the overall waste introduction into the surrounding environment; (4) if two or more of these measures are taken together, polluting potentials can be reduced much more sharply than would be the case with any single measure by itself.

Incompatibility of Industry and Agriculture

In general, cohabitation of agriculture and industry is not feasible. Although there are certainly some potential exceptions, such as the nuclear agra-industrial complexes conceptualized by the Atomic Energy Commission [1], there are two basic reasons why this generality remains valid.

First, both agriculture and industry prefer the same type of flat or gently sloping well-drained land. Since only the surface is used, they cannot occupy the same land concurrently. Therefore, they are essentially competitors for it. Second, the solidity and relative nonpermeability, not to mention the immobility, of the terrestrial substrate (i.e., its two-dimensionality) have the effect of inhibiting environmental assimilation of industrial by-products. Many industrial by-products can be deleterious to agriculture in any concentration, and since the terrestrial environment does tend to inhibit environmental assimilation of these by-products, many industrial operations pose constant pollution threats to nearby agriculture. So not only do food production and industry tend to compete for terrestrial space, but extra buffering space frequently must be provided between industrial and agricultural endeavors if the latter is to remain safe from the discharges of industry.

The Case for Siting Industry Offshore

The four-dimensional character of the ocean environment renders it infinitely more capable of dispersing and/or assimilating wastes than any comparable land area, a capability augmented by its vast size. This characteristic assumes vital importance at a time when industrial societies find their land environments growing intolerable. Both increasing population densities and growing industrialization are exceeding the capacities of terrestrial areas to absorb human-generated wastes. The result is increasing pressure on industries to either minimize waste output or locate elsewhere, far from residential or agricultural areas. The technical literature abounds with designs and budgets for minimizing industrial wastes. All are extremely expensive. Locating industry away from population centers, as some propose, has its drawbacks. Transportation links are lengthened, and the "natural" attractions of any undeveloped areas selected as the new hosts for relocated industries are endangered by the same two-dimensional waste-dispersal problem. As we have seen in Chapter 12, technology that will allow offshore siting of industry is advancing well; thus, it seems likely that future industry will move offshore as the results of economic analyses point more and more in this direction.

The offshore petroleum industry demonstrates that it is indeed a matter of economics. Petroleum recovery is attracted offshore by resource availability (i.e., oil) rather than driven there by the economic manifestations of environmental protection pressures. And both the incipient offshore power industry and the new designs for offshore transportation terminals imply that, from the economic point of view, the term "resource" may be applied to waste sinks, deep water, and other environmental characteristics, as well as to recoverable resources such as petroleum. On land, industry is attracted by resources such as labor, inexpensive property, convenient access to transportation terminals, and hospitable political climates, as well as to natural-resource availability. The situation is not qualitatively different with offshore siting: proximity to population centers, convenient and economic transportation, and, above all, release from environmental protection pressures seem quite likely to attract several industrial components of society to offshore locations in the not-too-distant future. And there is likely to be a *positive feedback loop* or self-stimulating effect involved in this movement; as on land, the existence of offshore industry of one type will stimulate the development of offshore industries of other kinds.

It is hoped, however, that as this movement occurs societies will benefit from past mistakes made in the terrestrial environment; that they will not concentrate industries so as to exceed even the greater assimilation capacity of marine areas; that they will seek to design synergistic relationships between industries so as to minimize the production of waste by-products; and that they will eliminate discharge of very long lived deleterious wastes that can gradually accumulate until they do irreparable harm.

The Case for Associating Offshore Industry and
Open Sea Mariculture

With the foregoing in mind, it is suggested that there are four fundamental reasons why offshore industry and open sea mariculture potentially are synergistic: (1) the practical four-dimensionality of hydrospace permits close spatial associations, so long as other factors do not prohibit them; (2) industrial structures in the marine environment attract and foster the growth of marine organisms, rather than prohibiting biological productivity as they do on land; (3) many by-products of industry are either vitally necessary to mariculture (heat and several organic and inorganic "wastes," particularly) or they can be readily processed into products useful in mariculture; if neither of these, many can be diffused through, and assimilated by, the dynamic oceanic environment without inhibiting mariculture; (4) offshore platforms and facilities are expensive, and the larger the number of productive activities an offshore facility can support, the more economically attractive it can be.

Contemplating the technological developments of the last four or five decades, and recognizing that technological development rates tend to increase exponentially, in light of the previous arguments it is not difficult to see open ocean mariculture–industrial complexes of impressive size as distinct possibilities within the next four or five decades. Moreover, opportunities for the first steps in such a direction already exist, and there seems little doubt that other opportunities are not far distant. Taking these opportunities in chronological order, in this chapter we attempt to envision the initial phases that certainly could mark the beginnings of sizable and profitable mariculture–industrial complexes of the not-too-distant future. Reasonably near term opportunities for associating open sea mariculture with offshore industry appear to lie with offshore petroleum operations and offshore power generation, in that order. After presenting ideas for these two possibilities, we return to a consideration of atoll-based mariculture. In each examination we attempt to consider all the main aspects, including species, platforms, husbandry concerns, and potential sources of energy.

MARICULTURE IN ASSOCIATION WITH OFFSHORE
PETROLEUM OPERATIONS

In the United States, offshore petroleum platforms are common off the coasts of the Gulf states and, to a lesser extent, the California coast. In both areas the platforms typically are located in waters that range in depth from 10 to 100 m and that are frequently fairly high in nutrients. Whereas platform legs offer a place for the attachment of cages and rafts, the possible use of a platform as an operations base suggests that floating and bottom-mounted containers also might be located beneath and in the vicinity of the platform; all these might be serviced from the platform either directly or by boat. When drilling operations have terminated and semiautomatic production is in progress, a platform may contain sufficient excess deck area for the installation of an on-site harvest processing facility. Such a facility probably would be required on only a few platforms; since multiple structures frequently are erected in any given area, one processing platform might serve a number of its neighbors, just as one factory ship now serves an entire fishing fleet.

As a matter of fact, experiments in petroleum-platform-based mariculture have already begun. The Atlantic Richfield Company, Texas A & M University, the National Marine Fisheries Service, and the Ralston Purina Company have entered into cooperative experiments to determine the feasibility of raising shrimp and oysters from some of Atlantic Richfield's platforms. So far the shrimp experiments are inconclusive, but the oyster experiments show high growth rates. The inevitable problems of competition, predation, and fouling plague this initial work, but there is certainly reason to expect that such problems will be brought under control. As a minimum, the work done thus far offers encouragement that edible marine crops can be grown in association with offshore petroleum production, and that natural nutrient concentrations in waters around such platforms, at least in the Gulf area, can be sufficiently high to allow rapid growth [2]. If this thrust proves successful, how might it develop?

Gulf of Mexico as a Potential United States Mariculture Site

Physical and Chemical Characteristics of Gulf Waters. Clearly, the natural and human-induced physical and chemical characteristics to be found in the immediate environment of offshore petroleum production platforms will vary widely, and so this subject can be treated here only on a very general level.

Several hundred platforms already dot the broad continental shelf of the Gulf of Mexico off the Gulf states of the southeastern United States. Since the western half of the Gulf features weak current activity and low tidal fluctuations, and since effluents from many large urban–industrial complexes flow into the area, one might expect significant nutrient levels to occur there. The recognized strength of the Gulf's fisheries industry combines with a variety of other evidence to confirm this expectation, if not necessarily verifying the reason for it.

Except for occasional hurricanes, the waters in the region of the petroleum platforms are generally calm, and currents usually are weak and variable outside of the eastern reaches. Oxygen concentrations appear adequate to sustain a high level of bioproductivity. Salinity is on the high end of the normal oceanic range, except near the mouth of the Mississippi River where it declines to about 26 parts per thousand (ppt). As one would expect, surface temperature variations are somewhat wide by open ocean standards, ranging from $25°C$ in winter to $29°C$ in summer.

Although there are concerns over hurricanes and localized contamination, it appears that the very broad continental shelf of the Gulf on which the petroleum platforms are located offers an attractive mariculture potential.

Mariculture Prospects for Gulf Platforms. The seeming attractivenss of oysters and peneid shrimp for culture associated with the Gulf platforms has already been mentioned. The problems encountered with the shrimp thus far are associated with cages and feeding mechanisms. The problems connected with oysters are in concentrating and harvesting mechanisms. It is suggested, then, that the Gulf platforms offer an appropriate opportunity to begin experiments with oyster attachment and crustacean containment devices such as those speculated upon in Chapter 10. Compared to the annual research and development budgets of major petroleum companies, the investment in this sort of experimentation would be trivial. Moreover, there is no doubt in our minds that the work would receive support from the federal government and from local academic institutions, just as the experiments already underway do.

It is widely recognized that the current American market demand for mussels is minimal. Nonetheless, as pointed out in Chapter 8, these mollusks remain the most efficient marine producers of harvestable animal protein. As the prices of agriculturally produced protein soar higher in the United States, perhaps our devotion to traditional meats will wane and our willingness to try new foods will increase. It could be, then, that the day of the mussel or mussel protein concentrate (MPC) market may be not so far off, if the organisms can be reared in abundance and marketed at attractive prices. Perhaps it is not too soon to begin developing mussel-culture technology along with technology for other shellfish.

Pompano culture is already receiving attention in the Gulf, although so far it unfortunately relies on the capture of juveniles. Other species that seem potentially attractive for culture here are a variety of other jacks, groupers, and red snapper. If one combines the present and future market potential of these finfish with the realization that the Japanese submersible-cage concept is made to order for the Gulf environment, two suggestions come to mind: (1) to extend reproduction physiology experiments to all of these finfish, and (2) to begin American research and development on the Japanese cage designs, employing petroleum platforms as experimental bases. Here again, initial investments would be modest and the potential returns, in terms of generally valuable mariculture and environmental knowledge as well as concrete economic results, could be quite significant.

Concomitantly, we might investigate the effectiveness of Klima's fish tents and of artificial reefs (Chapter 10). An additional minor investment could associate a variety of these structures with several offshore petroleum platforms to determine their long-range effects. To this work might be added research with the various other fish-attraction and harvesting techniques suggested in Chapter 10.

Experiments to increase primary production through the application of urban wastes and synthetic fertilizers might be added to the preceding projects. Beyond this is the development of pelletized feeds for use in the open ocean realm. If all were combined into an integrated and time-phased research program employing neighboring platforms, the annual research investment would still be nominal, and the possibility exists that the results might be spectacular in terms of both expanding the development of technology and achieving concrete economic results. As a minimum, the offshore petroleum industry would learn a great deal about the real and potential effects, both positive and negative, of its operations in the marine environment, and mariculturists would gain some practical knowledge of marine polyculture. As a maximum, the oil industry might find itself at the leading edge of a highly remunerative diversification.

Unfortunately, the Gulf platforms appear to offer largely unattractive sites for experimentation with conservative energy (see Chapter 13). Tidal differences are small; wave action usually is slight; waters in which the platforms stand are too shallow to offer significant thermal gradients; and geothermal sources, as far as we know now, are not attractive in this area. The possibility of wind power generation does appear to exist for some platforms; but due to the threat of hurricanes, any wind power machines sited in the area would have to be designed for high wind survivability. The one remaining possibility would be experimentation with direct solar radiation systems, such as the Helios–Poseidon concept described in Chapter 13. Small-scale experimental models of such systems might well be appropriate for association with platforms standing in deeper waters. Here again, the hurricane threat demands that the solar collectors either be exceptionally resistant to winds and waves, or be readily removable (or retractable) when hurricanes are expected. However, since severe environmental conditions are not uncommon elsewhere, collector survivability in high winds and waves is one of several critical obstacles in the development path of such systems. Perhaps, then, the Gulf would be an appropriate testing ground.

California Coast as a Potential United States Mariculture Site

Physical and Chemical Characteristics of Southern West Coast Waters. The petroleum platforms, existing and planned, off the coast of California are located in a typical eastern boundary current and its shoreward countercurrent, between two major urban complexes. Effluents from these and lesser urban areas in the region, combined with the naturally high nutrient levels on the west coasts of continents (discussed in Chapter 4), result in high potential bioproductivity. Particularly in the spring and summer months, the potential is realized regularly and sometimes results in massive dinoflagellate blooms— the red tides. These blooms could be catastrophic to mariculture interests; they produce harmful ectocrines (see Chapter 5), and also reduce dissolved oxygen concentrations markedly. The combination of the two effects could be disastrous to a large-scale intensive-culture operation. In addition, the proximity of major urban–industrial complexes poses the constant threat of direct contamination. In spite of these two concerns, the natural productivity of edible fish and shellfish in the region is high and varied. Consequently, there is little doubt that pilot-scale experiments in associating mariculture with offshore platforms in this area would produce valuable results. One is driven to caution, however, and to suggest that operations should remain at the pilot scale for several years in order that any obviously undesirable trends might be detected before significant capital investments are made.

Other characteristics of the California platform environment are relatively high salinity by open ocean standards, usually adequate dissolved oxygen, mild currents of fluctuating direction, freedom from intense storms, the occasional arrival of large swells, and normal tidal ranges from 1 to slightly more than 2 m.

Mariculture Prospects for California Platforms. There are no obvious reasons why the same oyster, shrimp, and mussel culture experiments suggested for Gulf platforms could not be conducted from California platforms as well. The average water temperatures are slightly lower, but the range of annual variance is slightly less, too. Red tides, as mentioned, could be a serious roadblock, but small-scale experiments carried on during the occurrence of several red tides would provide quantitative measures of their potential threat to large-scale offshore operations. Since the tides frequently occur at least once and sometimes two or three times each summer along the southern California coast, we might hope that two or three years' experimentation could provide realistic estimates of the danger to commercial-scale operations, as well as some insight into means of ameliorating their effects.

The spiny lobster, *Panulirus interruptus;* three coveted species of abalone, *Haliotis* spp.; the large Pismo clam, *Tivela stultorum;* and the scallop, *Pecten circularis,* all are indigenous to the area occupied by the California platforms. With one notable exception [3], no concerted culturing efforts have been directed yet toward any of these species so far as we know; but the Japanese have experimented with some encouraging results on other species of *Haliotis* and *Pecten,* and clam culture is already practiced in the U.S. East Coast states. As far as *Panulirus* is concerned, it appears to be similar in diet and overall natural history to its East Coast counterpart *Homarus* sp., and a good many studies have been directed toward its reproduction and growth. A long pelagic larval stage is one serious problem. Spiny lobsters bring a good price but do not now enjoy quite the same high market demand as *Homarus;* yet neither are they so fiercely armed. It may be that they would prove less prone to cannibalism and more amenable to intensive culture than *Homarus.*

In Chapter 10, suggestions were made about possible devices for culturing clams, scallops, and abalone in the open sea environment. The California platforms appear to offer an ideal opportunity for U.S. experimentation with these devices, using indigenous species of high market value. Research into culturing *Panulirus* in open sea benthic cages and in shelter-containing enclosures, such as that suggested in Chapter 10, might be an attractive addition to this work at some point in the future, but present problems with rearing the larvae should be attacked first.

Yellowtail (*Seriola* sp.) and albacore (genus and species) are large carnivorous finfish that migrate through California waters in large numbers, and are in high market demand. The Japanese already are culturing yellowtail in open water enclosures and ponds as well as pursuing artificial spawning [4]. To our knowledge no culture work whatsoever has been attempted with albacore. Future experiments associated with California platforms might combine work on culturing techniques for these two forms with the research on open sea enclosures. It is already well established that smaller fish are attracted in number to the oil platforms [5]. This, then, might be an excellent opportunity to experiment with various methods of stimulating even higher concentrations of baitfish in the vicinity of the platforms so that they might serve as seminatural foods for the large carnivores. Klima's fish tents, artificial reefs, conditioning to auditory stimuli, and light attraction are possible means.

Although the California platforms so far are fewer in number, they appear to offer opportunities for beneficial research and development of marine monoculture and polyculture systems similar to the potential of the Gulf platforms. For added advantage, moreover, if research were conducted simultaneously in both regions, important similarities and differences in the effectiveness of culturing techniques no doubt would come to light. This would help guard against premature rejection of some trial culturing techniques, and would also provide additional insight into general marine ecosystem dynamics.

In the realm of conservative energy, both winds and swells are fairly constant along the southwest coast. Furthermore, existing platforms are not far from Scripps Institution of Oceanography, where wave-pump experiments are underway. It may be that few of the California platforms are in water sufficiently deep to allow the use of wave pumps, since the present experimental wave-pump designs involve tubes over 30 m in length. Yet later designs may depart from the present approach sufficiently to be applicable to shallower waters.

Westerly breezes, although not strong, blow reliably during daylight hours, and winds of destructive force are rare or nonexistent, excepting only the local phenomenon known as the Santa Ana (when strong, hot, dry winds blow from the inland desert out to sea at speeds from 30 to 50 knots). Thus, the California platforms might offer a good location for experiments with modest wind-power machines just large enough to serve the mariculture operation itself.

MARICULTURE AND OFFSHORE POWER

Judging from present signs, power production will be the next major, formerly shoreside, industry (after petroleum) to move to sea. Because of the vast quantities of cooling water required by our ever-larger electrical power plants, these are already moving to the coasts. But the quantities of heated water they disgorge may even now be endangering coastal ecosystems; so it seems entirely possible that coastal-zone environmental preservation regulations may make offshore siting economically attractive by comparison in the not-too-distant future. As mentioned in Chapter 13, Offshore Power Systems, Incorporated already is planning offshore nuclear installations for the U.S. East Coast.

Whatever the various forms of conservative-energy systems that may be developed eventually, there is little doubt that the early ocean-based systems will employ nuclear energy to generate electrical power. Assuming no restrictive environmental regulations, these offshore nuclear plants may take in surface water and discharge it at the usual temperature elevation of $3°C$. This process will result in a plume of heated, low-density water spreading over the surface away from the plant. If the area is already nutrient-rich, some increase in bioproductivity will occur. Judging from experiments by the British Whitefish Authority [6], the increase could be quite significant. However, in warmer climates this heated water may be damaging [7]. But suppose that the intake cooling water consisted of cold, deep, nutrient water. First, as Gunderson and Bienfang [8] point out, power-plant cooling-water-volume requirements might be reduced as much as fivefold because of the deep water's low temperature, and the discharge water could still be at ambient surface temperature. From the point of view of the electrical utility, this fact might result it long-term savings. Further savings due to reduced maintenance requirements may be projected when one recognizes that both dissolved oxygen and fouling-organism larvae are much less concentrated in the deep water than in surface water. To tap the deep-water layer, however, requires refinement of deep-pipe technology, and we must admit that the initial capital investment might be high. Nevertheless, such a system certainly seems conceivable. How might it combine with mariculture then?

If a nuclear power plant drawing on deep water for coolant were located outside of strong surface currents, it would warrant the consideration of a

more or less uncontained sea-farm, utilizing the primary production stimulated by the output waters. Open sea oyster- and mussel-growing stations
along the lines of those mentioned in Chapter 10 might convert the primary
production directly into protein for human consumption. Shrimp, clams, and
scallops grown in subsurface devices developed from some of the suggestions
in that chapter could augment the oyster and mussel production.

It seems quite likely that small herbivorous fish would congregate in the
same densities at such a site as they do around natural upwellings, and with
them would come the carnivores. The addition of fish-attracting structures
might further increase the population density. If the discharge water were
somewhat warmer than the surrounding water, this might prove an additional
attractant as well as a growth stimulant, especially in winter at higher
latitudes. Where would the resulting marine ecosystem establish equilibrium?
How would the growth of beneficial plankton be stimulated and harmful
plankton be suppressed? How would harmful predation be controlled? How
long would it take such an ecosystem to become productive? What about
disease? What would be the effect of a temporary stoppage in pumping?
These and many other questions can be answered finally only by experience.
Yet there is little doubt that the bioproductivity increases would be nothing
short of spectacular. If our society has the will to begin experiments such as
this, we have seen in earlier chapters that there is every reason to believe that
the technology to support them would be forthcoming in a short time.

ATOLL-BASED MARICULTURE SYSTEMS

For better or worse, the peoples of the Pacific islands demand admission to
the 20th century. If no other means for acquiring the necessary economic
base are provided, they will follow Hawaii into the tenuous trap of tourism.
As the Pacific islands develop economically, we suggest that two dependency
conditions should be avoided if at all possible: (1) energy dependence on
fossil fuels, and (2) sole economic dependence on tourism. Lest there be a
misunderstanding, we do not propose that tourism be ignored; we suggest,
rather, that it not be viewed as a long-term panacea by itself. Nor can
near-term reliance on fossil fuels for power generation be ignored; but the
development of conservative-energy power systems should be the long-term
goal.

In 1971 the Oceanic Institute conceived what it termed "A Total Atoll
Production System" [9]. In this concept, lagoon mariculture was merged
with an agricultural system developed at the University of Arizona. The
agricultural system, designed for employment in arid coastal regions, combines diesel-generator power production, water desalination, and intensive
greenhouse agriculture. The growth rates, sizes, and quality of the vegetables
produced in this system are very high for the most part; and the electricity

and fresh water the system is capable of producing can be adequate to serve a modest nonindustrial community [10]. This agriculture system is in operation in Mexico. Thus far, the total atoll production concept has not been acted upon.

Bearing these ideas in mind, we are led to contemplate the possibility, for the Pacific island atolls, of combining the University of Arizona's agriculture concepts with mariculture, water desalination, power production, and even reasonable support of tourism. Could such a concept be developed along an avenue that would avoid dangerous addiction to external economic support?

Atoll Environment

The general features of Pacific atolls were delineated in Chapter 12 and the characteristics of their sea environment described in Chapters 4 and 5. Here we offer no more than a brief recapitulation.

Atolls and high and low islands in the Pacific are the tips of volcanic mountains rising steeply from the deep-sea bed. Where one of these mountains has subsided beneath the surface, coral growth activity maintains a ring of islets surrounding one or more lagoons. Where the tip of a mountain remains above the surface, high or low islands surrounded by coral reefs exist. In all cases, deep water is nearby.

Since coral grows only in warmer waters, relatively high water temperatures characterize the atoll marine environment. Surface waters typically are low in nutrients, although the underlying rock frequently is rich in phosphorus. Highly constant tropical trade winds, narrow tidal ranges, clean air, and abundant sunshine are the other usual environmental features.

Atoll Mariculture Prospects

At the outset it must be recognized that atoll mariculture will differ in at least one respect from mariculture based on man-made platforms: whereas the latter demands, for the most part, high-volume production systems only, the former admits consideration of low-volume cultures as well. The products of low-volume cultures on islands and atolls could provide a valuable component of the local diet for any indigenous population, as well as an attraction to visitors.

Nutrition. In Chapter 7 we discussed the question of stimulating plankton production with inorganic nutrients. In an atoll system there appear to be at least three prospective sources from which inorganic nutrients might be derived: commercial fertilizers, phosphorus-rich water from deep wells, and artificial upwelling. Of the three, the latter seems the most attractive, since the St. Croix experiments mentioned in Chapter 7 indicate its feasibility. Were the nutrient-rich deeper water to be pumped into the lagoon it would not require heating to remain on the surface; consequently, only low hydraulic pumping power would be required. However, the rate of influx of cold water into the lagoon could not be overly high without depressing biological activity. Should a required nutrient-input rate exceed the tolerable rate of cold-water input, it might be necessary to supplement the deep water with artificial fertilizers or to warm it before injecting it into the lagoon. During most days, warming might be accomplished by routing the water through wide, thin, solar-heated tanks before discharging it into the lagoon. But if reliance is placed on the direct utilization of solar radiation, a backup system for cloudy days will be necessary. This presumably could take the form either of an artificial fertilizer supply or of storable energy that could be used to heat the water. If some backup system were not provided, any intensive-production ecosystem that became dependent upon a high rate of nutrient input would be risking starvation during cloudy weather.

If a suitable, and presumably adjustable, nutrient-input rate is established, a high rate of plankton production should occur in the lagoon. With this the culturing of secondary producers for direct human consumption becomes possible. Oysters, mussels, clams, shrimp, and the herbivorous finfish are all possible candidates. Whether a lagoon would be devoted entirely to a mono-culture of one form, or whether a more complex polyculture or series of monoculture subsystems would be appropriate, would depend on a variety of factors, including the character of the lagoon, the species intended for culture and their interactions, and the needs of the human population of the atoll. Where an atoll would be a mariculture station only, it would probably tend toward monoculture; if an atoll or island supported a significant population, plus perhaps other industry, series monoculture or polyculture would proba-bly be more appropriate.

Species Selection. Opinions vary concerning which species are appropriate for atoll-based mariculture. In the recent Pacific Island Mariculture Confer-ence [11] no less than 32 kinds of organisms were examined with varying degrees of enthusiasm. These are

- Green sea turtle, *Chelonia* sp.
- Hawksbill turtle, *Eretmochelys imbricata*
- Cardinal fishes, family Apogonidae
- Eels, *Anguilla* sp.
- Goatfishes, family Mullidae
- Groupers, family Serranidae
- Jacks, family Carangidae
- Mahimahi (dolphinfish), *Coryphaena hippurus*
- Milkfish, *Chanos chanos*
- Mollies, *Poecilia sphenops* (for baitfish)
- Mullets, family Mugilidae
- Siganids, family Siganidae
- Tilapia, *Tilapia* sp.

- Brine shrimp, *Artemia salina*
- Coconut crab, *Birgus latro*
- Malaysian prawn, *Macrobrachium rosenbergii* (fresh water)
- Mangrove crab, *Scylla serrata*
- Shrimp, *Penaeus* sp.
- Northern lobsters, *Homarus* sp.
- Spiny lobsters, *Panulirus* sp.
- Various clams, families Lucinidae, Arcidae, Veneridae, and Cardiidae Psammobiidae
- Giant clams, *Tridacna* sp. and *Hippopus* sp.
- Mussels, *Mytilus smaragdinus* and *Perna canaliculatus*
- Octopus, *Octopus* sp.
- Giant oyster, *Crassostrea echinata*
- Japanese oyster, *Crassostrea gigas*
- Pearl oyster, *Pinctada* sp.
- Pink oyster, *Crassostrea mordax*
- Rock oyster, *Crassostrea glomerata*
- Winged pearl oyster, *Pteria* sp.
- Tritons, *Charonia* sp.
- Trochus, *Trochus niloticus*

Accepting all these forms as possibilities, the best crop composition for any given atoll system would result from creative design attuned to atoll characteristics. When a mariculture operation expands beyond the subsistence scale to become a significant component of an atoll economy, world market demand rather than local tastes alone will influence crop selection. One product to be considered for export might be *Artemia*. Brine shrimp are ideal food for the larvae and young of a number of carnivorous forms. Moreover, if they could be cultured and harvested in volume they could form a valuable component of prepared feeds. To the extent that mariculture in any form becomes important, the world market for *Artemia* will certainly increase.

Feed for carnivorous crops might be derived from imported pelletized feeds or from herbivores produced locally. The latter approach could involve either direct feeding by carnivores on the young herbivores (polyculture) or the harvesting of young herbivores for controlled feeding to a carnivorous culture (series monoculture).

If young herbivore production reached sufficient volume, their controlled release into the waters beyond the lagoons could result in a significant concentration of pelagic finfish, such as mahimahi and jacks, and of demersal predators such as groupers. Concentrations might be stimulated further by the addition of fish attractants, such as those with which Klima has experimented. The populations thus attracted might be harvested directly as an ancillary crop, and they might also comprise a valuable sport fishery for those islands catering to tourism.

The two sea turtles listed should be considered as a potential source of additional income for island populations. The females return to their hatching site to lay after reaching maturity and could be harvested selectively after laying their eggs. The natural laying sites could be protected until hatching occurred, or the eggs could be artificially incubated, and the young turtles, which apparently grow rapidly, could be raised in the lagoon until they were large enough to be reasonably predation-resistant, and then released. Alternatively, prepared feeds for these animals are being developed [11], so it may soon be feasible to culture them to maturity in the larger lagoons as one component of a polyculture system.

Another possibility, of course, is lobsters and crabs. If lagoons offered (or could be provided with) adequate shelter, and if population densities were low enough to reduce cannibalism to an acceptable level, these crustaceans might form valuable, detritus-removing components of a polyculture system. Going a step farther, if high-density rearing devices are developed, the deeper lagoons might prove an excellent place to locate them. There might even be a possibility that well-enclosed and nutrified lagoons could serve as synthetic "pelagic" regions for the larval stage of *Panulirus*.

Energy. Much more often than not, atolls are attractive sites for the development of conservative-energy systems. Even though tidal fluctuations are comparatively low, tidal flows through lagoon passages may measure several knots during more than half the daily cycle. Abundant sunshine makes the idea of solar stills for fresh water seem attractive and raises the distant vision of the Helios–Poseidon concept described in Chapter 13. The characteristic proximity of deep cold water might enable a modest ocean thermal gradient machine to serve the power needs of an atoll society, its tourism industry, and its mariculture industry. In addition, most atolls lie in the trade-wind belts, allowing the probable feasibility of wind machines to meet power needs. Combine these energy sources, either singly or in some integrated combination, with hydrogen production, and it seems that the dependency of atoll communities on waning fossil-fuel reserves is not necessarily mandatory.

THE NEXT STEPS

We should not be surprised to find major air and sea transportation terminals and noxious industry following power production into the oceans. The pressures either to remove them from the terrestrial living environment or to impose ever more restrictive controls on their interactions with that environment show every sign of continuing and increasing. As they do, offshore siting must begin to appear less and less economically prohibitive, as is the case with power production. Since these activities will require power, and since economies of scale as well as synergistic benefits can be effected through associations, one might expect the gradual evolution of offshore industrial complexes. This, of course, seems likely to lead toward larger and larger resident populations within these offshore complexes. And so the concept of floating cities, being pursued in Hawaii and Japan recently, might be more realistic than it appears at first glance.

Indeed, should transportation facilities and industry other than power plants begin to move seaward, we shall necessarily become ever more familiar with the marine environment. Then we may expect to see the eventual development of open ocean mariculture systems far more sophisticated than any of the concepts we have offered for consideration in this study. The basic engineering technology is available; it requires only purposeful development and refinement. If there are roadblocks other than inertial thinking and economic conservatism, they are our limited biological knowledge and sluggish legal system.

REFERENCES

1. Beall, S. E., Jr. 1970. Reducing the environmental impact of population growth by the use of waste heat. Presented at AAAS Symp. (Dec.), Pt. 1. 18pp.
2. Thompson, R. R., B. D. Baxter, and J. C. Parker. 1972. Cooperative environment projects, High Island Block 24L, Offshore, Texas. OTC paper 1676.
3. Bailey, J. H. 1973. Test-tube abalone. *Sea Frontiers 19*(3).
4. Milne, P. H. 1972. *Fish and shellfish farming in coastal waters.* Fishing News (Books) Ltd, London. 208 pp.
5. Carlisle, J. G., C. H. Turner, and E. E. Ebert. 1964. Artificial habitat in the marine environment. *Calif. Dept. Fish and Game Fishery Bull. 124.*
6. Nash, C. E. 1969. Thermal aquaculture. *Sea Frontiers 15*(5):168–276.
7. Jokiel, P. L., and F. L. Coles, 1974. Effects of heated effluents on corals off Kahoe Point, Oahu. *Pacific Science,* (Jan.) (in press).
8. Gundersen, K., and P. Bienfang. 1972. Thermal pollution: use of deep, cold, nutrient-rich sea water for power plant cooling and subsequent aquaculture in Hawaii. In FAO, *Marine pollution and sea life.* Fishing News (Books) Ltd., London. 624 pp.
9. The total atoll production system pilot project. 1971. *Oceanic Inst. Proposal OI-71-56,* (Mar.).
10. Hodges, C. N., and C. O. Hodge. 1971. An integrated system for providing power, water, and food for desert coasts. *Hort. Science 6*(1):30–33.
11. Proc. Pacific Island Mariculture Conf. 1973. Hawaii Inst. Marine Biol., University of Hawaii.

15

A NATIONAL MARICULTURE PROGRAM

J. A. Hanson

Considering the prospects and possibilities of the preceding chapter, is it worthwhile to consider launching a national-scale, mariculture research and development program? The answer unequivocally is yes.

What, in truth, might be the benefits of such a program? First, of course, there is the obvious potential of generating needed—and wanted—animal protein at competitive prices for domestic and foreign markets. It does not appear to be only a question of (national) self-sufficiency. If one or a few nations develop clear mariculture superiority, they will almost certainly enjoy a continuing world-market advantage as less alert nations continue to deplete natural fisheries. It would seem that any industrialized nation, watching the accelerating depletion of its natural resources and the concurrent increase in difficulties in maintaining a desirable world trade balance, could scarcely remain insensitive to this potential, which may well increase in importance as a direct function of the failure to control human populations and a concomitant rise in the cost of adequate nutrition.

Second, mariculture and marine environmental protection are both facets of the entire global environment question. Virtually all the knowledge developed by mariculture research will be valuable also in learning how to protect the marine environment from irreversible degradation, both locally and globally.

Third, and tied closely to the previous two considerations, those nations which are first into open sea mariculture operations will without doubt be the setters of precedent with respect to those changes in international maritime law which must necessarily accompany such developments. Technological, economic, and political advantages (and disadvantages) are inseparable in today's and tomorrow's world scene.

Fourth, U.S. foreign policy is based to a considerable extent upon its image as a wealthy and munificent world power, ready with technological and material wealth to assist less fortunate neighbors. To continue this foreign policy is to continue this image, and the image demands both attractive technology and material wealth.

Fifth, the rate of migration of industrial functions offshore, discussed in

Chapter 14, will be controlled by what is economically feasible. To the extent that mariculture for association with offshore industrial activities can be developed, it can share the costs and provide a share of the income. Thus, it can bring closer the date at which movement offshore of a given industrial function is economically feasible.

Finally, the commitment of federal and state governments to any clearly beneficial research-and-development thrust seldom fails to stimulate companion commitments from industry. Industrial participation in a national mariculture research-and-development program would not only accelerate its rate of progress, but would bring about immediate stimulation in local, regional, and national economic pictures.

If the nation were to agree with the foregoing assertions and to act upon its agreement, what form might the resulting program take? The task would be one of large-scale system management. As such it would be comparable to other national-scale research-and-development programs, for example, nuclear weaponry, the fast breeder reactor program, the Polaris program, and, of course, the NASA space-exploration program. The similarity of open sea mariculture development to other national technological thrusts is greatest with the space program. Both have multiple goals; both must develop knowledge of a new environment simultaneously as they develop technology to operate in that environment; and both must chart new biological territory as they proceed. But the development of mariculture would be at least an order of magnitude less expensive, and the economic stimuli that would be generated directly by committed research-and-development funds could be focused more broadly and flexibly.

In this final chapter we present recommendations for four broad research-and-development avenues aimed at an initial open sea mariculture capability. They are (1) marine environmental research, (2) biological research, (3) technological research and development, and (4) socioeconomic research. These proposals are followed by suggestions concerning appropriate roles for government, industry, and the academic community.

MARINE ENVIRONMENTAL RESEARCH

An understanding of the marine environment as a whole has developed steadily but not explosively. Most of us no longer hold that the world is flat and that winds, waves, currents, and storms are blessings and punishments of the gods. Instead, we now know fundamentally how the ocean–atmosphere heat engine is constructed, where the energy that drives it originates, and how it operates. Yet we have not reached a degree of refinement in our knowledge that allows us to predict very much ahead of time and with any precision what the machine will do over the long term, or at any given place at any given time. To phrase it succinctly, we have a generally accurate model of its structure but, as yet, only a characterization of its dynamics.

Our ignorance of ocean–atmosphere dynamics is not total by any means, and progress is being made. Tides can be predicted with some accuracy, as can tidal currents in selected regions. Research on intense cyclonic activity has been carried to the point where few lives now are lost as a result of inadequate storm warnings, and property damage is being reduced more and more. The ability to predict the arrival of extreme seas at specific points is improving, as is understanding of the generation and propagation of tsunamis. In a few regions, knowledge of major current dynamics is refined to the point that usefully accurate predictions can be made. Beyond these areas though, understanding of ocean–atmosphere dynamics consists largely of general characterizations.

Environmental Conservation

At the majority of oceanic locations, the direction and speed of currents throughout the water column can be predicted only within broad ranges. Our picture of the water layers of the oceans, the dynamics of their interactions, and the flow of matter and energy among them is certainly quite general. If these subjects were known adequately, we could predict the fate of the pesticides, heavy metals, radionuclides, and nutrients introduced into the marine environment at given locations and rates.

It seems that knowledge of the radiant energy and thermal exchange dynamics of the ocean–atmosphere system is not much past the characterization stage also. How delicately balanced is this system? To what extent can its dynamics be modified by combinations of atmospheric pollutants, marine thermal discharges, artificial upwellings, and interruption of incident solar radiation before the system moves beyond its present stability thresholds? Admittedly, some of the technology with which human societies conceivably might modify the equilibrium of this system is still in its conceptual stages, and the natural energy flux that operates this machine is so great that it is hard to conceive of human efforts modifying its operation. But it does not seem out of place to suggest that it is necessary to understand the dynamics of this system before technology, upon which civilizations have or may become dependent, exceeds its limits of stability—or, conversely, to gain reasonable assurance that such an event cannot occur.

Finally, there is the question of the nutrient-rich Intermediate Water. To what extent might the nutrients in these waters be depleted before unfortunate manifestations appear? What would be the effects of large-scale mixing between these waters and surface waters? Could the nutrients be replaced by introducing urban wastes beneath the thermocline? If so, what sort of pretreatment, if any, would be necessary?

Clearly, "ecological scare talk" is out of place here, particularly if it is based upon as-yet-undeveloped technological concepts. Yet the global atmosphere is already showing definite signs of man-induced changes, as is the marine food web. Furthermore, many technological concepts that may impact the dynamics of the ocean–atmosphere system may not be so far from implementation. Therefore, it seems far from irresponsible to suggest that for a variety of reasons, in addition to the needs of open sea mariculture, concerted research programs aimed at explaining the dynamics of the ocean–atmosphere heat engine are in order.

Predictive Modeling

Those who lean toward system modeling and simulation dream of someday developing a family of deterministic mathematical models that, when adequately interfaced and properly sequenced in some very large and very fast computer system, could reproduce the dynamic interactions of the significant factors of the ocean–atmosphere heat engine far faster than real time, and so predict usual and unusual phenomena. Perhaps someday this dream will be realized, but it is unlikely to happen in the near future. There are two approaches one might take to realizing such a dream: the first might be called inductive, and the second, deductive.

The inductive approach calls for the collection of detailed data over extensive time periods and at frequent time and geographic intervals throughout the global oceans and atmosphere. Theoretically, this mass of data could be analyzed using existing mathematical techniques to establish correlations that could lead to the construction of deterministic (quantitative cause and effect) models of fair validity. Practically, it cannot yet be done. The data do not exist, and it would take decades (or possibly centuries) and billions of dollars to gather an adequate data base—barring a major breakthrough in environmental sensing and recording technology. Furthermore, if the data base did exist today, current computer technology could not handle its monstrous volume.

The deductive approach would begin with existing theories of ocean–atmosphere dynamics, and would develop and refine these, gradually interfacing them as computerized submodels while spot-checking (calibrating) their accuracy against empirical data. Eventually, a complete model of the engine would emerge. But the mathematics of today allow only probabilistic models, which are better certainly than broad theory and superstition. But are such models an adequate base upon which to make decisions that could affect the ocean–atmosphere energy balance? For the time being, they might have to be.

Environmental Research Priorities

Facing up to the realities and narrowing the purview back down again to oceanic dynamics of direct concern to mariculture, we find we are not the first to recommend three specific research focuses. These are

1. *To understand the fate of urban wastes, pesticides, heavy metals, and radionuclides introduced into the oceans at present and postulated locations*

and rates. This alone is a prodigious order, which involves understanding the interactions among major physical, chemical, and biological factors of the oceanic system. Undoubtedly, initial attacks must be theoretical, and so, probabilistic. Nonetheless, every practical measure should be taken to verify theory-based predictions.

2. *To understand the dynamics of natural and artificial upwellings of Intermediate Water.* This would include empirical research on natural upwellings well beyond the efforts that have already taken place, and, based partly on these findings, the development of theoretical models of artificial upwellings of various types—thermally induced, mechanically induced, pneumatically induced, contained, and uncontained. Here again, the order is a huge one encompassing major elements of physical, chemical, and biological oceanography.

3. *To expand and improve our current programs aimed at predicting extreme marine conditions.*

Although these are only three of many possibilities, they are the problems that appear to be of critical importance to mariculture and associated offshore activities, and together they are of high ecological significance as well. In addition, the challenges they represent are enormous; in fact, should substantial progress be made toward the achievement of these goals, breakthroughs could be involved that would open the way to simulations of the entire ocean—atmosphere system.

Since the tasks are basic research, the prime responsibility for them appears to fall into the realm of oceanographic institutions of proven excellence. But the tasks are sufficiently broad that it is not difficult to conceive of valuable roles for other academic institutions, industry, the military establishment, and nonmilitary federal organizations.

BIOLOGICAL RESEARCH

Organism Selection

First contemplate the thousands of marine organisms for which man has found uses. Then contemplate what is known about the reproduction, development, growth, nutrition, and general ecology of each. Such a picture reveals that in this realm ignorance vastly exceeds knowledge and that, if one wishes to embark on intelligent mariculture, considerable organismic selectivity is in order. So, initially, it would probably be wise to concentrate the biological research efforts of a national mariculture program on no more than a few dozen organisms, or groups of organisms so closely related that findings for one species in the group likely would be applicable to most of its members.

The criteria for selecting organisms upon which to concentrate research and development are several and were covered in Chapter 8. They include such considerations as (1) abundance of present knowledge, (2) world market demand, (3) productivity potential, and (4) polyculture potential. Unfortunately, neither the criteria themselves nor assessments of their satisfaction are readily quantified. Thus, the nomination and selection of species to best meet these criteria are matters of authoritative debate. In Chapter 8, Nash's views on the matter are apparent. Before any decision to concentrate national efforts might be made, these views should be subjected to ratification and modification by a broadly qualified panel. Even then, a program should not doggedly hold with the organisms selected initially. If extreme difficulties are encountered, a candidate probably should be dropped; or if new knowledge increases the apparent attractiveness of a previously rejected organism, its addition should be considered seriously.

Given an initial list of organisms upon which to concentrate research efforts, we suggest that biological research thrusts could fall into the seven categories discussed next. Unfortunately, this breakdown is artificial, as is the case in any categorization, and obvious overlaps occur. Nonetheless, it is more useful than no clear organization at all.

Primary Production

Many attractive open sea mariculture concepts hinge upon primary production, as does the marine food web itself. For mariculture purposes primary production will frequently require artificial stimulation. Furthermore, there is no disagreement that some phytoplankton affect productivity in successive trophic levels more advantageously than others: they exhibit more rapid biomass growth, better nutrient qualities, and lower toxicity characteristics. There is some disagreement, however, on whether or not one should, in the interests of mariculture, attempt to control species compositions in a plankton culture, particularly in the open ocean. One argument stresses the magnitude of the natural physical, chemical, and biological forces and concludes that the equilibrium which can be maintained practically, in terms of planktonic species composition, will be dominated by the interaction of these forces, certainly in the open oceans and probably even in very large containers such as atoll lagoons. Another argument points to some evidence that minor variations in inorganic constituents may produce marked variations in species compositions of phytoplankton cultures and hypothesizes that, as a result, the energy necessary to control species compositions in large cultures could be small.

Observing that natural primary-production phenomena in upwellings definitely seem adequate for mariculture purposes, one also is reminded that even uncontrolled introduction of urban wastes into the oceans seems to cause productive responses more often than deleterious ones. Thus, initial primary-production research should focus upon the avoidance of known deleterious organisms in artifically induced primary production. The work should seek to isolate the conditions which cause blooms of undesirable plankters so that these conditions can be avoided in mariculture systems. A second and concurrent line of research might continue to chart more exactly the population concentrations at which some planktonic organisms cross the threshold from benign or useful to toxic.

A third line of research—one that has already begun—is the indexing and quantification of the amino acids, vitamins, and so on, that are characteristics of various planktonic groups. Associated with this work could be a fourth line, the determination and quantification of the feeding preferences of plankton feeders and the conditions under which these preferences and different feeding modes are manifested.

These four lines of research are all related, yet apparently not so closely that they could not be pursued by separate institutions as long as adequate communications were maintained.

Nutrition

For those species selected for intensive research and development, two major and very practical questions must be answered. These, in order of logical priority, are

1. What feeding systems are adequate to sustain and produce acceptable survival and growth rates throughout larval and postlarval stages?

2. What feeding systems will produce the most favorable economic conversion rates (ECR's; see Chapter 7) in cultured populations?

Clearly, the answers to these two questions imply a variety of research topics. The most important are

- Nutritional requirements of each organism
- Nutritional characteristics of all candidate nutrients
- Digestion and assimilation of candidate nutrients by selected organisms
- Acceptability of candidate nutrients in both raw and blended form
- Toxicity and disease characteristics of candidate feeds

Taking into account the varying nutritional requirements of successive development stages, a matrix with the development stages of a few dozen organisms on one axis and nutritional research topics on the other could be drawn. In the resulting intersections several hundred specific research projects would appear. This is all right if one is interested in keeping people busy on useful research projects; but not if one is interested in achieving practical mariculture results as early as possible.

It is the nature of basic science to explain how and why phenomena occur. It is the goal of mariculture to achieve predictably consistent results. The two are not identical. Aquacultural history shows, on the one hand, a multitude of nutritional physiology research projects producing few practical results, and, on the other hand, straightforward, "green-thumb" culturing attempts producing profitable results. Therefore, knowing we invite heated argument, we recommend that the first steps in nutrition research be straightforward trial-and-error experimentation on an organism-by-organism basis. The objective of each experimental program would be to find or develop feeds for all developmental stages of the organism of interest that will result in *acceptable* ECR's. With this objective achieved, the nutritional factor will no longer be a roadblock to practical culture. Once other obstacles have been removed and practical culture achieved, research on refining feeds and nutritional systems should be initiated. It is this latter phase that invites detailed research in the physiology of nutrition.

Nutrition research is, of course, complementary to larval-rearing research, just as larval-rearing research is complementary to research on induced spawning. Should an initial selection of organisms for concerted mariculture research be made, it will certainly contain a variety adequate to allow a number of qualified institutions each to assume responsibility for research on one or a few species most appropriate to their geographic location and resources. We suggest, then, that a single institution should assume responsibility for nutrition, hatchery, and husbandry research on any given organism. Although some interinstitutional complementary research on a given organism might prove desirable in some cases, it is recommended that this be held to a minimum.

Hatchery Research

An organism cannot be considered a serious candidate for large-scale culturing until its juveniles can economically be made available in large numbers. Typically, this implies some human control over spawning, and the ability to bring high percentages of the spawned larvae through to the postlarval stage, ready for grow-out.

For many organisms this requirement implies achieving the ability to induce spawning artificially; it does not for forms that spawn spontaneously and frequently. Some organisms have a simple larval development sequence that requires a short time to complete. In others the sequence of metamorphosis is moderately to highly complex. In all species, the survival and development rates are sensitive to the complex interactions of nutrition, physical and chemical environment, and disease.

Following the attitudes expressed in discussing nutrition research, we advance the opinion that the objective of hatchery research is to reliably supply large numbers of postlarval organisms, as opposed to thoroughly charting the physiology of larval development. Thus, the groups responsible for hatchery research would be advised to experiment directly with all reasonable methods and combinations of methods for inducing spawning (where spawning inducement appears desirable) and for larval rearing of selected species. Such experimentation would begin with methods already proved useful for other species, and advance toward intuitive approaches as necessary. Recourse to extended basic physiology research would be made only when all other ideas had been exhausted, and at this point elimination of the organism from the candidate list might be considered. As with nutrition research, once an economical and reliable hatchery technique has been developed, the hatchery roadblock to practical culturing will be removed. And again, after commercial-scale culturing has become feasible, additional physiological research might point the way to higher and more economical postlarval yields.

Husbandry

Nutrition is far from being the sole factor determining maturation and growth rates. Temperature and light are known to be important, and there is evidence that a variety of other "normal" physical and chemical factors usually are important as well. For example, salinity has been shown to be important to salmonids. Chemical toxicity can be tolerated in some concentrations by any organism without adversely affecting maturation and growth, but tolerances vary among organisms, and sensitivity is dependent on other environmental factors. Disease and parasitism can probably be considered always deleterious; these are discussed separately. The question for each organism selected for concerted research is, "For a given nutritional system, what combination of other environmental factors results in optimum economic conversion rates?"

Again, we are of the opinion that these questions can be answered most effectively by first experimenting directly with a variety of combinations that are practical to maintain in commercial-scale operation. The research strategy might be to hold all but one parameter constant, at some presumed mid-value, until an optimum value for the experimental variable is determined. Then, holding the former variable constant at that optimum value and all but one of the other parameters constant at their original values, vary the next selected parameter to determine an optimum value. This could be continued until initial approximations of optima for all parameters were achieved. Iterations of this procedure might serve to refine the initial values appreciably, or they might not: one additional iteration should serve to reveal the answer. This research method is described here as serial for the sake of simplicity; probably it would not have to be serial in practice. Rather, given adequate facilities, simultaneous runs representing variabilities in different parameters could serve to shorten the time required for each iteration.

This work clearly would complement nutritional and hatchery research, and so it seems best undertaken, in each case, by the same institution that pursues the other two objectives for a selected organism.

Disease, Competition, and Predation

Epizootic disease is a constant threat to any high-density animal population. Virtually every commercial aquaculture venture today continues to suffer losses from outbreaks of disease and parasitism. Even though highly economical and effective hatchery and husbandry technology may be developed, the commercial success of large-scale aquaculture will remain problematic until common forms of disease are controlled.

Once a group of marine species has been selected for concerted mariculture research, what should be the strategy for a broad attack on the disease and parasite problem? Most often, one type of disease organism is capable of infecting multiple types of host species; infrequently, a disease or parasite organism is specific to a single host. However, effective preventions and treatments frequently vary according to the host organism of concern, and host species also vary in their sensitivities to a given disease. One might visualize a matrix with the selected mariculture species on one axis and disease organisms that infect them on the other. When knowledge becomes adequate, descriptions of the disease organisms and their life cycles would appear on the disease-organism axis; and symptoms, prevention, and treatment appropriate to each host organism would appear in the intersections.

This picture suggests a two-pronged attack on the disease problem: one prong would be directed toward the natural history of disease-causing organisms and effective means of interrupting the life cycles; the other prong would be directed toward the symptomology, prevention, and treatment of the diseases they cause in each of the host organisms selected for concerted research. The first prong probably would best fall to one, or a very few, highly qualified institutions. The second seems properly the responsibility of the institution conducting the hatchery, nutrition, and husbandry research on each potential host species.

Predation is a problem different from parasitism since parasites themselves are not considered valuable. But some predators are considered desirable animals so long as they do not feed excessively at the open sea mariculture trough. For these, nonlethal control measures should be sought. In many cases, predators also may expand their populations with prey abundance. Generally, this would be unwelcome in a cultured crop, even though the predators might be "desirable" organisms in moderation or elsewhere. Viewing the problem along another dimension, predation may occur in the form of chronic infestation or in the form of "raids" by groups of highly mobile organisms. For the former, methods of population control must be found; for the latter, means of minimizing the incidence of occurrence as well as the effects are in order.

With adequate interinstitutional communication, we suggest that research on predation control could be conducted by institutions other than those conducting research on organisms for culture. For slow-moving predators, such as sea stars, that tend to infest a cultured crop, research could probably concentrate as much on natural history and interruption of life cycles as on direct control measures. For more mobile predators that could arrive swiftly in numbers and severly damage a cultured crop in a short time, research should be concentrated on direct control measures of the sort mentioned in Chapter 9.

Behavioral Research

Many important aspects of stock control hinge on responses to a single stimulus and to combinations of stimuli; efficient reproduction, nutrition, protection, concentration, and harvesting, for example, all depend to some extent on understanding the behavior of cultured marine organisms. The stimuli include visual patterns, light intensity and periodicity, auditory patterns, chemicals, electrical fields, magnetic fields, and conditioning histories. Adequate stock control depends, then, upon understanding how the stock will respond to given stimulus patterns and upon the ability to produce effective patterns at will.

Once basic biological obstacles to hatchery and husbandry technology have been (or are near to being) overcome for any organism, it will be the time to extend research on that organism into behavioral phenomena important to efficient culturing and harvesting. The work should be focused first on organismic response patterns already recognized, or suspected, from previous work, with that focus being modified by what is technologically possible. With some imaginative thinking, understanding of the behavioral characteristics of cultured organisms and human technological capabilities should develop together so as to form an efficient component of husbandry systems for each candidate organism.

Behavioral research would seem a logical extension of physiological research; as such, it should be a follow-on responsibility for the institutions that accepted responsibility for the basic biological work on each organism. The question of technological development is treated later.

Symbiotic Possibilities

Up to this point, the treatment of organisms on an individual basis has been implied, indicating (falsely) that only monoculture was under consideration. Since this most clearly is not the case, it is now time to consider research activities designed to advance polyculture knowledge in terms of true multiproduct polyculture systems and in terms of single-product systems that gain effectiveness from symbiotic phenomena.

Basic hatchery and husbandry research undoubtedly will reveal numerous symbiotic possibilities for most of the organisms investigated. Additionally, free communication between researchers in a national program would, if encouraged, lead to realistic proposals for experimental polyculture systems.

As soon as adequate progress had been made with a representative subset of the organisms selected for concerted research, it would be time to stimulate such polyculture-system proposals and, subsequently, to select the dozen or so polyculture-system experiments that showed highest promise.

These experiments should be conducted in the most appropriate marine locations, such as coastal waters, offshore petroleum platforms, atolls, and coastal and offshore power production stations. At this point, component research and development paths would begin to converge on the early mariculture systems suggested in Chapter 14. Monoculture experiments with selected organisms in these locations easily may have begun well in advance of this time, of course, as part of the research attendant to individual organisms.

To review, research on symbiotic potentials would include the following.

• Multiproduct, multi-trophic-level, polyculture systems
• Biological controls for disease and parasitism
• Biological control of marine fouling
• Possibilities for biological predation control

In addition, and conceivably in advance, if technological development allows, experiments in natural responses to artificial upwellings should be initiated. In these experiments, uncontained upwellings would be maintained in selected locations over extended time periods to determine the species composition and productivity of the ecosystems that would result naturally. This, and the symbiotic research already mentioned, would merge as favored organisms and culture techniques were gradually introduced into the ecosystems already associated with the upwellings.

At this stage, true system development projects would have come into being. Every project would involve multiple institutions, each pursuing efforts in its own realms of expertise, in cooperation with other involved institutions and under the overall authority of a single project office.

TECHNOLOGICAL RESEARCH AND DEVELOPMENT

In Chapter 14 we noted that the development of open sea mariculture in many instances is closely tied to the development of other forms of offshore technology. The ties are so close, in fact, that the rate at which commercial-scale offshore mariculture might be developed very probably will be strongly affected by the rate at which other forms of offshore technology develop. It would be, then, clearly in the best interests of a national mariculture program to encourage and support other forms of offshore technology, and to make mariculture objectives, requirements, and benefits well known to these potential allies. Those responsible for a national mariculture development program should establish and maintain broad-based, noise-free, two-way communication links with the offshore petroleum industry, with the incipient offshore power industry, with advance thinkers who even now are shaping concepts for other forms of offshore industry, and (for atolls) with the government of the Trust Territory of the Pacific. It is most likely that the interests of all can best be served by mutual cooperation.

If basic offshore engineering research, such as floating platforms and energy-source development, can be pursued external to the national mariculture program, we suggest four technological research focuses to address initially within the program. These are containment and attraction, concentration and harvesting, feeding technology, and artificial upwellings.

Containment and Attraction

In Chapter 10 we examined devices for containing and attracting marine organisms in the open sea and Milnes' recent book [1] treats these subjects as they apply to coastal waters. With this information at hand, it is suggested that a complete national mariculture program would include projects aimed at expanding present knowledge in the following realms.

1. Large-scale survivable and maintainable finfish enclosures for use in the open sea.

2. Artificial-reef technology and ecology in tropical, subtropical, and temperate waters.

3. Surface and mid-water fish attraction structures.

4. Artificial substrates and containers for rearing and harvesting benthic mollusks and crustacea.

5. Bubble curtains.

6. Fouling-resistant materials for use in marine enclosures and related structures.

For the first five projects the first milestone would be to develop (where necessary) realistic design concepts and then to assess their probable productivity potential. In some cases, bubble curtains for example, at least moderate-scale empiricism would be required to achieve even a rough feasibility assessment. When satisfied with respect to the potential feasibility and productivity of each concept, the national mariculture program then could select further to determine which projects should be given priority.

The artificial-reef project and the surface and mid-water attraction project should be pursued by three or more well-coordinated institutions that are appropriately located. Finfish enclosures, mollusk substrates, crustacean containers, bubble curtains, and fouling-resistant materials are projects that should each be spearheaded by a separate institution with the cooperation of appropriate industrial and governmental organizations.

Concentration and Harvesting

For concentration and harvesting, we have described in Chapter 10 what we feel should be done. Present methods should suffice for initial mariculture systems, however, and so we recommend support for research and development in this area at a continuing moderate level until initial mariculture system successes firmly define precise operational requirements for greatly improved technology.

Feeding Technology

Mechanisms for preparing, storing, and dispersing feeds in supplementary and provisional mariculture systems will be required for the implementation of many open sea mariculture concepts. With attention to present ideas about operational requirements, conceptual designs should be developed and evaluated for practicality fairly early in the development of a national mariculture program. Then, as the first mariculture operations that need this support enter experimental stages, initial feeding-system designs could be finalized, built, and tried. Early experience would provide a basis for ongoing refinements in subsequent designs. Ocean engineering departments of several institutions could participate in these developments, initially at modest levels.

Artificial Upwellings

Theoretically, upwellings of nutrient-rich Intermediate Water can be induced by thermal, chemical, mechanical, and pneumatic techniques. Among the concepts advanced so far are

- Pumping for nuclear power plant cooling or for use in the condenser side of ocean thermal gradient machines
- The "perpetual salt fountain" concept, first advanced by Strommel et al.
- Pneumatic bubble generators
- Injection of heated water at depth

As a component of any national mariculture program, we suggest that a single institution possessing highly competent ocean engineering and oceanography departments with a record of working well together should accept responsibility under a modest grant or contract for developing a comprehensive artificial-upwelling research-and-development plan. The plan should include basic research requirements, plans for small-scale experiments, and projections for larger-scale pilot studies. It should also emphasize cooperative roles for governmental and private industrial organizations, and attend to the need to maintain successful pilot upwellings for an extended period in order that the biological research mentioned earlier in this chapter may be conducted.

SOCIOECONOMIC RESEARCH

If mariculture is to become important in the national economy, it must not only become technologically practical, but must also fit into the legal and economic structures that will exist in the future. For a prospective national mariculture program, this means that first the legal and economic needs of mariculture must be defined; then the current legal and economic structures should be described in terms of their probable impact on mariculture. With this information in hand, it should be possible to decide where mismatches are most likely to exist and then to initiate steps to alleviate the conflicts, both by modifying mariculture and, because we are speaking of the future, by modifying laws and possibly economic concepts.

Legal and Political Questions

The legal and political questions were examined at some length in Chapter 3 and the problems pointed out. It remains only to summarize here the steps we feel should be taken with respect to local, state, national, and international legal structures.

Tenancy and Property Rights. Local and state wetlands laws of all coastal states require examination and cataloging. Where laws are inimical to mariculture, more workable alternatives should be suggested. For open sea mariculture the question of tenancy and property rights on the high seas and the issue of boundary definitions of state and national territorial waters demand a great deal of examination and eventual resolution. To be sure, final international agreements in these realms will be made under United Nations auspices. Nevertheless, if a national mariculture program is initiated, the intent implied thereby dictates that open sea mariculture interests and requirements find their way into U.S. law-of-the-sea policy positions as quickly as possible. Two fruitful research projects are implied here; each could be the responsibility of a different qualified institution.

Environmental Protection Versus Mariculture. At present, certain "environmental protection acts" are inimical to mariculture interests. Yet, in theory, intelligent mariculture and environmental protection are highly compatible; current conflicts between the two most probably result from shortsighted and piecemeal legislation. We suggest that an in-depth study based upon the characteristics of mariculture on the one hand, and upon the fundamental requirements of marine environmental protection on the other, could result in the drafting of landmark legislation that would serve the best interests of both. This work should be the responsibility of a single highly qualified institution and enthusiastically supported by appropriate federal and state agencies.

Conflicts Among Hydrospace Users. If, as suggested, a systems approach is taken to developing mariculture, it will be necessary to consider operational compatibilities among developing mariculture and other uses of the marine environment. Careful designs and well-thought-out agreements can avoid the majority of conflicts, if conflict potentials are recognized in time; but statutory guidelines for conflict resolution will be required also. A project aimed at first characterizing the hydrospace usage requirements of present and planned marine activities, then determining potential conflicts, and, finally, suggesting workable resolutions would be highly beneficial. Such a project should be conducted by a single institution with very broad scientific and technical resources supporting a highly qualified legal staff.

The Economic Question

Adequate federal spending can, by itself, support open sea mariculture. Although there might be some who would support this approach, we are not among them. The role of the federal government should, indeed, be that of stimulating mariculture development; but if mariculture techniques are not assimilated by industry and carried on to become profitable endeavors, a mistake will have been made. We suggest, then, that a valuable fourth analysis would be aimed at defining specifically the socioeconomic conditions which would have to be satisfied in order for mariculture to attract the economic and technological support of industry. This should fall to the responsibility of a highly respected business-management institution.

A ROUGH FRAMEWORK

With interim open sea mariculture objectives having been suggested in Chapter 14, and research-and-development thrusts aimed at these objectives having been discussed in this chapter, it behooves us now to do our best to complete the picture. As yet no attention has been directed to schedule, organization, and funding. In this section we attempt in a general way to fill these three gaps.

Scheduling

Although a realistic scheduling exercise is not something we would presume to accomplish alone, it is possible here to at least summarize and sequence the research and development that has been recommended, and also to make some informed estimates of the time frames required. This has been done in Figure 15.1, in which a nominal decade is assumed as the planning period.

Green-thumb monoculture approaches have already commenced on petroleum platforms in the Gulf of Mexico and in association with coastal power plants. We recommend that these be intensified with selected organisms, and augmented in the second and third years with work on California platforms and atolls. As biological obstacles are removed, these experiments would evolve into more formalized polyculture development projects.

In the field of marine environmental research, a reasonably heavy emphasis beginning in the second year of the program is recommended for a high rate of early progress in research on the fate of wastes and toxins, and on upwelling dynamics. Knowledge in these two areas will be highly valuable,

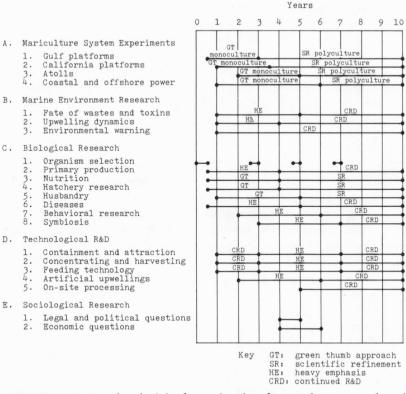

FIGURE 15.1 *Rough schedule for a decade of mariculture research and development.*

whether or not other mariculture research produces the desired results, so research may begin early. Work on environmental warning is already receiving a good deal of attention, and we recommend only that it be continued at an adequate level.

In the field of biological research, the first task would be the selection of the organisms upon which attention is to be concentrated. Continuing intermittent bars on line C-1 in Figure 15.1 indicate that the original selection is likely to be modified by research results. Our opinion that in many cases the shortest road to interim success lies in intelligent, goal-oriented, green-thumb approaches is manifested in Figure 15.1, which frequently shows green-thumb approaches first, followed by more scientific refinement of methods once an organism has been shown to be suitable for culture.

There should be a good idea of which organisms might be cultured before devoting significant funds to culturing technology. So, in general, suggested heavy emphasis in technological research and development are scheduled to coincide with expected early results from the biological research.

Since the effectiveness of the sociological research is dependent to some extent on national attitudes, it is recommended that emphasis on this area be delayed until the program has developed some maturity and public recognition.

This suggested schedule is intended only as a point of departure from which a more definitive scheduling project might develop. The more arguments it generates, the better, for they will indicate a real interest in the possibility of a national mariculture program.

Organization

Quite clearly, if an organized, national-level mariculture program is to be, it will fall to the federal government to take the initiative. Just as clearly, if industry does not support the program and eventually assume responsibility for carrying it forward, the program will not have proved essentially viable. The criteria for organization, then, are (1) effective involvement of state and federal agencies, (2) effective involvement of industry, and (3) a manageable structure, with effective communications that encompass the full breadth of applicable science and technology.

The development of a national mariculture program organization containing all the necessary constituents would be a sizable task. In Figure 15.2 we offer a generalized framework, which, if fleshed out with existing governmental, industrial, and scientific organizations, seems to meet the preceding criteria.

In this proposed structure all organizations involved might contribute to the staffing of the scientific and technical advisory board. The interagency steering committee would consist of members of the participating state and federal agencies; it would be responsible for advising the lead agency on matters of policy, regulation, and legislation. Industrial participants would be selected on the basis of their applicability to and interest in the program. The research-and-development work would be carried on by a multitude of institutions, which very likely would include industrial participants as well as academic, private, and public scientific organizations. Whatever federal lead agency is selected, it would be obliged to establish a mariculture program office staffed with qualified research-management personnel.

Funding

Obviously, the background required to estimate the funding requirements of a national mariculture program with any degree of confidence whatever does not yet exist. Consequently, any attempt to make such an estimate here is in danger of being considered irresponsible. However, there does exist a good deal of experience with each category of research that would be involved in a national mariculture program, and with the funding under which such research presently is conducted. Using this as a basis, an off-hand estimate of funding requirements, although it cannot be advertised as sup-

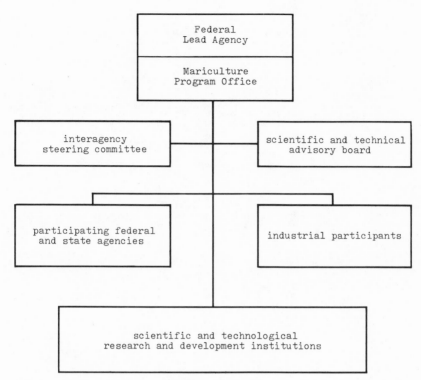

FIGURE 15.2 *Organizational framework for a national mariculture program.*

portable, might be helpful in placing a national mariculture program in approximate budgetary perspective. This we attempt in Table 15.1, in which guesses are made as to the funding that could produce significant results in each of the research-and-development categories suggested. All these estimates are arguable in the specific; yet few investigators in the field would argue with the assertion that funding along the lines suggested most certainly would result in a decade of rapidly advancing mariculture technology. In other words, there is no serious doubt that the funding suggested here would carry mariculture a long way toward reality.

Yet another perspective from which the funding question may be considered is to estimate the magnitude of support that would signify a real federal commitment sufficient to stimulate complementary commitments by nonfederal institutions. With this idea in mind, coupled with the necessity to allow time for a program of this breadth to gain momentum, we suggest consideration of an appropriation of $18 million, $25 million, and $32 million, respectively, for the first 3 years of a national mariculture program.

TABLE 15.1 *Rough Estimate of Funding Requirements for a National Mariculture Program*

Research	Annual budget requirements (in millions of dollars)	Total for 10-year period ($ millions)	Average annual total ($ millions)
A. Systems research			
1. Gulf platforms	$1 for first 3 yr and $1/yr for next 7	$ 8.0	$ 0.8
2. California platforms	$1 for first 2 yr and $1/yr for next 6	7.0	0.7
3. Atolls	$3/yr for first 3 yr and $5/yr for next 5	34.0	3.4
4. Coastal and offshore power	$1/yr for first 4 yr and $3/yr for next 5	19.0	1.9
		68.0	6.8
B. Marine environment research			
1. Fate of wastes and toxins	$1/yr for first 4 yr and $0.5/yr for next 5	6.5	0.6
2. Upwelling dynamics	$1/yr for first 3 yr and $0.5/yr for next 6	6.0	0.6
3. Environmental warning	$0.25/yr for 10 yr	2.5	0.2
		15.0	1.5
C. Biological research			
1. Organism selection	$0.1/yr for 10 yr	1.0	0.1
2. Primary production	$1/yr for 10 yr	10.0	1.0
3. Nutrition	$2/yr for 10 yr	20.0	2.0
4. Hatchery research	$4/yr for 10 yr	40.0	4.0
5. Husbandry	$4/yr for 10 yr	40.0	4.0

6. Disease, etc.	$4/yr for 10 yr	40.0	4.0
7. Behavioral research	$1/yr for 10 yr	10.0	1.0
8. Symbiosis	$6/yr for 4 yr and $3/yr for next 4	36.0	3.6
		197.0	19.7
D. Technological research and development			
1. Artificial upwellings	$2/yr for first 4 yr and $0.5/yr for next 4	$10.0	$1.0
2. Containment and attraction	$2/yr for first 4 yr and $0.5/yr for next 5	10.5	1.05
3. Concentration and harvesting	$2/yr for first 4 yr and $0.5/yr for next 5	10.5	1.05
4. Feeding technology	$2/yr for first 4 yr and $0.5/yr for next 5	10.5	1.05
5. On-site processing	$2/yr for first 4 yr and $0.5/yr for next 5	10.5	1.05
		41.5	4.15
E. Sociological research			
1. Legal and poligical questions	$0.5	0.5	0.05
2. Economic questions	$0.5	0.5	0.05
		1.0	0.1
		$322.5	$32.25

Toward the end of the second year would be the time to evaluate prospects for funding the remaining 7 years.

CONCLUSIONS

As this final chapter of a 2-year study is being completed, in the summer of 1973, we contemplate a nation with an annual gross national product of around $1 trillion; yet its balance of trade with other countries is largely unfavorable; its food prices are rising alarmingly; and its dwindling natural resources are causing it to become increasingly dependent on foreign markets or to seek resources in the seas. Overlapping this is a strong internal demand to clean up the environment, which has the effect of hampering industrial development, driving up prices to the consumer, and further degrading the U.S. world market position. We certainly do not suggest that mariculture development will cure all national ills; but it appears now to be a stride in the right direction. Is it worth committing about three one-millionths of this nation's annual gross national product, or about $0.16 per capita per year, in an attempt to take this stride?

REFERENCE

1. Milne, P. H. 1972. *Fish and shellfish farming in coastal waters.* Fishing News (Books) Ltd. London.

INDEX